T0310071

Biomaterials Science: Processing, Properties and Applications II

Biomaterials Science: Processing, Properties and Applications II

Ceramic Transactions, Volume 237

Edited by
Roger Narayan
Susmita Bose
Amit Bandyopadhyay

A John Wiley & Sons, Inc., Publication

Published by John Wiley & Sons, Inc., Hoboken, New Jersey.
Published simultaneously in Canada.

No part of this publication may be reproduced, stored in a retrieval system, or transmitted in any form
or by any means, electronic, mechanical, photocopying, recording, scanning, or otherwise, except as
permitted under Section 107 or 108 of the 1976 United States Copyright Act, without either the prior
written permission of the Publisher, or authorization through payment of the appropriate per-copy fee to
the Copyright Clearance Center, Inc., 222 Rosewood Drive, Danvers, MA 01923, (978) 750-8400, fax
(978) 750-4470, or on the web at www.copyright.com. Requests to the Publisher for permission should
be addressed to the Permissions Department, John Wiley & Sons, Inc., 111 River Street, Hoboken, NJ
07030, (201) 748-6011, fax (201) 748-6008, or online at http://www.wiley.com/go/permission.

Limit of Liability/Disclaimer of Warranty: While the publisher and author have used their best efforts in
preparing this book, they make no representations or warranties with respect to the accuracy or
completeness of the contents of this book and specifically disclaim any implied warranties of
merchantability or fitness for a particular purpose. No warranty may be created or extended by sales
representatives or written sales materials. The advice and strategies contained herein may not be
suitable for your situation. You should consult with a professional where appropriate. Neither the
publisher nor author shall be liable for any loss of profit or any other commercial damages, including
but not limited to special, incidental, consequential, or other damages.

For general information on our other products and services or for technical support, please contact our
Customer Care Department within the United States at (800) 762-2974, outside the United States at
(317) 572-3993 or fax (317) 572-4002.

Wiley also publishes its books in a variety of electronic formats. Some content that appears in print may
not be available in electronic formats. For more information about Wiley products, visit our web site at
www.wiley.com.

Library of Congress Cataloging-in-Publication Data is available.

ISBN: 978-1-118-27332-6
ISSN: 1042-1122

Printed in the United States of America.

10 9 8 7 6 5 4 3 2 1

Contents

Preface

This volume is a collection of twenty-six research papers from the Next Generation Biomaterials and Surface Properties of Biomaterials symposia, which took place during the Materials Science & Technology 2011 Conference & Exhibition (MS&T'11) in Columbus, Ohio on October 16–20, 2011.

These symposia focused on several key areas, including biomaterials for tissue engineering, ceramic biomaterials, metallic biomaterials, biomaterials for drug delivery, nanostructured biomaterials, biomedical coatings, and surface modification technologies.

We would like to thank the following symposium organizers for their valuable assistance: Paul Calvert, University of Massachusetts; Kalpana Katti, North Dakota State University; Mukesh Kumar, Biomet Inc; Kajal Mallick, University of Warwick; Sharmila Mukhopadhyay, Wright State University; Vilupanur Ravi, California State Polytechnic University, Pomona; Varshni Singh, Louisiana State University; and Thomas Webster, Brown University.

Thanks also to all of the authors, participants, and reviewers of this Ceramic Transactions proceedings issue.

We hope that this issue becomes a useful resource in the area of biomaterials research that not only contributes to the overall advancement of this field but also signifies the growing roles of The American Ceramic Society and its partner materials societies in this rapidly developing field.

ROGER NARAYAN, *UNC/NCSU Joint Department of Biomedical Engineering*
SUSMITA BOSE, *Washington State University*
AMIT BANDYOPADHYAY, *Washington State University*

MECHANICAL AND MICROSTRUCTURAL CHARACTERIZATION OF 45S5 BIOGLASS® SCAFFOLDS FOR TISSUE ENGINEERING

E. A. Aguilar-Reyes; C. A. León-Patiño; B. Jacinto-Díaz;
Universidad Michoacana de San Nicolás de Hidalgo, Morelia, Michoacán, México

L.-P. Lefebvre
National Research Council Canada, Industrial Materials Institute, Boucherville, Canada

ABSTRACT

Bioglass® 45S5 is the most widely investigated bioactive glass. It was recently demonstrated that this glass is better integrated into bone when the structure is porous. The National Research Council Canada has recently developed a process to produce metallic and ceramic foams using a powder metallurgy approach. Preliminary tests showed that Bioglass® foams can be produced with this process. These foams are produced by foaming a powder mixture composed of Bioglass® particles, a binder and a foaming agent. Highly porous Bioglass® foams (64-79% porosity) were successfully synthesized from dry powder blends. The porosity of the resulting foams was open and the pore size was essentially between 175 and 880 μm in diameter, as required for bone ingrowth. During sintering, crystallization took place and $Na_6Ca_3Si_6O_{18}$ (primary phase) and $Na_2Ca_4(PO_4)_2SiO_4$ (secondary phase) were observed. The mechanical compressive strength ranged from 1.7 to 5.5 MPa.

INTRODUCTION

Bioactive glasses have generated significant interest due to their excellent bioactivity and biocompatibility. The 45S5 glass (composition: $24.5Na_2O$-$24.5CaO$-$6P_2O_5$-$45SiO_2$ in wt. %) has been by far the most widely investigated bioactive glass [1]. This glass has been used clinically in the form of suspension or dense structures for dental applications and middle ear prosthesis [2].

Porous structures have been proposed in the late 60's to enable orthopedic implant fixation without cement [3,4]. By the mid-1970s, many studies provided evidences of biologic fixation through bone ingrowth into the porosity [5]. During the last decades, important research and development activities were conducted to develop materials with the optimum structure, composition and properties. Porous metals and metallic foams have been extensively used since then as coating or scaffold in load bearing applications.

Porous ceramics (HA, CaP) have been used for bone repair and reconstruction. These ceramics are used to promote the regeneration of damage tissue as template for cell interaction and new tissue ingrowth. The minimum pore size required to regenerate mineralized bone is generally considered to be ~50 μm. Structures with pores larger than 150 μm are usually preferred as they provide an ideal environment for bone reorganization and vascularization [6]. One limitation comes from their brittleness and low strength that limited their use in location not subjected to high initial loads.

Some researchers have recognized the interest for the development of porous Bioglass structures. Different groups have worked on the development of these materials [7-12]. One of the main challenges is to develop structures than can support some load [11, 13] since the work performed on this material (45S5 scaffolds) do not meet the minimum requirements of load.

The objectives of this work were to synthesize 45S5 Bioglass® foams using a novel powder metallurgy approach [14] and characterize the structure and mechanical properties of the foams. The process used has been developed for the production of open cell metallic and ceramic foams having different composition, porosity level, structure and properties. The process consists in dry-mixing metallic or ceramic powders with a solid polymeric binder and a foaming agent. The mixture is then

1

molded into the desired shape and heat treated in a three-step thermal treatment. During the first step (foaming), the binder is melted to create a suspension charged with the inorganic particles. During this step, the foaming agent decomposes and releases a gas that expands the structure and creates the interconnected porosity. The second step (debinding) is done to decompose the polymeric binder at moderate temperature. The third step (sintering) is done to provide the mechanical strength to the material. The structure of the foams can be modified by changing the formulation or varying the foaming and sintering conditions.

MATERIALS AND EXPERIMENTAL PROCEDURE

The 45S5 Bioglass® was prepared from a mixture of SiO_2 (99.6%, Aldrich), CaO (98%, Aldrich), Na_2CO_3 (100%, J.T. Baker), and P_2O_5 (≥ 98%, Aldrich) powders in the stoichiometric amounts to obtain the final composition ($24.5Na_2O-24.5CaO-6P_2O_5-45SiO_2$ wt. %) by the conventional melt-quenching method using a fused silica crucible. The glass was then grounded and sieved to obtain particle size less than 63 μm. The powder was observed with a JEOL JSM-6100 scanning electron microscope (SEM) and the particle size distribution was analyzed by laser diffraction using a Beckman Coulter LS 13 320 analyzer.

Bioglass® foams were produced with the Bioglass powder using the process described in the introduction. The foams were machined into small cylinders ($D = 15.2$ mm; $L = 8$ to 13 mm) after foaming. The cylinders were then debinded by thermal decomposition and sintered at different temperatures (950, 975, 1000 and 1025°C) in air to consolidate the material.

The density ρ_{foam} of the foams was measured using the mass and dimensions of the sintered cylinders. The porosity p was calculated with the formula:

$$p = 1 - \frac{\rho_{foam}}{\rho_{solid}} \tag{1}$$

where $\rho_{solid} = 2.7$ g/cm^3 is the theoretical density of 45S5 Bioglass® [15]. The measurement was done on 10 foam cylinders from each experimental condition. The structure of the foams was observed with a JEOL JSM-6100 scanning electron microscope and a Metris X-Tek HMX ST 225 CT System (Nikon). Pore size distribution was evaluated by image analysis on 2-D sections extracted from the μCT reconstructions using a Clemex image analyser. The glass powder and sintered foams were characterized by X-ray diffraction (XRD) to determine their crystalline nature. The foams were first ground into powder and then analyzed using a Bruker AXS D8 Discover diffractometer. The acquisition was performed on the 20-90° 2θ range using 0.01° step (2 s per step). The foams were machined to obtain cylindrical samples ($D = 9$ to 12 mm; $L = 3$ to 8 mm) used for the mechanical test. The strength of the foams, evaluated on 5 specimens per series, was characterized using an MTS 5 kN testing machine. The cross-head speed was 2.5 mm/min. The compressive strength was defined as the maximum on the stress-strain curves.

RESULTS AND DISCUSSION

The resulted powder is irregular and has a particle size smaller than 63 μm (average = 25 μm; $d_{10} = 1.8$ μm; $d_{50} = 19.2$ μm and $d_{90} = 57.4$ μm). The distribution was multimodal, probably associated with the batch grinding process used.

Figure 1 shows a typical foam cylinder produced in this study. The material has sufficient mechanical strength to be easily handled. The structure is uniform across the section of all foams (Figure 1 and 2). The foam has a 3-D interconnected porous network. SEM micrographs of the foams sintered at different temperatures (Figure 3) show that the general structure of the foam is slightly affected by the sintering temperature. Images taken at higher magnification (Figure 4) reveal, however, that the structure of the wall of the pores is significantly denser when the sintering temperature increases. Indeed, increasing sintering temperature significantly affects solid state diffusion and influence the bonding between the particles. At low sintering temperature, the particle can still be observed in the wall of the pores while they are not visible after treatment at higher temperatures (Figure 4b). Consequently, sintering temperature has an impact on the densification and porosity of foams. Indeed, porosity decreases from 79% down to 64% when sintering temperature increases from 950 to 1025°C; a similar behavior was observed by Bretcanu et al. [16], they showed a variation of shrinkage as function of sintering temperature with a maximum densification of the specimens (cubic samples) after 950°C with an approximate value of shrinkage of 36%. While not clearly visible on the SEM micrographs, densification during sintering also affects the pore size distribution. Broad pore size distribution of materials sintered at various temperatures has been observed. For instance, in the material sintered at 1025°C, the pores are essentially (i.e. 61%) comprised between 175 μm (D_{10}) and 520 μm (D_{90}), corresponding to pore size range optimal for bone ingrowth. The average pore size decreases when the sintering temperature increases.

Figure 1 45S5 Bioglass® foam sintered at 1000°C for 1h.

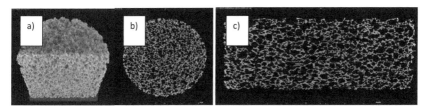

Figure 2 a) 3D and b) 2D perpendicular and c) 2D longitudinal sections extracted from μCT reconstructed images collected on a foam sintered at 1000°C.

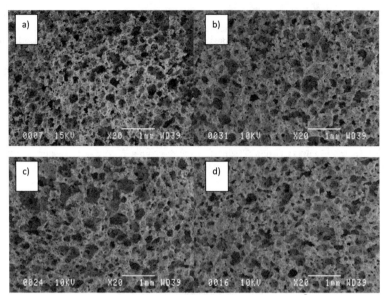

Figure 3 SEM micrographs showing the structure of 45S5 Bioglass® foams sintered at a) 950 b) 975 c) 1000, and d) 1025°C for 1h.

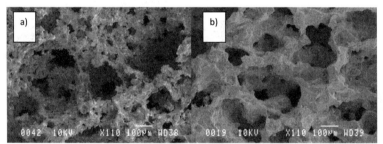

Figure 4 Higher-magnification SEM micrographs showing the structure of 45S5 Bioglass® foams sintered at a) 950, and b)1025°C in air for 1h.

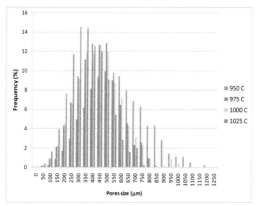

Figure 5 Pore size distribution of foams sintered at various temperatures.

Figure 6 shows the X-ray diffractograms of the Bioglass® powder and foams sintered at different temperatures. The spectrum of 45S5 Bioglass® powder shows that the initial powder is amorphous. The spectra of the sintered foams shows that the material has crystallized during sintering. The peaks observed corresponds to $Na_6Ca_3Si_6O_{18}$ and $Na_2Ca_4(PO_4)_2SiO_4$ crystalline phases. These crystalline phases have been observed in previous studies on sintered bioactive glasses [16-19]. Xin et al. [20] sintered porous body of 45S5 bioglass at 1000°C for 2 h and obtained $Na_2Ca_2Si_3O_9$ as a main phase, and $Ca_3(Si_3O_9)$ as a second phase; SBF immersion results showed that any of these phases inhibited the formation of Hydroxyapatite (HA) crystals when the sintered material was tested in vitro. A similar behavior was obtained by Chen et al. [11] who obtained $Na_2Ca_2Si_3O_9$ as a main phase after in vitro tests and showed biocompatibility by the formation of HA on the surface of foams. The diffractograms of the glass sintered at various temperatures show essentially the same phases.

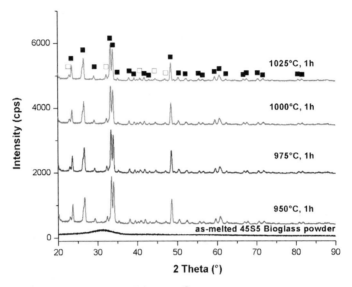

Figure 6 XRD spectra of as-melted 45S5 Bioglass® powder and foams sintered at 950, 975, 1000, and 1025°C for 1h. Main phase $Na_6Ca_3Si_6O_{18}$ and second phase $Na_2Ca_4(PO_4)_2SiO_4$ are marked by (■) and (□), respectively.

Compression tests were carried out on foams sintered at various temperatures. An example of the stress-strain curve obtained is presented in Fig 7. The curves obtained are representative of those generally obtained with ceramic and glass foams. They are highly corrugated due to their brittle nature. The compression strength of the foam, defined as the maximum on the compression curves is affected by the sintering temperature and increases from 1.7 to 5.5 MPa when sintering temperature increases from 950 to 1025°C (Figure 8). The increase of the properties with sintering can be associated with the density of the materials. The increase is also associated with the consolidation of the struts of the foams. As shown in Figure 4, the sintering of the particle is much better as the sintering temperature increases. This provides less porosity in the struts between the particles and structure with higher mechanical strength.

The values obtained in this study are significantly higher than those reported by Chen et al. [11] on highly porous (~90%) Bioglass® foams (i.e. 0.3 – 0.4 MPa). This may come from the structure and porosity of the materials investigated that was significantly different. On the other hand, the values compare well with those obtained by Shih-Ching et al. [12] (7.2 - 5.4MPa) on foams with lower porosity (43-47%).

Table 1 Procedures used to manufacture randomly porous scaffolds for tissue engineering

Material	Process	Sintering	Compressive Strength (MPa)	Pores sizes (μm)	Porosity (%)	Ref
45S5 BG	Replica	1000°C, 1h	0.27-0.42	510-720	89-92	11
45S5 BG	Mix with an aqueous solution, compressed, then calcined	1025°C, 1h	5.4-7.2	420 in length 100 in breadth	45.9-47.2	12
HA	Replica	1300°C, 3h	0.21-0.41	420-700	82-86	21
HA	Gel casting and replica combo	1350°C, 2h	0.55-5.0	200-400	70-77	22

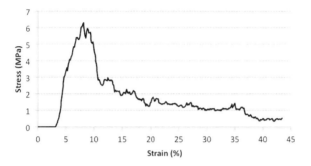

Figure 7 Typical compressive stress-strain curve of the 45S5 Bioglass® foam sintered at 1025°C for 1h.

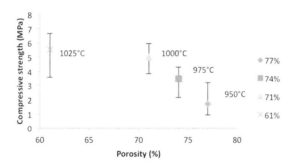

Figure 8 Mechanical behavior-porosity for 45S5 Bioglass® foams sintered at various temperatures. The bars in the figure correspond to standard deviations of a total of ten specimens.

CONCLUSIONS

Highly porous Bioglass® foams (64 to 79 % porosity) were successfully synthesized from dry powder blends. The porosity of the resulting foams is open and the pore size is appropriate for bone ingrowth (100 – 1000 μm). After sintering, the particles are well bonded together. During sintering, crystallization takes place and $Na_6Ca_3Si_6O_{18}$ (primary phase) and $Na_2Ca_4(PO_4)_2SiO_4$ (secondary phase) are observed, as reported by other researchers [17-19]. Foams produced present good mechanical strength (1.7 - 5.5 MPa) comparable with the cancellous bone (2-12 MPa). Future studies should focus on the evaluation of the biocompatibility of the foams.

ACKNOWLEDGMENTS

The authors would like to acknowledge Mario Laplume (NRC) and Maxime Gauthier (NRC) for their help with the fabrication of the specimens, Éric Baril (NRC) for the X-ray diffraction analysis, Fabrice Bernier (NRC) for the X-ray tomography acquisition and image analysis and Manon Plourde (NRC) for the compression test. The authors would also like to thank to CONACyT Mexico for financial support.

REFERENCES

1. Hench L.L., Splinter R.J., Allen W.C., Greenlee T.K., Bonding mechanisms at the interface of ceramic prosthetic materials, *J. Biomed. Mater. Res.*, 5 (6), pp. 117-141, 1971.
2. Wang C., Kasuga T., Nogami M., Macroporous calcium phosphate glass-ceramic prepared by two-step pressing technique and using sucrose as a pore former, *J. Mater. Sci.: Mater. Med.*, 16 (8), pp. 739-744, 2005.
3. Cameron HU, Pilliar RM, McNab I. The effect of movement on the bonding of porous metal to bone. J Biomed Mater Res., 1973;7(4):301–11.
4. Bobyn JD, Pilliar RM, Cameron HU. The optimum pore size for the fixation of porous surfaced metal implants by the ingrowth of bone. *Clin Orthop Relat Res.* 1980;150:263–70.
5. M.Spector, Historical Review of Porous-Coated Implants, J.Arthroplasty, Volume 2, Issue 2, 1987, Pages 163-177.
6. Hulbert SF, Young FA, Mathews RS, Klawitter JJ, Talbert CD, Stelling FH. Potential of ceramics materials as permanently implantable skeletal prostheses. *J Biomed Mater Res* 1970; 4 (3): 433-56.
7. Oana Bretcanu, Claire Samaille, Aldo R. Boccaccini, Simple methods to fabricate Bioglass® - derived glass- ceramic scaffolds exhibiting porosity gradient, *J Mater Sci*, 2008, 43:4127-4134
8. Junmin Qian, Yahong Kang, Zilin Wei, Wei Zhang, Fabrication and characterization of biomorphic 45S5 bioglass scaffold from sugarcane, *Materials Science and Engineering C*, 2009, 29:1361-1364
9. Sanjukta Deb, Ramin Mandegaran, Lucy Di Silvio, A porous scaffold for bone tissue engineering/45S5 Bioglass® derived porous scaffolds for co-culturing osteoblasts and endothelial cells, *J Mater Sci*: Mater Med, 21:893-905
10. V. Cannillo, F. Chiellini, P. Fabbri, A. Sola, Production of Bioglass® 45S5 – Polycaprolactone comnposite scaffolds via salt-leaching, *Composites Strusctures* 92, 2010, 1823-1832
11. Chen Q. Z., Thompson I.D., Boccaccini A.R., 45S5 Bioglass®-derived glass-ceramic scaffolds for bone tissue engineering, *Biomaterials*, 27, pp. 2414-2425, 2006.
12. Shih-Ching W., Hsueh-Chuan H., Sheng-Hung H., Wen-Fu H. Preparation of porous 45S5 Bioglass®- derived glass-ceramic scaffolds by using rice husk as a porogen additive, *J. Mater Sci: Mater Med*, 20, pp. 1229-1236, 2009.
13. Chen Q. Z., Boccaccini A. R., Poly (D,L-lactic acid) coated 45S5 Bioglass® -based scaffold: Processing and characterization, *Journal of Biomedical Materials Research Part A* 2006, 77:445-457
14. Lefebvre L.P., Thomas Y., Method of making open cell material, US patent No. 6660224, 2003.
15. Hench L.L., Wilson J., Surface-active biomaterials, *Science*, 226, pp. 630–6, 1984.

16. Bretcanu O., Chatzistavrou X., Paraskevopoulos K., Conradt R., Thompson I. and Boccaccini A.R., Sintering and crystallisation of 45S5 Bioglass® powder, *J. Eur. Ceram. Soc.*, 29, pp. 3299–3306, 2009.
17. Lefebvre L., Chevalier J., Gremillard L., Zenati R., Thollet G., Bernache-Assolant D., Govin A., Structural transformations of bioactive glass 45S5 with thermal treatments, *Acta Mater.*, 55, pp. 3305–3313, 2007.
18. Clupper D.C., Mecholsky Jr. J.J., LaTorre G.P., Greespan D.C., Sintering temperature effects on the in vitro bioactive response of tape cast and sintered bioactive glass-ceramic in Tris buffer, *J. Biomed. Mater. Res.*, 57, pp. 532-40, 2001.
19. R. A. Martin, H. Twyman, D. Qiu, J. C. Knowles, R. J. Newport, A study of the formation of amorphous calcium phosphate and hydroxyapatite on melt quenched Bioglass_ using surface sensitive shallow angle X-ray diffraction, *J Mater Sci: Mater Med* (2009) 20:883–888
20. Renlong Xin, Qiyi Zhang, Jiacheng Gao, Identification of the wollastonite phase in sintered 45S5 bioglass and its effect on in vitro bioactivity, *Journal of Ion-Crystalline Solids* 356 (2010) 1180–1184
21. Hae-Won Kim, Jonathan C. Knowles, Hyoun-Ee Kim, Hydroxyapatite porous scaffold engineered with biological polymer hybrid coating for antibiotic vancomycin release. *Journal of materials science: materials in medicine* 16 (2005) 189-195.
22. Hassna Rehman Ramay, Miqin Zhang, Preparation of porous hydroxyapatite scaffolds by combination of the gel-casting and polymer sponge methods, *Biomaterials* 24 (2003) 3293–3302

NEXT-GENERATION ROTARY ENDODONTIC INSTRUMENTS FABRICATED FROM SPECIAL NICKEL-TITANIUM ALLOY

William A. Brantley, Jie Liu, Fengyuan Zheng, Scott R. Schricker and John M. Nusstein
College of Dentistry, Ohio State University
Columbus, OH, USA

William A.T. Clark and Libor Kovarik
Department of Materials Science and Engineering, Ohio State University
Columbus, OH, USA

Masahiro Iijima
School of Dentistry, Health Sciences University of Hokkaido
Ishikari-Tobetsu, Japan

Satish B. Alapati
College of Dentistry, University of Illinois at Chicago
Chicago, IL, USA

ABSTRACT
    Next-generation rotary endodontic instruments with improved clinical performance are being fabricated from special pseudoelastic NiTi wire (termed M-Wire) that has a nanoscale martensitic structure. In this article we report etched microstructures, Vickers hardness measurements, and SEM observations of as-manufactured and clinically-used GT® Series X™ rotary instruments, with comparisons to rotary instruments fabricated from conventional dental pseudoelastic NiTi alloy. The next-generation NiTi instruments have significantly higher Vickers hardness and much less evidence of clinical wear compared to conventional NiTi instruments. Etched microstructures show a martensitic structure, which is consistent with our previously reported STEM observations. Differential scanning calorimetric analyses have shown that the manufactured instruments have similar phase transformation behavior to starting M-Wire blanks. The properties and clinical performance of the next-generation NiTi rotary instruments arise from their special martensitic structure. More TEM studies are needed to elucidate details of this complex microstructure and effects of clinical use.

INTRODUCTION
    Since the first rotary nickel-titanium endodontic instrument (ProFile®, Tulsa Dental) was marketed in 1993, these instruments have been widely used in dental practice due to the special properties of the near-equiatomic NiTi alloy: pseudoelastic behavior and much lower modulus of elasticity compared with stainless steel.[1] [In the dental literature, the term *superelastic* is generally used instead of *pseudoelastic*.] Although NiTi rotary instruments have the ability to prepare curved root canals with facility, occasional instrument fracture does occur, and this may bring serious consequences.[2] The fracture of NiTi rotary instruments during clinical use may be due to cyclic loading or a single episode of sudden overload.[3-5]
    Considerable research has been focused on improving the cyclic fatigue resistance of NiTi rotary instruments using surface modifications, such as ion implantation,[6,7] physical vapor deposition[8] and cryogenic treatment[9]. These methods improved the fatigue resistance and cutting efficiency of

11

NiTi instruments by increasing surface hardness.[b-y] Therefore, the surface hardness of a rotary instrument may be considered as closely related to its fatigue resistance and cutting efficiency.

A study of clinically-used conventional NiTi rotary instruments has shown that the mean Vickers hardness number for the tip region ranged from 313 to 324,[10] and these values were similar to the previously reported[1] Vickers hardness of a shape memory NiTi orthodontic wire. The mean Vickers hardness of the used instruments was lowest at the tip region compared to the intermediate and shank regions.[10]

Wear resistance of NiTi rotary instruments can be qualitatively evaluated from scanning electron microscope (SEM) observations of the surfaces.[9,11] Manufacturer defects in as-received instruments or defects created during canal preparation may serve as crack initiation sites, and crack propagation induced by stress concentration at defects will significantly decrease fatigue resistance,[12] leading to separation. Therefore, it is important that instruments are resistant to damage and exhibit wear resistance during clinical use.

Recently, a rotary NiTi instrument (GT® Series X™), machined from special M-Wire[13] blanks processed by a proprietary thermomechanical technique (Sportswire LLC, Langley, OK), has been marketed (Dentsply Tulsa Dental, Tulsa, OK). Laboratory studies (Sportswire LLC) have shown that M-Wire has a higher ratio of tensile strength to upper pseudoelastic plateau stress, compared with pseudoelastic wire used to manufacture conventional rotary instruments. The manufacturer claims that the GT® Series X™ instruments have significantly increased resistance to cyclic fatigue compared to conventional rotary instruments, and this is supported by a recent laboratory study.[14] However, no data on the hardness and wear resistance of these new rotary instruments have been reported.

The purpose of this study was to compare effects of clinical use on hardness of GT® Series X™ rotary instruments and NiTi rotary instruments fabricated from conventional pseudoelastic wire and to obtain qualitative information about clinical wear resistance from SEM observations. Additional study of the etched microstructures and the NiTi phase transformation behavior using differential scanning calorimetry (DSC) provided complementary information about the M-Wire instruments.

EXPERIMENTAL PROCEDURES

New GT® Series X™ instruments were provided by Dentsply Tulsa Dental. GT® Series X™ instruments and ProFile® instruments that had experienced 7 – 8 clinical uses were obtained from the Dental Faculty Practice of the College of Dentistry, Ohio State University. All used instruments had no visible evidence of permanent torsional deformation. In addition to the instruments, segments from one batch of M-Wire (provided by Sportswire LLC) similar to that utilized to manufacture GT® Series X™ rotary instruments and one batch of 35°C Copper Ni-Ti® wire (Ormco, Orange, CA) employed for orthodontic treatment were also selected for study.

Five randomly selected samples in each group of instruments were cut perpendicularly to the long axis with a water-cooled diamond saw. Three adjacent segments were obtained from each sample instrument for study: the tip region, the adjacent intermediate segment, and the adjoining shank region. Each segment was 4 – 5 mm in length. The M-Wire blanks and 35°C Copper Ni-Ti® archwires were also cut into segments approximately 5 mm in length.

These segments were embedded in acrylic metallographic resin and subjected to a standard sequence of metallographic preparation. A polishing machine (Metaserv 2000, Buehler UK Ltd, Coventry, West Midlands, England) was used to grind and polish the surfaces of the mounted specimens with sandpaper (Carbimet Paper Discs, 240 and 600 grit, Buehler, Lake Bluff, IL), followed by alumina slurries (Alpha Micropolish, Buehler) having 6 μm, 1 μm and 0.5 μm particle sizes.

Vickers hardness measurements were performed at room temperature on randomly selected, representative, as-received and used instruments with a 300 g load and 15 s dwell time (Micromet 2100, Buehler). Five indentations were made at the center and at equidistant adjacent locations in each segment for each instrument. By observing the sample at lower magnification under the optical microscope associated with the microhardness tester, it was ensured that the distance between each indentation was greater than at least ten indentation widths, so that the effect of previous loading on the hardness of the adjacent loaded area would be minimized. Diagonal lengths were measured with this optical microscope and converted to Vickers hardness number (VHN).[12]

Three comparisons were made: (1) means of hardness measurements in the three regions of the new or used GT® Series X™ instruments; (2) means of hardness measurements in the same corresponding single region of the new and used GT® Series X™ instruments; and (3) means of hardness measurements in all three regions of the used GT® Series X™ instruments and used ProFile® instruments. Statistical comparisons employed one-way ANOVA, followed by the Tukey multiple range test, with $P < 0.05$ for statistical significance.

An optical microscope (PME, Olympus) provided excellent views of the etched microstructures of the GT® Series X™ instruments. Specimens were resin-mounted and polished with the series of sandpaper and alumina abrasives (Buehler), ending with a 0.5 μm slurry of the latter. Polished surfaces were immersed in a solution of 3 mL hydrofluoric acid, 5 mL nitric acid and 20 mL acetic acid, which had been found[15] to be an effective etchant for the NiTi microstructures.

RESULTS

Figure 1 shows mean values and standard deviations of VHN for as-received and used size 30/.04 taper GT® Series X™ instruments. Mean values for tip, intermediate and shank regions were 374, 380 and 392, respectively. No significant difference ($P < 0.05$) was found when the same segments were compared for as-received and used instruments. The Vickers hardness at the shank region was significantly higher than for the tip region and intermediate segment of as-received GT® Series X™ instruments ($P < 0.05$), which was attributed to the manufacturing process. Vickers hardness values for GT® Series X™ instruments were significantly higher than for segments from used ProFile® instruments, which had mean values less than 350.[15] The GT® Series X™ instruments also had significantly higher VHN than the approximate value of 350 found for M-Wire blanks.[15]

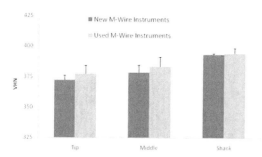

Figure 1. Vickers hardness for as-received and used GT® Series X™ size 30/.04 taper instruments.

Figure 2 is a representative SEM image showing manufacturing defects on the surface of an as-received GT® Series X™ instrument. Debris, metal rollover at the edges of the radial lands, and parallel grooves from the machining process were evident on all as-received instruments, and similar results have been observed for conventional NiTi rotary instruments.[16,17]

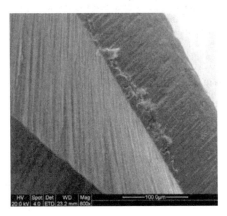

Figure 2. Secondary electron image of a representative as-received GT® Series X™ instrument.

Figure 3 is a representative SEM image of a used GT® Series X™ instrument. Compared with used ProFile® instruments in a previous study,[16] the used GT® Series X™ instruments have shallower areas of surface deformation and fewer embedded dentin chips.

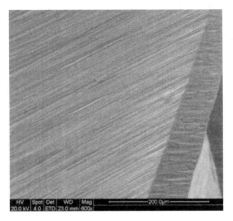

Figure 3. Secondary electron image of a representative used GT® Series X™ instrument.

Figure 4 presents an optical microscope image of the etched microstructure of a representative as-received GT® Series X™ instrument, which has the appearance of a classic martensite structure, very similar to that previously[13] observed for the starting M-Wire blanks. There were no substantial differences between the microstructures along the axes or cross-sections of the as-received or used instruments. Further careful examination is needed to verify the possibility that some differences exist in the microstructure for the tip region of the instruments.

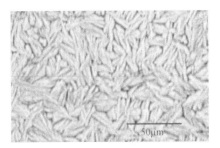

Figure 4. Etched microstructure for a representative GT® Series X™ instrument.

The martensitic-type structure of the starting M-Wire is complex, as shown below in the STEM (Tecnai TF-20) photomicrograph (Figure 5) obtained at room temperature, which has higher resolution than the STEM images in our previous[13] publication. This nanoscale architecture should provide effective strengthening mechanisms that account for the high hardness of these wires, which increases further when the GT® Series X™ instruments are fabricated by machining the blanks.

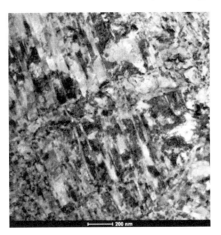

Figure 5. High-resolution STEM image of a cross-section region of M-Wire.

DISCUSSION

The objectives of this study were to compare effects of clinical use on the Vickers hardness of GT® Series X™ rotary instruments and NiTi rotary instruments fabricated from conventional pseudoelastic wire and to obtain qualitative information about clinical wear resistance of GT® Series X™ instruments from SEM observations. The surface hardness of NiTi rotary instruments has been previously shown to contribute significantly to their cutting ability and wear resistance during canal instrumentation.[6-9,17-19] In the present research,[15] it was found that GT® Series X™ instruments have significantly higher hardness than conventional ProFile® instruments.

No significant difference (P < 0.05) was found for the mean hardness of used GT® Series X™ instruments, compared with as-received instruments. The absence of an evident effect of clinical use is consistent with a previous study[10] of used conventional rotary NiTi instruments. The Vickers hardness of the M-Wire blanks employed to manufacture the GT® Series X™ instruments was significantly higher than that for the batch of pseudoelastic 35°C Copper Ni-Ti® orthodontic wire (approximately 300). However, the Vickers hardness of the M-Wire blanks was significantly lower than that for the as-received GT® Series X™ instruments (P < 0.05). Therefore, the M-Wire blanks work hardened substantially during the thermomechanical processing procedure, and further work hardening occurred when these wire blanks were machined to fabricate the GT® Series X™ instruments. Further STEM study of the ultrastructure of M-Wire instruments is planned.

There was no substantial difference in the austenite-finish (A$_f$) temperature of approximately 50°C for the GT® Series X™ instruments compared to that previously[13] reported for M-Wire blanks. Figure 6 shows a representative differential scanning calorimetry (DSC) plot for a specimen consisting of the three segments of a used size 20/.04 taper instrument. The two-step heating transformation from martensite to R-phase, followed by transformation from R-phase to austenite (completed at the A$_f$ temperature) was not as distinctly resolved as previously[13] found for M-Wire blanks, although there is some evidence of both transformations, which were clearly observed[15] for other instrument specimens. Figure 6 suggests that at body temperature (37°C) the instrument is a mixture of martensite and austenite (and perhaps R–phase).

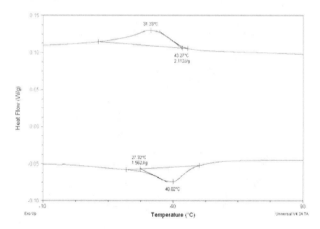

Figure 6. DSC heating (lower) and cooling (upper) curves for a used GT® Series X™ instrument.

Complementary analyses of specimens using the technique of Micro-X-ray diffraction[20] have shown that the GT® Series X™ instruments contain martensitic NiTi as well as perhaps both R-phase and austenitic NiTi.[15] The complexity of these analyses will be discussed in a separate publication.

While the present Vickers hardness measurements have provided much information about the GT® Series X™ instruments, use of a nanoindenter may yield more detailed information within the different regions of these instruments. The hardness at the instrument tip may be volume-sensitive due to the small diameter, which ranges from 200 to 400 μm. With the much smaller tip and lower loading force (< 1 g), a nanoindenter might yield a reduced effect of material volume on hardness, unless substantial submicron variations in microstructure exist for the GT® Series X™ instruments.

As would be predicted from the significantly higher Vickers hardness, superior clinical wear resistance was qualitatively observed for the used GT® Series X™ instruments, with fewer surface defects and less deformation of the radial lands, compared with used conventional ProFile® instruments[16]. The effect of surface defects on conventional NiTi rotary instruments has been discussed.[21-23] These instruments are reported to separate as a result of cyclic fatigue, in which crack initiation occurs on the surface, followed by transgranular crack growth. The excellent clinical wear resistance of the GT® Series X™ instruments observed in the present study is attributed to their increased surface hardness. Detailed information about wear resistance should be obtained in a future study by examining the same areas on each instrument with the SEM before and after clinical use.

CONCLUSIONS

The GT® Series X™ M-Wire rotary endodontic instruments have significantly higher Vickers hardness compared with rotary instruments made from conventional pseudoelastic NiTi wire. The increased hardness, which results from a nanoscale martensitic-type structure, may contribute to better wear resistance and cutting efficiency for these instruments. The shank region of new instruments has the highest hardness compared to the tip and middle regions. No effect of clinical use on the Vickers hardness of the M-Wire instruments was found. These instruments displayed qualitatively better wear resistance, along with less permanent deformation of the surface features after 7 – 8 times of clinical use, compared with the conventional ProFile® rotary instruments. Further STEM study of the complex ultrastructure is needed, along with electron diffraction analyses to provide unambiguous phase identification. Use of a nanoindenter is recommended to obtain more detailed information about local variations in hardness along the axis and over the cross-section of the as-received and used instruments.

ACKNOWLEDGMENT

We thank Dentsply Tulsa Dental for providing rotary instruments and Dr. William Ben Johnson for providing M-Wire segments.

REFERENCES

[1]W.A. Brantley, Orthodontic wires, In: W.A. Brantley and T. Eliades (editors), Orthodontic Materials: Scientific and Clinical Aspects, Stuttgart: Thieme, pp. 77-103 (2001).

[2]P. Parashos and H.H. Messer, Rotary NiTi Instrument Fracture and Its Consequences, J. Endod., **32**, 1031-43 (2006).

[3]P. Parashos, I. Gordon, and H.H. Messer, Factors Influencing Defects of Rotary Nickel-Titanium Endodontic Instruments after Clinical Use, J. Endod., **30**, 722-5 (2004).

[4]S.B. Alapati, W.A. Brantley, T.A. Svec, J.M. Powers, J.M. Nusstein, and G.S. Daehn, SEM Observations of Nickel-Titanium Rotary Endodontic Instruments that Fractured During Clinical Use, J. Endod., **31**, 40-3 (2005).

[5]A.P. Spanaki-Voreadi, N.P. Kerezoudis, and S. Zinelis, Failure Mechanism of ProTaper Ni-Ti Rotary Instruments During Clinical Use: Fractographic Analysis, Int. Endod. J, **39**, 171-8 (2006).

[6]D.-H. Lee, B. Park, A. Saxena, and T.P. Serene, Enhanced Surface Hardness by Boron Implantation in Nitinol Alloy, J. Endod., **22**, 543-6 (1996).

[7]E. Rapisarda, A. Bonaccorso, T.R. Tripi, G.G. Condorelli, and L. Torrisi, Wear of Nickel-Titanium Endodontic Instruments Evaluated by Scanning Electron Microscopy: Effect of Ion Implantation, J. Endod., **27**, 588-92 (2001).

[8]E. Schäfer, Effect of Physical Vapor Deposition on Cutting Efficiency of Nickel-Titanium Files, J. Endod., **28**, 800-2 (2002).

[9]T.S. Vinothkumar, R. Miglani, and L. Lakshminarayananan, Influence of Deep Dry Cryogenic Treatment on Cutting Efficiency and Wear Resistance of Nickel-Titanium Rotary Endodontic Instruments, J. Endod., **33**, 1355-8 (2007).

[10]S.B. Alapati, W.A. Brantley, J.M. Nusstein, G.S. Daehn, T.A. Svec, J.M. Powers, W.M. Johnston, and W. Guo, Vickers Hardness Investigation of Work-Hardening in Used NiTi Rotary Instruments J. Endod., **32**, 1191-3 (2006).

[11]C. Eggert, O. Peters, and F. Barbakow, Wear of Nickel-Titanium Lightspeed Instruments Evaluated by Scanning Electron Microscopy, J. Endod., **25**, 494-7 (1999).

[12]G.E. Dieter. Mechanical Metallurgy (3rd ed), New York: McGraw-Hill, pp. 329-32, 375-6, and 394-412 (1986).

[13]W.A. Brantley, J. Liu, W.A.T. Clark, L. Kovarik, C. Buie, M. Iijima, S.B. Alapati, and W.B. Johnson, Characterization of New Nickel-Titanium Wire for Rotary Endodontic Instruments. In: R.J. Narayan, P.N. Kumta, and W.R. Wagner (editors), Advances in Biomedical and Biomimetic Materials, Ceram. Trans., **206**, 49-57 (2009).

[14]C.M. Larsen, I. Watanabe, G.N. Glickman, and J. He, Cyclic Fatigue Analysis of a New Generation of Nickel Titanium Rotary Instruments. J. Endod., **35**, 401-3 (2009).

[15]J. Liu, Characterization of New Rotary Endodontic Instruments Fabricated from Special Thermomechanically Processed NiTi Wire [Ph.D. Thesis], Columbus: Ohio State University (2009).

[16]S.B. Alapati, W.A. Brantley, T.A. Svec, J.M. Powers, and J.C. Mitchell, Scanning Electron Microscope Observations of New and Used Nickel-Titanium Rotary Files, J. Endod., **29**, 667-9 (2003).

[17]G. Chianello, V.L. Specian, L.C.F. Hardt, D.P. Raldi, J.L. Lage-Marques, and S.M. Habitante, Surface Finishing of Unused Rotary Endodontic Instruments: A SEM Study, Braz. Dent. J., **19**, 109-13 (2008).

[18]U.-M. Li, M. Iijima, K. Endo, W.A. Brantley, S.B. Alapati, and C.-P. Lin. Application of Plasma Immersion Ion Implantation for Surface Modification of Nickel-Titanium Rotary Instruments, Dent. Mater. J., **26**, 467-73 (2007).

[19]E. Rapisarda, A. Bonaccorso, T.R. Tripi, I. Fragalk, and G.G. Condorelli, The Effect of Surface Treatments of Nickel-Titanium Files on Wear and Cutting Efficiency, Oral Surg. Oral Med. Oral Pathol. Oral Radiol. Endod., **89**, 363-8 (2000).

[20]M. Iijima, H. Ohno, I. Kawashima, K. Endo, W.A. Brantley, and I. Mizoguchi, Micro X-ray diffraction study of superelastic nickel–titanium orthodontic wires at different temperatures and stresses, Biomaterials **23**, 1769-74 (2002).

[21]L. Borgula, Rotary Nickel-Titanium Instrument Fracture: An Experimental and SEM Based Analysis, Melbourne: University of Melbourne, School of Dental Science (2005).

[22]G.S.P. Cheung, B. Peng, Z. Bian, Y. Shen, and B.W. Darvell, Defects in ProTaper S1 Instruments after Clinical Use: Fractographic Examination, Int. Endod. J., **38**, 802-9 (2005).

[23]G.S.P. Cheung, Z. Bian, Y. Shen, B. Peng, and B.W. Darvell, Comparison of Defects in ProTaper Hand-Operated and Engine-Driven Instruments after Clinical Use, Int. Endod. J., **40**,169-78 (2007).

PREPARATION OF NANOPHASE HYDROXYAPATITE VIA SELF PROPAGATING HIGH TEMPERATURE SYNTHESIS

Sophie C. Cox and Kajal K. Mallick

Warwick Manufacturing Group
School of Engineering
University of Warwick
Coventry CV4 7AL
United Kingdom

ABSTRACT

Nano hydroxyapatite (nHAP) is known to exhibit enhanced biological response when interacting with cells. Phase pure HAP nanoparticles were synthesized by self-propagating high temperature synthesis (SHS) using citric acid and urea as fuels and aqueous precipitation with calcium nitrate and ammonium hydrogen phosphate. The effects of the critical factors governing the synthesis protocols such as pH, reaction and calcination temperatures on the stoichiometry, crystallinity, crystal size, and shape of the nano HAP powders are reported. The results indicated that the crystal sizes varied with some agglomeration in the nanoregime of 10-80 nm, which is comparable to the size of natural nanocrystalline HAP found in human bones and teeth. The influence of grain size, sintering temperature and density of bodies using nHAP as precursor is discussed. Microstructure, crystal structure, morphology, surface area of nHAP powders and dense bodies were investigated using differential thermal analysis and thermal gravimetry analysis (DTA-TG), X-ray diffraction (XRD), scanning electron microscopy (SEM), transmission electron microscopy (TEM), and particle size analysis.

INTRODUCTION

Despite four decades of biomaterials research and development, synthetic bone graft substitutes (BGS) are still largely inferior to allo- and autografts due to their lack of osteoinductive and osteogenic properties[1]. BGS currently are yet to be optimised for load bearing applications within the body, however there are many promising non-structural products available for clinical use[2]. Current BGS materials lack the high functionality of bone tissue. It has only recently been recognised that in order to achieve the necessary biofunctionality, novel biomaterials need to be bioinspired in order to perform at a similar level to that of natural bone[3, 4].

Bone tissue is best described as a nanocomposite, which can be defined as a multiphase material in which the majority of the dispersed phase components have one or more dimensions of the order of 100 nm or less[5]. It is this natural nanostructure that nanotechnology aims to emulate for different tissue engineering applications. There is hope that the use of nanophase materials may address some of the shortcomings of current grafts, such as donor site morbidity[6, 7], limited harvesting sites and donors[8] and cost[9]. Natural bone consists primarily of a soft protein-based hydrogel template (collagen and non-collagenous proteins) reinforced by a hard inorganic mineral phase of which hydroxyapatite $(Ca_{10}(PO_4)_6(OH)_2$, HAP) is the chemical and crystallographic template. Approximately, 70 wt% of the bone matrix contains nanocrystalline HAP which are 20-80 nm long and 2-5 nm thick[10, 11].

Since synthetic HAP mimics the natural mineral component of mammalian bones and teeth it is at the forefront of explored bioceramics. HAP has been intensively investigated and widely used as a bone replacement material in the areas of: orthopaedics, dentistry and maxillofacial[12-14]. In addition to being non-toxic, its main advantage over other biomaterials is its excellent biocompatibility and also its bioactive behaviour, which allows it to integrate into living tissue by the same processes active in the remodelling of natural healthy bone[7, 15]. Significant research has been devoted to the development of the physical and chemical preparation of HAP as well as methods to control its morphology. Clinically, it is used in the form of powders, granules, coatings, dense and porous blocks, non-viral carriers for plasmid DNA gene delivery, and in various biocomposites[16-20].

Despite its many advantages, micron-sized HAP due to its typically high stability exhibits a relatively low bioresorbability, an undesirable characteristic for a BGS, and brittle characteristics[21, 22]. The nanometer structures and molecules found in bone tissue clearly indicate that bone-forming cells are accustomed to interacting with nanometer morphology surfaces[23], thus suggesting the importance of investigating the use of nanophase HAP (nHAP) for use as a BGS[24-26].

Nanocrystalline HAP powder exhibits a greater surface area[27], which can improve the fracture toughness of the HAP ceramic by reducing the temperature required during sintering[28]. Moreover, research has shown that nanosized ceramics can exhibit significant ductility before failure contributed by the grain-boundary phase. It has also been reported that brittle ceramics with nanograin dimensions can withstand a greater plastic strain, of up to 100%[29].

Recent investigations of nanophase materials have illustrated their potential for bone repair. More specifically, nanobiomaterials offer a unique approach to overcome the shortcomings of many conventional materials. It has been reported that nHAP is expected to have a greater bioactivity than coarser micron sized HAP[30]. Since osteoblasts discriminate nanoscale from conventional micron HAP there is a great possibility that osteoblasts could react sensitively toward different nanometric characteristics of HAP. For example, increased osteoblast adhesion on nano grained materials in comparison to conventional micron grained materials has been reported[31, 32]. Enhanced osteoblast proliferation in vitro and long-term functions have also been reported on ceramics with grain and fibre sizes less than 100 nm[32, 33]. It has been shown that osteoblasts can generate approximately 60% more new cells when they are exposed to HAP that contains nanometer-scale features, compared with HAP containing micrometre-size particles[34]. Additionally, modified osteoclast behaviour has been documented on nanophase ceramics[35] and increased bone formation, in comparison to conventional apatites, has been demonstrated on metals coated with nHAP in vivo[36]. Improved biological performance when nHAP is used as a precursor, for instance nano-hydroxyapatite/collagen (nHAC) composites has been reported and developed for bone tissue engineering[37]. Experimental results show that biomimetic nHAC tissue scaffolds can be fabricated and exhibit enhanced bioactivity, high osteoconductive activity and controlled resorption rates[38].

It is possible to improve the mechanical properties of HAP ceramics by controlling important parameters of powder precursors such as particle size and shape, particle distribution and agglomeration[14]. Studies have shown that the addition or use of nHAP in biocomposites has been found to improve their mechanical strength[39]. It is clear that nanophase ceramics represent a unique and promising class of orthopaedic and dental implant formulations with improved biological, particularly osteointegrative and biomechanical properties[35, 40]. These improved biomechanical properties suggest that there may be a potential use for nanophase bioceramics to be used as a component of a load bearing biocomposite for use as a structural BGS.

However, despite their promising potential, difficulties with controlling the particle size distribution, agglomeration and surface area have arisen during the synthesis of HAP particles[5]. This has created a need to search for and study advanced methods so that the synthesis of HAP nanocrystals can be accurately be controlled. In recent years, a number of fabrication methods for the controllable synthesis of nano-materials have been reported including precipitation[41, 42], sol-gel processing[43-48], solid state reactions[49, 50], microemulsion techniques[15, 51-53], hydrothermal reaction[54, 55], mechanochemical synthesis[56, 57], microwave-hydrothermal process[58] and vibro-milling from natural bone. The size of particles obtained from these methods varies from tens to hundreds of nanometres. Moreover, the control of the shape of resultant particles is a problem, frequently pin-like or irregular particles are produced.

Recently, there has been a growing interest in producing ceramics *via* self-propagating high temperature synthesis (SHS), or otherwise known as self-propagating combustion synthesis (SPCS) due to its potential to produce sub-micron sized calcium phosphates. An average particle size of 0.45μm was reported when urea was used as a fuel in SHS to produce HAP[59]. Furthermore, synthesis of HAP and other calcium phosphate crystallites in the range of 50-65 nm were reported using single and mixed fuels, namely citric and succinic acid[60]. The motivation of the presented work is to synthesise and compare nHAP powders produced using: an aqueous precipitation method, urea based SHS and citric acid based SHS. Each of the powders will be characterised with XRD, particle size analysis using the Scherrer equation, SEM, TEM, and DTA-TG.

EXPERIMENTAL

Aqueous precipitation
    Stock solutions of calcium nitrate ($Ca(NO_3)_24H_2O$ > 99 %, Sigma-Aldrich, UK) and ammonium phosphate (($NH_4)_2HPO_4$ >98%, Sigma-Aldrich, UK) were formed separately by dissolving 5.90 g and 1.98 g into 300 ml and 125 ml of deionised water respectively. The two solutions were then combined under stirring conditions (250 rpm). The pH value of the solution was adjusted to 10 by adding ammonium hydroxide solution (NH_4OH, 28-30 %, Sigma-Aldrich, UK) dropwise at room temperature and under stirring conditions (250 rpm). An extra 5 mmol of calcium nitrate was then added. The mixture was stirred vigorously (300 rpm) and after two hours the pH was readjusted to 10 using ammonium hydroxide solution. This final mixture was stirred moderately for 12 hours.
    The formed precipitate was filtered three times using a 1 L Buckner flask to remove any alkalinity and a small amount of the sample was retained for thermal analysis. The remaining filtered precipitate was oven dried for 24 hours in a furnace preheated to 80 °C. After drying the precipitate was finely ground using a pestle and mortar. Sintering was then performed at: 600, 800, 900 and 1100 °C, for 1 hour. Samples using this method shall be referred to as AP followed by their batch/sample number, for example AP B1.1 refers to sample 1 (sintered at 600 °C) from Batch 1 synthesised *via* aqueous precipitation (Figure 1).

SHS using urea as a fuel
    5.82 g of calcium nitrate and 1.9 g of ammonium phosphate were added to 100 ml of deionised water, resulting in the formation of an opaque white solution. This solution was made clear by adding approximately 0.5ml of concentrated nitric acid (70%, Sigma Aldrich, UK) while stirring. 12 g of powdered urea [$(NH_2)_2CO$, ≥ 98% purity, Sigma, UK] was added and the final solution stirred for 5 minutes at room temperature.
    The solution was poured into an alumina crucible and placed in a furnace preheated to 500 °C were it combusted after approximately 15 minutes. The products of combustion were lightly ground using a pestle and mortar to form fine powder and then calcined for 1 hour at various temperatures: 600, 800, 900 and 1100 °C, as shown in Figure 2. Samples synthesised using this method will be referred to as USHS followed by their batch/sample number, e.g. USHS B1.3 denotes batch one sample two (sintered at 900 °C).

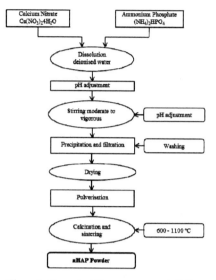

Figure 1: Flow chart for aqueous precipitation

SHS using citric acid as a fuel

  Calcium nitrate (4.8 g) and citric acid (4 g HOC(COOH)(CH₂COOH)₂, Sigma, UK) were added to 100 ml of deionised water and stirred for 10 minutes. The pH of the solution was adjusted to 10 using ammonium hydroxide solution. 3 g of ammonium phosphate was dissolved separately in 25 ml of deionised water and added dropwise to this solution while   stirring at 250 rpm, thus resulting in the formation of a white precipitate. The precipitate was dissolved using concentrated nitric acid (70%, Sigma Aldrich, UK) to a pH value of 1 resulting in a clear solution. The uncovered glass beaker containing the final solution was placed on a magnetic hot plate with continuous stirring and a temperature of 70 °C was maintained for 2 hours.

  The homogeneous solution mixture was transferred to an alumina crucible and placed into a preheated furnace at 450 °C. Combustion occurred after approximately 10 minutes with a continuous evolution of gases. The brown precursor obtained was calcined at 900 °C for 2 hours to remove the combustion products. Samples were subsequently sintered at 900 and 1100 °C for 1 hour (Figure 3). Samples synthesised *via* this method is referred to hereafter as CSHS followed by their batch/sample number, e.g. CSHS B1.4 refers to sample four (sintered at 1100 °C) of Batch 1 using citric acid SHS.

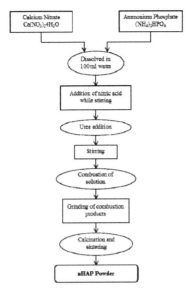

Figure 2: Flowchart for SHS using urea as a fuel

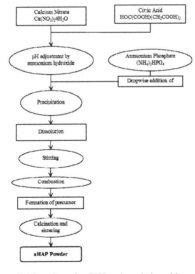

Figure 3: Flowchart for SHS using citric acid as a fuel

Powder characterisation

The crystalline phase and crystallite size of the nHAP was determined using powder X-ray diffraction (XRD) on a Bruker D5000 diffractometer in Bragg-Brentano geometry with a monochromatic CuK_α radiation (λ = 1.54056 Å), using aluminium sample holders. Patterns were matched to reference data on the JCPDS database. The crystallite size (L) was estimated for each run using the Scherrer equation[61]:

$$B(2\theta) = \frac{K\lambda}{L\cos\theta}$$

where K = Scherrer constant ($\approx$ 0.9), λ = X-ray wavelength, B = peak width and θ =Bragg angle.

The thermal behaviour of the precipitate after the filtering stage of the AP method prior to drying was determined by simultaneous differential thermal analysis and thermogravimetry (DTA-TGA) (STA1500 TA Instruments, West Sussex, UK). 20 mg of sample was heated from 20 – 1200 °C at a constant ramp rate of 10 °C min^{-1} in flowing air.

Scanning electron microscopy (SEM) and transmission electron microscopy (TEM) of the samples was carried out using a Zeiss Supra55 FEGSEM and a Jeol 2000FX, respectively to study and compare the particulate morphology for each method.

RESULTS AND DISCUSSION

Comparison of experimental methodologies

The solution produced via the AP was controlled via pH, adjusted twice for each batch produced. This ensured that the solution was homogeneously mixed at a molecular level resulting in a single phase HAP precipitate. This precipitate remained single phase up to a temperature of approximately 800 °C, but at higher sintering temperatures was transformed into a powdered mixture of HAP and beta-tricalcium phosphate (β-TCP).

The success of the SHS process is dependent on the intimate mixing of constituents using a suitable fuel, in this case urea or citric acid, in an aqueous medium and a vigorous exothermic redox reaction between the fuel and an oxidizer, namely nitric acid. Nitrate solutions typically decompose at temperatures below 700 °C with the evolution of gases, such as NO_2, NO, N_2O_5[62].

The gaseous products of combustion for USHS are expected to consist of nitrous oxides as urea is known to decompose into biuret ($C_2H_5N_3O_2$, which decomposes itself >300 °C), cyanuric acid (HCNO) and ammonia (NH_3) when heated to 200 °C. When the ambient temperature reaches 500 °C the gaseous mixture spontaneously ignites, increasing the local temperature of the dried foam, formed via dehydration, to approximately 1300 °C[62]. This process was completed between 15 and 20 minutes.

The success of CSHS depends on the formation of a gel precursor before ignition, reported to occur at 185 °C[60], in conjunction with the evolution of large amounts of nitrogen dioxide gas, and the formation of a black or dark brown coloured precursor. The CSHS experimental method used here formed a solution before ignition. The dispersion of Ca and P ions within the solution, compared to a gel, is unpredictable and thus leads to the formation of a multiphase powder after calcination is completed.

Phase formation via combustion cannot be controlled because of the spontaneous ignition of the fuel and oxidizer. Thus, the volatile molecular dispersion of Ca and P ions within the solution leads to the formation of multiphase calcium phosphates when SHS is used as a synthesis method. The high temperature of the reaction front makes it difficult to ensure all the fuel is oxidized. If any of the constituents is left unreacted undesirable carbonaceous component may remain resulting in the formation of $CaCO_3$which was found to be present in USHS B3 following combustion.

Crystal analysis (XRD)

The XRD patterns for AP 2 (Figure 4) confirmed presence of single-phase HAP (JCPDS: 09-432) at lower sintering temperatures and the peak broadening indicates a low crystallinity and small crystallite size, see Figure 4 (a). A mixture of HAP and β-TCP were found in samples sintered at higher temperatures as shown in Figures 4 (b) and (c). The onset of this phase transformation begins at approximately 800 °C (Figure 7).

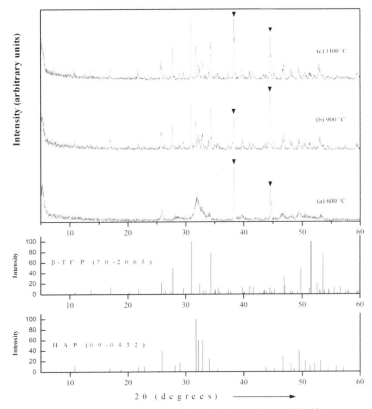

Figure 4: Phase formation for AP B2 (▼refers to Al sample holder)

A combination of different calcium phosphates (HAP, α-TCP and β-TCP) were confirmed for USHS B3 (Figure 5). This batch used 24 g of powder that resulted in the formation of $CaCO_3$ after combustion, which originated from the remaining unreacted urea. Batch 1 used only half the amount of fuel indicating that all reacted during combustion hence no calcite was found in sample USHS B1.1.

XRD patterns for the CSHS samples are shown in Figure 6. Only two samples were analysed as the products from the combustion reaction did not burn off until the samples were sintered above 900

°C. Both CSHS B2.3 and CSHS B2.4 were matched to calcium pyrophosphate, $Ca_2O_7P_2$ and no other phase transformation was observed between 900 and 1100 °C (Figure 6).

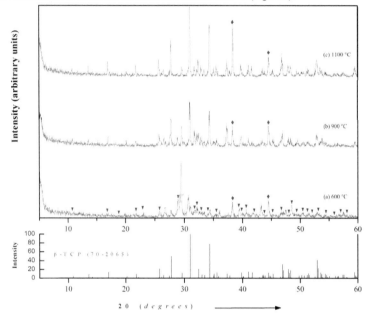

Figure 5: Phase formation for USHS B3 (▼refers to HAP and ♦ to Al sample holders)

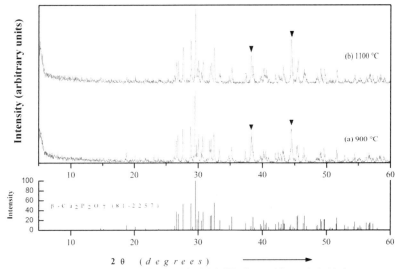

Figure 6: Phase formation for CSHS B2 (▼ refers to Al sample holder)

Thermal analysis

The DTA-TG curves for the samples synthesised by AP from 20 to 1200°C shows two successive steps of weight loss. The endotherm below 170 °C is attributed to the loss of hydroxyl ions in water from the sample and the exothermic slope with an onset at 800°C indicating the small weight loss due to the dehydroxylation of HAP to β-TCP during phase transformation.

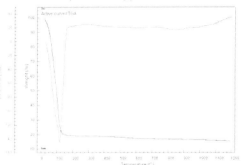

Figure 7: DTA-TGA trace of precipitate formed by AP method

Particle size analysis

Table 1 presents a summary of the phases present in each sample, chracterised *via* XRD and an estimate of crystallite sizes determined by Scherrer equation[61] using data shown in Figures 4 – 6, respectively.

Synthesis Method	Sample number	Phase(s) present	Estimated average crystallite size (nm)
AP	B2.1	HAP	13
	B2.3	HAP, Whitlocklite, Calcium hydrogen phosphate hydroxide	18 (HAP)
	B2.4	HAP, β-TCP	18 (HAP), 18 (β-TCP)
USHS	B3.1	HAP, α-TCP, β-TCP, $CaCO_3$, $Ca_2O_7P_2$	18 (HAP), 19 (β-TCP)
	B3.3	β-TCP, $Ca_2O_7P_2$	18 (β-TCP), 18 ($Ca_2O_7P_2$)
	B3.4	β-TCP	130
CSHS	B2.3	$Ca_2O_7P_2$	20
	B2.4	$Ca_2O_7P_2$	49

Table 1 Phase identification and estimation of crystallite size

The range of average particle size, studied *via* SEM varied between $0.12 - 0.6$ µm ($\pm$ 0.04 µm), $0.2 - 20$ µm ($\pm$ 0.08 µm) and $0.24 - 5.6$ µm ($\pm$ 0.15 µm) for AP, USHS and CSHS respectively. The typical morphology of the powder particles synthesised *via* AP and CSHS were consistent throughout for each sample analysed. However, two distinctive morphologies were observed for USHS samples (Figure 8). Both small and near globular (200 – 400 nm) and large agglomerated rectangular particles (12 – 20 µm) were found to be present within all USHS samples (Figure 9), with an increasing proportion of rectangular particles present with increasing sintering temperature. Both SHS methods produced more rectangular shaped particles in comparison to the globular synthesised *via* AP method.

Figure 8: SEM micrograph showing typical morphology of the particles as synthesised *via* (a) AP B1.4, (b) USHS B1.4 and (c) CSHS B1.4

Figure 9: SEM micrograph showing particle morphologies present in USHS (a) Small globular (USHS B3.1) and (b) Large agglomerated rectangular (USHS B3.4)

TEM was performed on samples AP B2.1 and AP B2.4. The typical morphology of the as synthesised particles can be seen in Figure 10. The range of particle sizes were between 11.6 and 44.4 nm (average

26.24 nm) and 30.7 and 96.2 nm (average 58.7 nm) for samples AP B2.1 and AP B2.4, respectively. Greater agglomeration was observed for sample AP B2.4 due to the exposure to higher sintering temperature where sizes of agglomerated particles varied between 100.3 – 1780 nm (average 594 nm).

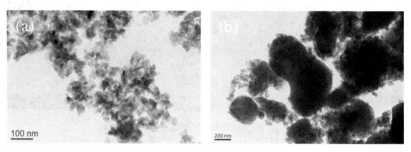

Figure 10: TEM comparison of particle sizes (a) AP B2.1 and (b) AP B2.4

CONCLUSIONS

The experiments conducted have shown that it is possible to synthesis phase pure nHAP of 10 – 80 nm crystallites *via* aqueous precipitation at ambient temperature. In comparison, multiphase calcium phosphates were formed when urea and citric acid SHS was used as the synthesis method due to the uncontrolled reaction front developed during the ignition process. The conversion of nHAP to other types of calcium phosphate occurred above 800 °C. This is attributed to the difficulty in maintaining the homogeneous conversion of thermal dynamically unstable ionic species during the combustion process. Increasing the sintering temperature was found to increase agglomeration of particles in all methods investigated, thus increasing the average particle sizes to 0.1 – 20 μm.

ACKNOWLEDGEMENTS

This study was supported by Warwick Postgraduate Research Scholarship (WPRS) and the Institute of Materials, Minerals and Mining (IOM3). The authors are grateful to Martin Davis, Stephen York and Richard Walton of Departments of Engineering, Physics and Chemistry respectively, University of Warwick for their assistance with characterisation of the samples.

REFERENCES

[1]F. Chai, G. Raoul et al., Biomaterials as bone substitutes: classification and contribution, *Revue de stomatology et de chirurgie maxilla-faciale* (2011)

[2]K. A. Hing, Bone repair in the twenty-first century: biology, chemistry or engineering?, *Philosophical Transactions of the Royal Society of London Series a-Mathematical Physical and Eng. Sci.*, **362 (1825)**: 2821-2850 (2004)

[3]S. C. Leeuwenburgh, J. A. Jansen et al., Trends in biomaterials research: an analysis of the scientific programme of the World Biomaterials Congress 2008, *Biomater.*, **29 (21)**: 3047-3052 (2008)

[4]C. Ortiz and M. C. Boyce, Materials science – bioinspired structural materials, *Sci.*, **319 (5866)**: 1053-1054 (2008)

[5]D. Williams, The relationship between biomaterials and nanotechnology, *Biomater.*, **29 (12)**: 1737-1738 (2008)

[6]A. Bigi, E. Boanini et al., Hydroxyapatite gels and nanocrystals prepared through sol-gel process, *J. Solid State Chem.*, **177 (9)**: 3092-3098 (2004)

[7]S. V. Dorozhkin, Nanosized and nanocrystalline calcium orthophosphates, *Acta Biomaterialia*, **6 (3)**: 715-734 (2010)

[8]Y. Masuda, K. Matubara et al., Synthesis of hydroxyapatite from metal alkoxides through sol-gel technique, *Nippom Seramikkusu Kyokai Gakujutsu Ronbunshi - J. Ceram. Soc. Japan*, **98 (11)**: 1255-1266 (1990)

[9]P. Kessler, M. Thorwarth et al., Harvesting of bone from the iliac crest – comparison of the anterior and posterior sites, *Brit. J. Oral and Maxillofacial Sur.*, **43 (1)**: 51-56 (2005)

[10]L. J. Zhang and T. J. Webster, Nanotechnology and nanomaterials: promises for improved tissue regeneration, *Nano Today*, **4 (1)**: 66-80 (2009)

[11]S. C. Leeuwenburgh, I. D. Ana et al., Sodium citrate as an effective dispersant for the synthesis of inorganic-organic composites with a nanodispersed mineral phase, *Acta Biomaterialia*, **6 (3)**: 836-844 (2010)

[12]M. Akao, H. Aoki et al., Mechanical-properties of sintered hydroxyapatite for prosthetic applications, *J. Mater. Sci.*, **16 (3)**: 809-812 (1981)

[13]L. L. Hench, Bioceramics: from concept to clinic, *J. Amer. Ceram. Soc.*, **74 (7)**: 1487-1510 (1991)

[14]S. Best and W. Bonfield, Processing behaviour of hydroxyapatite powders with contrasting morphology, *J. Mater. Sci. Mater. in Medicine*, **5 (8)**: 516-521 (1994)

[15]M. P. Ferraz, F. J. Monteiro et al., Hydroxyapatite nanoparticles: a review of preparation methodologies, *J. Appl. Biomater. Biomech.*, **2 (2)**: 74-80 (2004)

[16]P. N. Kumta, C. Sfeir et al., Nanostructured calcium phosphate for biomedical applications: novel synthesis and characterisation, *Acta Biomaterialia*, **1 (1)**: 65-83 (2005)

[17]D. Olton, J. H. Li et al., Nanostructured calcium phosphates (NanoCaPs) for non-viral gene delivery: influence of the synthesis parameters on transfection efficiency, *Biomaterials*, **28 (6)**: 1267-1279 (2007)

[18]Q. Fu, M. N. Rahaman et al., Freeze-cast hydroxyapatite scaffolds for bone tissue engineering applications, *Biomed. Mater.*, **3(2)** (2008)

[19]C. E. Pedraza, D. C. Bassett et al., The importance of particle size and DNA condensation salt for calcium phosphate nanoparticle transfection, *Biomater.*, **29 (23)**: 3384-3392 (2008)

[20]G. L. Converse, T. L. Conrad et al., Mechanical properties of hydroxyapatite whisker reinforced polyetherketoneketone composite scaffolds, *J. Mech. Behaviour Biomed. Mater.*, **2(6)**: 627-635 (2009)

[21]R. Murugan and S. Ramakrishna, Aqueous mediated synthesis of bioresorbable nanocrystalline hydroxyapatite, *J. Crystal Growth*, **274 (1-2)**: 209-213 (2005)

[22]B. Cengiz, Y. Gokce et al., Synthesis and characterisation of hydroxyapatite nanoparticles, *Colloids and Surfaces a-Physicochemical and Eng. Aspects*, **322 (1-3)**: 29-33 (2008)

[23]Y. Zhao, Y. Zhang et al., Synthesis and cellular biocompatibility of two kinds of HAP with different nanocrystals morphology, *J. Biomed. Mater. Res. B Appl. Biomater.*, **83 (1)**: 121-126 (2007)

[24]C. R. Kothapalli, M. T. Shaw et al., Biodegradable HA-PLA 3-D porous scaffolds: effect of nanosized filler content on scaffold properties, *Acta Biomaterialia*, **1 (6)**: 653-662 (2005)

[25]S. Liao, W. Wang et al., A three-layered nano-carbonated hydroxyapatite/collagen/PLGA composite membrane for guided tissue regeneration, *Biomater.*, **26 (36)**: 7564-7571 (2005)

[26]L. J. Kong, Y. Gao et al., A study on the bioactivity of chitosan/nano-hydroxyapatite composite scaffolds for bone tissue engineering, *Euro. Poly. J.*, **42 (12)**: 3171-3179 (2006)

[27]R. Z. LeGeros, Biodegradation and bioresorption of calcium phosphate ceramics, *Clinical Mater.*, **14 (1)**: 65-88 (1993)

[28]K. C. Yeong, J. Wang et al., Fabricating densified hydroxyapatite ceramics from a precipitated precursor, *Mater. Letters*, **38 (3)**: 208-213 (1999)

[29]J. Karch, R. Birringer et al., Ceramics ductile at low-temperature, *Nature*, **330 (6148)**: 556-558 (1987)

[30]S. I. Stupp and G. W. Ciegler, Organoapatites – materials for artificial bone 1. Synthesis and microstructure, *J. Biomed. Mater. Res.*, **26 (2)**: 169-183 (1992)

[31]T. J. Webster, C. Ergun et al., Enhanced functions of osteoblasts on nanophase ceramics, *Biomater.*, **21 (17)**: 1803-1810 (2000)

[32]Z. L. Shi, X. Huang et al., Size effect of hydroxyapatite nanoparticles on proliferation and apoptosis of osteoblast-like cells, *Acta Biomaterialia*, **5 (1)**: 338-354 (2009)

[33]H. Liu and T. J. Webster, Nanomedicine for implants: a review of studies and necessary experimental tools, *Biomater.*, **28 (2)**: 354-369 (2007)

[34]T. J. Webster, K. Ellison et al., Increased osteoblast function on nanostructured materials due to novel surface roughness properties, *Thermec'2003*, Pts 1-5, **426 (4)**: 3127-3132 (2003)

[35]T. J. Webster, C. Ergun et al., Enhanced osteoclast-like cell functions on nanophase ceramics, *Biomater.*, **22 (11)**: 1327-1333 (2001)

[36]P. Li, Biomimetic nano-apatite coating capable of promoting bone ingrowth, *J. Biomed. Mater. Res. A*, **66 (1)**: 79-85 (2003)

[37]S. M. Zhang, F. Z. Cui et al., Synthesis and biocompatibility of porous nanohydroxyapatite/collagen/alginate composite, *J. Mater. Sci. – Mater. in Medicine*, **14 (7)**: 641-645 (2003)

[38]X. Y. Shen, L. Chen et al, A novel method for the fabrication of homogeneous hydroxyapatite/collagen nanocomposite and nanocomposite scaffold with hierarchical porosity, *J. Mater. Sci. – Mater. in Medicine*, **22 (2)**: 299-305 (2011)

[39]Q. Fu, N. Zhou et al., Effects of nano HAP on biological and structural properties of glass bone cement, *J. Biomed. Mater. Res. Part A*, **72A (2)**: 156-163 (2005)

[40]T. J. Webster, R. W. Siegel et al., Osteoblast adhesion on nanophase ceramics, *Biomater.*, **20 (13)**: 1221-1227 (1999)

[41]L. Bernard, M. Freche et al., Preparation of hydroxyapatite by neutralization at low temperature – influence of purity of the raw material, *Powder Tech.*, **103 (1)**: 19-25 (1999)

[42]A. C. Tas, Synthesis of biomimetic Ca-hydroxyapatite powder at 37 degrees C in synthetic body fluids, Biomater., 21 (14): 1429-1438 (2000)

[43]D. M. Liu, T. Troczynski et al., Water-based sol-gel synthesis of hydroxyapatite: process development, *Biomater.*, **22 (13)**: 1721-1730 (2001)

[44]D. M. Liu, Q. Z. Yang et al., Structural evolution of sol-gel-derived hydroxyapatite, *Biomater.*, **23 (7)**: 1679-1687 (2002)

[45]I. S. Kim and P. N. Kumta, Sol-gel synthesis and characterisation of nanostructured hydroxyapatite powder, *Mater. Sci. and Eng. B-Solid State Mater. for Ad. Tech.*, **111 (2-3)**: 232-236 (2004)

[46]T. A. Kuriakose, S. N. Kalkura et al., Synthesis of stoichiometric nano crystalline hydroxyapatite by ethanol-based sol-gel technique at low temperature, *J. Crystal Growth*, **263 (1-4)**: 517-523 (2004)

[47]F. Wang, M. S. Li et al., A simple sol-gel technique for preparing hydroxyapatite nanopowders, *Mater. Letters*, **59 (8-9)**: 916-919 (2005)

[48]B. H. Fellah and P. Layrolle, Sol-gel synthesis and chacracterisation of macroporous calcium phosphate bioceramics containing microporosity, *Acta Biomaterialia*, **5 (2)**: 735-742 (2009)

[49]R. A. Young and D. W. Holcomb, Variability of hydroxyapatite preparations, *Calcified Tissue Inter.*, **34**: S17-S32 (1982)

[50]P. Parhi, A. Ramanan et al., A convenient route for the synthesis of hydroxyapatite through novel microwave-mediated metathesis reaction, *Mater. Letters*, **58 (27-28)**: 3610-3612 (2004)

[51]G. K. Lim, J. Wang et al., Processing of hydroxyapatite *via* microemulsion and emulsion routes, *Biomater.*, **18 (21)**: 1433-1439 (1997)

[52]G. C. Koumoulidis, A. P. Katsoulidis et al., Preparation of hydroxyapatite *via* microemulsion route, *J. Colloid and Interface Sci.*, **259 (2)**: 254-260 (2003)

[53]Y. X. Sun, G. S. Guo et al, Synthesis of single-crystal HAP nanorods, *Ceram Inter.*, **32 (8)**: 951-954 (2006)

[54]H. S. Liu, T. S. Chin et al., Hydroxyapatite synthesised by a simplified hydrothermal method, *Ceram. Inter.*, **23 (1)**: 19-25 (1997)

[55]Y. J. Wang, S. H. Zhang et al., Hydrothermal synthesis of hydroxyapatite nanopowders using cationic surfactant as a template, *Mater. Letters*, **60 (12)**: 1484-1487 (2006)

[56]K. C. Yeong, J. Wang et al., Mechanochemical synthesis of nanocrystalline hydroxyapatite from CaO and CaHPO4, *Biomater.*, **22 (20)**: 2705-2712 (2001)

[57]W. L. Suchanek, P. Shuk et al., Mechanochemical-hydrothermal synthesis of carbonated apatite powder at room temperature, *Biomater.*, **23 (3)**: 699-710 (2002)

[58]Y. C. Han, S. P. Li et al., Synthesis of nanocrystalline hydroxyapatite powders by citric acid sol-gel combustion method, *Mater. Res. Bulletin*, **39 (1)**: 25-32 (2004)

[59]A. C. Tas, Combustion synthesis of calcium phosphate bioceramic powders, *J. Euro. Ceram. Soc.*, **20 (14-15)**: 2389-2394 (2000)

[60]S. Sasikumar and R. Vijayaraghavan, Solution combustion synthesis of bioceramic calcium phosphates by single and mixed fuels – a comparative study, *Ceram. International.*, **34 (6)**: 1373-1379 (2008)

[61]P. Scherrer, Bestimmung der Grösse und der inneren Struktur von Kolloidteilchen mittels Röntgenstrahlen, *Nachr. Ges. Wiss. Göttingen*, **26**: 98-100 (1918)

[62]H. Varma, K. G. Warrier et al., Metal nitrate-urea decomposition route for Y-Ba-Cu-O powder, *J. Amer. Ceram. Soc.*, **73 (10)**: 3103-3105 (1990)

LOW TEMPERATURE SINTERING OF Ti-6Al-4V FOR ORTHOPEDIC IMPLANT
APPLICATIONS

Kyle Crosby, Leon Shaw
Department of Chemical, Materials, and Biomolecular Engineering
University of Connecticut, Storrs, CT 06269

ABSTRACT
    Titanium alloys are widely used in orthopedic implants due to their high strength to weight
ratio and proven biological stability. However, coatings are often required to improve the bioactivity of
Ti-based implants. Another approach to enhancing the bioactivity is to fabricate functionally graded
Ti-6Al-4V (Ti)/hydroxyapatite (HA) implants, thus avoiding a sharp interface between the Ti core and
HA coating. To fabricate such graded implants, the sintering temperature of Ti must be 1000°C or
lower to prevent embrittlement of the Ti alloy. Here we investigate the creation of nanostructured Ti
powders via high-energy ball milling, while preventing their oxidation in the subsequent powder
pressing and sintering processes. It is found that methods for green body formation can affect the
sintered density while the use of nanostructured Ti powder can reduce sintering temperature
substantially. These results reveal that nanostructured Ti powders have great potential for achieving
functionally graded Ti/HA orthopedic implants.

INTRODUCTION
    In general, biomedical engineering is a merging of engineering design with the scientific
foundation of medicine in an effort to repair a damaged or missing biological component. As a whole,
biomedical engineering is focused on combining the problem solving capabilities of both fields to
improve our ability to diagnose and treat medical problems. One such form of expanding biomedical
technology is the biomedical implant. A biomedical implant is a device that is designed and
implemented to replace or repair a missing or damaged biological structure. An implant differs from a
medical transplant in the sense that a transplant indicates the use of existing biological tissue as
opposed to an implant which requires the use of foreign non-biological material. Biomedical implants
may be functional if they are coated with a drug delivery or bioactive material. Current widely used
biomedical implants include heart pacemakers, cochlear implants, hip prostheses, knee prostheses, and
dental implants. Powder metallurgy has been studied extensively for use in hip implant applications
where biocompatibility is desired without a sacrifice of load bearing mechanical integrity.
    The original design of the hip implant from a geometrical point of view has not changed much
since its inception, mainly because each implant is designed specifically to fit the exact shape and size
of the original bone material that has failed. The material choice has, however, undergone quite drastic
alterations recently, including the use of alloys rather than pure metal components, as well as the use of
a functional surface to further promote bone in growth and stymie infection. The major advancements
in this area include the shift from dense, stainless steel or pure titanium components toward α-β
titanium alloys (i.e. Ti-6Al-4V) and porous surface components which allow for bone ingrowth thus
securing the mechanical stability of the implant [1]. Another interesting progression in hip implant
materials is the use of a coating material that is bioactive [2]. The coating material of choice is
hydroxyapatite ($Ca_5(PO_4)_3(OH)$, HA), which itself comprises 50% of the natural bone material in
humans, thus making it nearly perfectly biocompatible. In order to develop composite Ti-6Al-4V + HA
prostheses, the material processing of each constituent must be studied to determine the best approach
for the simultaneous fabrication of the composite component [3]. This study is aimed at reducing the
Ti-6Al-4V powder metallurgy processing temperature to avoid the embrittlement of Ti-6Al-4V
induced by the HA additive during sintering.

EXPERIMENTAL

Ti-6Al-4V starting powder was purchased and used without any pretreatment (Advanced Specialty Metals, Nashua, NH). The as-received spherical powder was identified as -60+325 mesh, indicating particles between 45-250 μm. This as-received powder morphology was confirmed via scanning electron microscopy (SEM, FEI Electroscan ESEM 2020, Hillsboro, OR). In order to reduce the particle size of the as-received Ti-6Al-4V powder and to introduce defects which will help to increase the diffusion rate and the driving force for sintering, high-energy ball milling was applied. A SPEX mill, high energy, shaking style apparatus which has no moving internal parts, was utilized for this purpose. Stearic acid was used as a process control agent (PCA) during milling to prevent severe cold welding of the Ti-6Al-4V powder. The stainless steel SPEX mill vial used for this study was fabricated with an o-ring seal so that all ball milling was done under ultrahigh purity (UHP) argon by loading the SPEX vial within a glovebox. The milling time investigated ranged from 1 to 5 hours. Furthermore, for long milling time experiments (> 0.5 or 1 hour), ball milling was conducted via several short milling segments. There was a 15-minute break and thus cooling at room temperature between these short segments. These breaks between segments prevented overheating of the SPEX vial. The SPEX milling conditions investigated, including the duration of each milling segment, the total milling time and the weight percentage of the PCA with respect to the Ti-6Al-4V sample, are summarized in Table 1.

The ball-to-powder ratio (i.e. the charge ratio) was 40:1 for sample 110406 in Table 1 and 10:1 for all other samples. The 40:1 charge ratio involved the use of only two large 12.7 mm stainless steel grinding balls, while for the 10:1 charge ratio experiments samples were processed initially using two large 12.7 mm balls plus four small 6.35 mm balls (i.e. samples 110422 and 110503 in Table 1), but later the 10:1 was applied via four large 12.7 mm balls plus eight small 6.35 mm balls (i.e. samples 110510, 110516, 110525, 110527, 110602, 110607 in Table 1).

Table 1. Ti-6Al-4V SPEX milling conditions

Sample ID	Stearic Acid (wt%)	Milling Duration (hr)	Powder Details	Fig. 2 XRD ID
110406	2	0.5 + 0.5	Loose, easy collection	1
110422	2	1 + 1 + 1 + 1	Severe cold welding	2
110503	2	1	Light cold welding	3
110510	3	1 + 1 + 1 + 1	Severe cold welding	4
110516	4	1 + 1 + 1 + 1	Light cold welding	5
110519	4	2	Loose, easy collection	6
110525	4	1 + 1 + 1	Light cold welding	7
110527	4	1 + 1	Loose, easy collection	8
110602	4	1 + 1 + 1 + 1 + 1	Moderate cold welding	9
110607	5	1 + 1 + 1 + 1	Moderate cold welding	10

Green body pellets are prepared from the as-received Ti-6Al-4V powder as well as from the as-milled Ti-6Al-4V powders using a hardened steel die at a unixial pressure of 300 MPa. Sintering of the green body pellet was initially done at 1250°C for two hours under dynamic vacuum in an Al_2O_3 tube furnace. It was found that this sintering condition did not result in high density samples because of insufficient sealing of the Al_2O_3 tube and thus the partial oxidation of Ti-6Al-4V powder during sintering. To protect the sample from oxidation during the sintering process, samples were sealed in a quartz tube which was evacuated to <10 mTorr prior to necking the ends closed. To further prevent oxidation, some green pellets were wrapped in 0.025 mm thick tantalum foil (Alfa Aesar, Ward Hill,

MA) prior to evacuating and sealing the quartz tube. The quartz tube with samples sealed inside was then placed into the Al_2O_3 tube furnace, and sintering was carried out under a positive argon pressure.

X-ray diffraction (XRD) (Bruker D8 Advance, Madison, WI) studies were conducted for each SPEX milled sample to determine phase purity and monitor any compounds that may be forming during the milling process using $Cu_{K\alpha}$ radiation, 40 kW, 40 mA, 0.01°/step, and 4°/min scan rate. In addition to monitoring phase formation, XRD was also used to evaluate crystallite size via peak broadening through the use of the Scherrer formula,

$$\beta_g = k\lambda \, / \, D\cos\theta \qquad\qquad (1)$$

where β_g is the full width half maximum peak breadth, k is the shape factor (approximately 1.0), λ is the x-ray wavelength (Cu $K_{\alpha 1}$ = 1.541 nm), D is the crystallite size and θ is the Bragg angle [4]. The correction for instrumental broadening was conducted using the procedure described in [5] with the aid of coarse-grained silicon (Si) of 99.9% purity and the following equation,

$$\beta_g^{\,2} \, (2\theta) = \beta_h^{\,2} \, (2\theta) - \beta_f^{\,2} \, (2\theta) \qquad\qquad (2)$$

where β_g is the full width half maximum from the desired curve if there were no instrumental broadening, β_h is the full width half maximum from the milled sample, and β_f is the full width half maximum from the silicon standard. The pure silicon instrumental peak broadening quantity is subtracted out of the XRD scan for a given sample, while the internal strain broadening correction has not been applied presently.

Environmental SEM (Philips ESEM 2020) was used to examine the average particle size of the as-milled powders. The average particle size via SEM involves direct measurement of the powder particles using a line which has been calibrated based on the scale bar specific to the magnification of the image. Once the calibration is determined, approximately 15 particles are measured from each of 5 images of the same powder condition, such that the average particle size takes into account the distribution of particle sizes over more than 50 particles. This method does not show the true range of particle sizes, but it does accurately portray the trend which is experienced during ball milling.

The as-sintered density was measured directly using SEM images by taking the volume fraction of solid Ti-6Al-4V minus the volume fraction of porosity from a cross-sectioned and polished pellet. The Image J analysis software was used for this purpose. The SEM sample was prepared by cutting the as-sintered pellet down the middle using a diamond blade at a speed of 350 rpm, hot mounting the sectioned half, grinding with silicon carbide paper (400, 600, 800, 1200 grit) and polishing with colloidal silica (0.05 μm).

Thermogravimetric analysis (TA Instruments Q500, New Castle, DE) was applied to reveal the temperature at which the stearic acid processing agent burns off and/or other degassing reactions occur during sintering of the as-milled Ti-6Al-4V powders. The TGA experiments were performed after flowing argon for 60 minutes prior to heating and the heating rate was 5°C/min.

RESULTS AND DISCUSSION

The as-received Ti-6Al-4V powder morphology is shown in Figure 1, while the XRD spectrum for each SPEX milled Ti-6Al-4V condition appears in Figure 2. The SEM images used to determine the as-milled particle size is shown in Figure 3. The average crystallite size of the two major Ti-6Al-4V reflections (010) and (002) is derived from the XRD data in Figure 2 using the Scherrer formula. This data together with the particle size measurements from Figure 3 are used to show trends such as the change in both particle size and crystallite size as a function of milling duration, as well as the change in both particle size and crystallite size as a function of stearic acid concentration (wt%). This data can be seen in Figures 4 – 7 respectively.

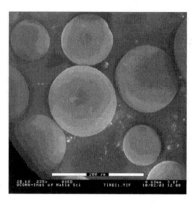

Figure 1. ESEM image of Ti-6Al-4V as received powder.

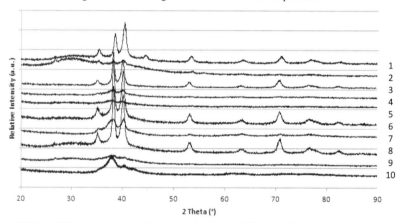

Figure 2. X-ray diffraction spectrum for each of the SPEX milling conditions described in Table 1.

As seen in the Powder Details column of Table 1, longer milling durations require additional PCA to prevent cold welding and allow for easy collection of the processed powder, a method derived from prior work by Shaw, et al. [6]. The XRD data in Figure 2 shows the as milled powder condition in terms of crystallinity; i.e. spectra with sharp peaks have large crystallites, while spectra with low broad humps have ultrafine crystallites. Particles with ultrafine crystallites are expected to have higher densification rates and be easier to densify than the counterparts with larger crystallites.

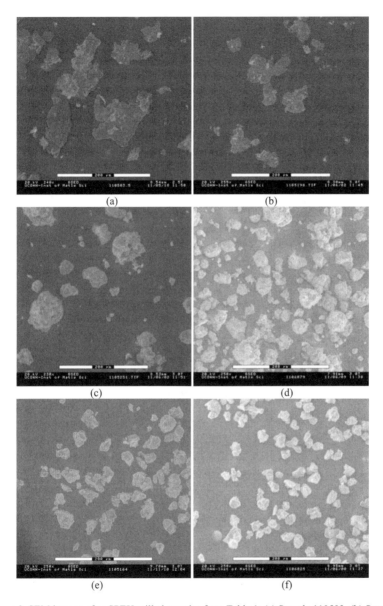

Figure 3. SEM images of as SPEX milled samples from Table 1; (a) Sample 110503, (b) Sample 110519, (c) Sample 110525, (d) Sample 110607, (e) Sample 110516, (f) 110602.

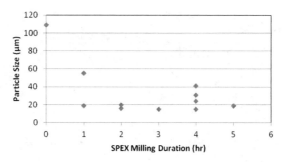

Figure 4. Average Ti-6Al-4V powder particle size (via SEM image analysis) as a function of SPEX milling duration.

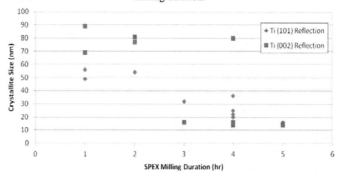

Figure 5. Average crystallite size of Ti-6Al-4V (101) and (002) reflections as a function of SPEX milling duration.

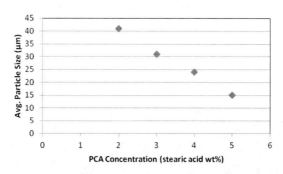

Figure 6. Average particle size (via SEM image analysis) of 1+1+1+1 hour as-milled Ti-6Al-4V powder as a function of the process control agent concentration.

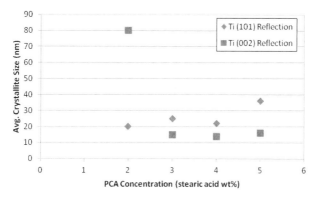

Figure 7. Average crystallite size of 1+1+1+1 hour as-milled Ti-6Al-4V (101) and (002) reflections as a function of the process control agent concentration.

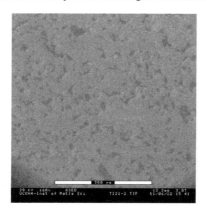

Figure 8. SEM image of the cross section of SPEX milled sample 110516, sintered at 1250°C for 2 hours under dynamic vacuum with insufficient sealing of the Al_2O_3 tube.

Figure 8 shows the cross-section image of the pellet sintered under dynamic vacuum, but with insufficient sealing of the Al_2O_3 tube. It is apparent that there is a large amount of porosity. The inadequate densification problem is attributed to the partial oxidation of Ti-6Al-4V particles during sintering. The oxide scale from the partial oxidation can prevent intimate Ti-6Al-4V to Ti-6Al-4V particle contact, thus preventing sintering from proceeding to completion. Prior to sintering the sample shown in Figure 4 has a green density of 3.12 g/cm^3 or about 70% of the theoretical, as calculated via direct caliper measurements. The sintered density is found to be 80% of the theoretical or about 3.56 g/cm^3. This result is generally poor as compared to prior work done by Shibo [7], who report as sintered density values greater than 94%.

Figure 9. Ti 6Al-4V as pressed pellets sealed in quartz tube and additionally wrapped in tantalum foil (right) for oxidation protection during sintering.

Thus, sealing via quartz tubes is employed for the subsequent sintering experiments. By sealing the samples in quartz tubing, which is stable up to 1650°C, the Ti-6Al-4V pellets which are extremely sensitive to oxygen are effectively secluded from an oxide forming environment throughout the sintering process. The addition of tantalum foil in the sintering protocol was done with the notion that the tantalum foil will act as an oxygen getter to even further protect the sample from oxidation during sintering because tantalum has a higher oxygen affinity than titanium. It was our notion that the tantalum would be equally effective as an oxygen getter when applied to the as-milled Ti-6Al-4V samples, and after many sintering trials that notion appears to hold valid. The quartz tube sealed samples are shown in Figure 9. As seen in the image, the quartz tube contains two samples with the one on the left having no tantalum protection but the one of the right being wrapped in tantalum foil. The excess powder seen inside the tube in Figure 9 is the result of the loose pellet on the left rolling in the tube during the glass sealing process. The as-pressed pellet crumbles slightly during the sealing process because the pellet is still in the green body state where it has low mechanical strength and thus the edges of the pellet have eroded leaving behind Ti-6Al-4V residue.

Sintering results using the modified quartz tube sealing procedure have proven to be effective when compared to Shibo's results mentioned earlier; i.e. our as-received Ti-6Al-4V reaches 90% density (Figure 10) using a larger starting particle size than Shibo, while our as-milled Ti-6Al-4V samples reach 97% density with only 2 hours of sintering time at 1250°C (Figure 11). These results are promising from a mechanical integrity point of view because Dewidar has shown the effect of sintering temperature on the compressive yield strength of Ti-6Al-4V alloys prepared by powder metallurgy at our sintering temperature [8]. Our results still fall short when compared to electro-discharge-sintering results presented by Lee which show a solid Ti-6Al-4V core that is > 99% dense [9]. However, the key factor to note is that the samples Lee has prepared have a large porosity when moving radially from the solid Ti-6Al-4V core, whereas our samples that are sintered using traditional powder metallurgy show uniform density throughout the sintered pellet. As per Shibo's results, the application of hot isostatic pressing is the only method to further increase the sintered density to > 99% for samples prepared using traditional pressing and sintering techniques. Overall the goal of fabricating a fully dense Ti-6Al-4V component will be more sensitive to reductions in the sintering temperature than to changes in the dwell time as per Bautista's results, thus our work will aim to continue to drive the sintering temperature to a minimum as required for sintering of composite implant components [10].

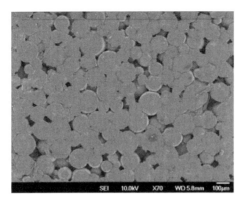

Figure 10. FESEM image of the as-polished cross section of an as-received Ti-6Al-4V sample that was sintered at 1250°C for 2 hours sealed in quartz and wrapped in Ta foil.

CONCLUSION

In this work we have shown that the application of the correct powder processing parameters are crucial to obtaining a material which is refined in terms of particle size and crystallite size as compared to the as-received commercial Ti-6Al-4V powder. Those processing parameters that are investigated here are the SPEX milling duration and the process control agent concentration (stearic acid wt%). The SPEX milling duration that results in the smallest particle size (< 20 μm) and smallest crystallite size (< 25 nm), while maintaining a sufficient yield, is shown to be 1 hour intervals for a total milling duration of 4 hours. The use of a 15-minute break between one-hour milling segments helps to prevent excessive heating of the SPEX vial, which helps to mitigate cold welding. It was found that for optimal effectiveness of SPEX milling of Ti-6Al-4V powders the use of 4 wt% stearic acid should be added to the powder charge when SPEX milling for a total of 4 hours in order to further mitigate cold welding. The application of longer milling durations does not further reduce particle and/or crystallite size because cold welding begins to become more prevalent at longer milling durations. Furthermore, the addition of a higher stearic acid concentration does not further reduce particle and/or crystallite size because the higher concentration allows for excess stearic acid to act as a lubricant such that the resulting powder has a more textured morphology rather than resulting in fine equiaxed particles and crystallites.

After witnessing sintering impedance, due to oxidation of the powder particles prior to particle coalescence during heating, it is shown that sealing of the Ti-6Al-4V pellets under vacuum in a quartz tube with tantalum foil wrapping has completely alleviated this issue. By reducing the particle size and crystallite size as described above, the sintering temperature necessary to produce highly dense Ti-6Al-4V bodies is lowered to 1250°C as compared to the as-received commercial powders which require temperatures above 1300°C to produce similar density when the compacts are sintered for only 2 hours. From this work it can be seen that there is room to further reduce the sintering temperature by producing green bodies with a higher starting density via higher uniaxial pressure, and/or by the application of longer sintering dwell times.

Figure 11. FESEM image of the as-polished cross section of a SPEX milled (1+1+1+1 hr with 4 wt% stearic acid) Ti-6Al-4V sample sintered at 1250°C for 2 hours sealed in quartz and wrapped in Ta foil.

ACKNOWLEDGMENTS
This research was sponsored by the U.S. National Science Foundation (NSF) under the contract number CBET-0930365. The support and vision of Ted A. Conway is greatly appreciated.

REFERENCES
[1] R. Nicula, F. Luthen, M. Stir, B. Nebe, E. Burkel, "Spark plasma sintering synthesis of porous nanocrystalline titanium alloys for biomedical applications," *Biomolecular Engineering*, 24, 564-567 (2007).

[2] X. Hu, H. Shen, Y. Cheng, X. Xiong, S. Wang, J. Fang, S. Wei, "One-step modification of nano-hydroxyapatite coating on titanium surface by hydrothermal method," *Surface & Coatings Technology*, 205, 2000-2006 (2010).

[3] K. Niespdziana, K. Jurczyk, J. Jakubowicz, M. Jurczyk, "Fabrication and properties of titanium-hydroxyapatite nanocomposites," *Materials Chemistry and Physics*, 123, 160-165 (2010).

[4] L. Azaroff, *Elements of X-ray Crystallography*, McGraw-Hill, New York, 552 (1968).

[5] Z-G. Yang, L. Shaw, "Synthesis of nanocrystalline SiC at ambient temperature through high energy reaction milling," *Nanostructured Materials*, Vol.7, No. 8, 873-886 (1996).

[6] L. Shaw, M. Zawrah, H. Villegas, H. Luo, and D. Miracle, "Effects of Process-Control Agents on Mechanical Alloying of Nanostructured Aluminum Alloys," *Metallurgical and Materials Transactions A.*, 34A, 159-170 (2003).

[7] G. Shibo, Q. Xuanhui, H. Xinbo, Z. Ting, D. Bohua, "Powder injection molding of Ti-6Al-4V alloy," *Journal of Materials Processing Technology*, 173, 310-314 (2006).

[8] M.M. Dewidar, J.K. Kim, "Properties of solid core and porous surface Ti-6Al-4V implants manufactured by powder metallurgy," *Journal of Alloys and Compounds*, 454, 442-446 (2008).

[9] W.H. Lee, C.Y. Hyun, "Fabrication of fully porous and porous-surfaced Ti-6Al-4V implants by electro-discharge-sintering of spherical Ti-6Al-4V powders in a one-step process,"*Journal of Materials Processing Technology*, 189, 219-223 (2007).

[10] A. Bautista, C. Moral, G. Blanco, F. Velasco, "Influence of sintering on the corrosion behavior of a Ti-6Al-4V alloy," *Materials and Corrosion*, 56, No. 2, 98-103 (2005).

CYTOTOXICITY EVALUATION OF 63S BIOACTIVE GLASS NANOPARTICLES BY
MICROCALORIMETRY

A. Doostmohammadi[1*]

A. Monshi[1]

M.H. Fathi[1]

O. Braissant[2]

A.U. Daniels[2]

1: Biomaterials Group, Materials Engineering Department, Isfahan University of Technology, Isfahan 84156-83111, Iran

2: Isfahan University of Medical Sciences, Isfahan 81746-73461, Iran

3: Laboratory of Biomechanics & Biocalorimetry, Coalition for Clinical Morphology & Biomedical Engineering, University of Basel, Faculty of Medicine, Basel, Switzerland

* Corresponding author: alidm14@ma.iut.ac.ir

ABSTRACT

The cytotoxicity evaluation of 63S bioactive glass nanoparticles with yeast and human chondrocyte cells was carried out using isothermal micro-nano calorimetry (IMNC). Bioglass nanoparticles were made via sol-gel method. Elemental analysis was carried out by XRF. Amorphous structure of the glass was detected by XRD analysis. Finally, the cytotoxicity of bioactive glass nanoparticles with yeast and cultured human chondrocyte cells was evaluated using IMNC. The results confirmed the viability and proliferation of human chondrocyte cells in contact with 63S bioglass nanoparticles. Also the results indicated that yeast model which is much easier to handle, can be considered as a good proxy and can provide a rapid primary estimate of the ranges to be used in assays involving human cells. All of these results confirmed that IMNC is a convenient method which caters to measuring the cell-biomaterial interactions alongside the current methods.

1. INTRODUCTION

Biocompatibility evaluation is one of the most important assessments to be performed prior to clinical use of biomaterials [1]. At the time being, there are a number of different methods in use for evaluating biocompatibility. These methods are time consuming and the results depend extremely on the human skill [2].

There is a need for a convenient method for screening of biocompatibility and cytotoxicity (as a criterion of biocompatibility), a method that can directly evaluate cell growth as well as cell adhesion to biomaterials surfaces. Furthermore, it should be possible to sample the medium for analysis of compounds released by the cells [3].

Isothermal micro-nano calorimetry (IMNC) is capable of measuring the heat production or consumption rate in the μW range with a calorimeter operating at nearly a constant temperature. Therefore, this technique allows direct and continuous monitoring of the metabolic activity of living cells [4].

As a result, IMNC can be extensively used for monitoring the growth and activity of microorganisms [5], human cells or cell lines [6, 7] and evaluating the cell-biomaterial interactions.

In this study, biocompatibility of a specific prepared bioglass was evaluated with IMNC. Bioactive glasses (SiO_2) glasses containing Ca and P) are well-known materials for use in implant applications, and have been shown to augment the formation of bone and other tissues [8]. Bioactive glasses in the system SiO_2-CaO-P_2O_5 obtained by sol-gel method present good characteristics of biocompatibility and bioactivity. They can be designed with controlled compositions and high specific surface area in order to be biodegradable [9].

Compared to particles of μm or larger sizes, bioactive glass nanoparticles may provide a means for more rapid release of Ca and P where this is desired [8].

Bioactive glass can be used in granular form to fill bone defects. It has the capacity to bond to the osseous tissue; moreover, bone tissue repair and growth can be enhanced by its osteoconductive properties [8, 9].

Bioactive bioceramics such as bioglass have gained access to great application successes in bone repairing [9, 10]. Bioactive glass could elicit a specific biological in-vivo response at the interface and attach to the tissues, such as soft tissue and bone, with a strong chemical bond. This is the reason that bioactive glasses have been used for many different applications [8-10]. Certain compositions of bioactive glasses containing SiO_2- CaO-P_2O_5 can bond to both soft and hard tissue without any intervening fibrous layer [8].

In this study, bioglass (63S) particles were made via sol-gel method. Characterization experiments were carried out on the material and it's biocompatibility with yeast and cultured human chondrocyte cells was evaluated using IMNC.

2. MATERIALS AND METHODS

2.1. Synthesis and preparation

2.1.1. Bioactive glass

Colloidal solutions (sols) of 63S composition (63 mol% SiO2, 28 mol% CaO, 9 mol% P2O5) were prepared by mixing distilled water, 2N Hydrochloric acid (Merck), tetraethyl orthosilicate (TEOS, Merck), triethyl phosphate (TEP, Sigma-Aldrich) and calcium nitrate (Merck) [10]. The initial procedure involved mixing TEOS (28 ml) and ethanol (40 ml, Merck) as an alcoholic media. Distilled water was added to solution and allowed to mix until the solution became clear. The H2O:(TEOS) molar ratio was 4:1. After 30 min, TEP (2.3 ml) was added to the stirring solution. After another 20 min, calcium nitrate (12 gr) was added. The solution was then stirred for an additional hour. The gel was heated (60 °C, 10 hours), dried (130 °C, 15 hours) and thermally stabilized (600 °C, 2 hours) according to established procedures [9, 10]. The produced gel was ground with a mortar and pestle to disagglomerate the particles. Finally the particles were sieved to make a distribution of particles of size less than 50 μm (L3-M50 50 μm stainless steel sieve & Sonic Sifter Separator, Advantech Manufacturing Co., New Berlin WI, USA). Bioactive glass particles were sterilized at 180°C for 1 h. phosphoric acid. They were rinsed again in sterile distilled water, and sterilized at 200 °C.

2.2. Characterization

In order to detect the phases and the elements present in the chemical compositions, the basic characterization experiments were carried out.

2.2.1. Elemental composition analysis

The elemental composition of bioactive glass particles was confirmed by X-ray fluorescence spectroscopy (XRF), (PW2404, PHILIPS) and energy dispersive X-ray analysis (EDX) technique (SUPRA 40 VP FE-SEM).

2.2.2. X-ray diffraction

X-ray diffraction (XRD) technique (Philips X'Pert-MPD system with a Cu Kα wavelength of 1.5418 A°) was used to analyze the structure of the prepared bioactive glass. The diffractometer was operated at 40 kV and 30 mA at a 2θ range of 20–80° employing a step size of 0.02°/s.

2.2.3. Particle Morphology and size by Scanning Electron Microscopy (SEM)

Particle samples were mounted on aluminum SEM pins and coated with Au/Pd. They were then observed with a scanning electron microscope (SUPRA 40 VP FE-SEM, Carl Zeiss AG, Germany) operated at an acceleration voltage of 20 kV. The size range of the particles was determined by measuring a statistically relevant numbers of particles from the electron micrographs, using the measuring tools provided for this purpose by Zeiss with their SEM.

2.3. Direct cytotoxicity evaluation with isothermal micro-nano calorimetry

Monitoring the cell growth in solid substrates is often difficult. However, in this context microcalorimetry offers a way to measure the growth and activity of the cells over time in the presence of solid compounds such as bioactive glass nanoparticles.

2.3.1. Yeast experiment

Yeast (*Saccharomyces cerevisiae*) were grown and maintained on YPD (Yeast Peptone Dextrose) medium. Overnight cultures were performed before the experiment. The culture was diluted ca 5000 times in order to have an optical density at 600 nm below 0.05. 3 ml of the diluted culture was added to increasing amounts of bioactive glass (0 to 50 mg/vial) in 4 ml calorimetric ampoules. The ampoules were sealed and placed in equilibration position in the microcalorimeter (TAM48, Waters/TA) for at least 15 minutes. Following equilibration, ampoules were lowered in measuring position equilibrated again for 45 minutes. After this second equilibration time, growth related heat-flow was recorded for ca 300 hours.

2.3.2. Chondrocyte experiment

Trypsinized human chondrocyte culture was performed in Dubelco's modified eagle's medium containing 10% fetal bovine serum, D-glucose (4.5 mg/mL), 0.1 mM nonessential amino acids, 1 mM sodium pyruvate, 100 mM HEPES buffer, penicillin (100 units/mL), streptomycin (100 μg/mL), and L-glutamine (0.29 mg/mL). In addition, the medium was added with growth factors (transforming growth factor (TGF-1, 1 ng/mL), fibroblast growth factor-2 (FGF-2, 5 ng/mL), and platelet-derived growth factor type bb (PDGF-bb, 10 ng/mL)). The cultures were aseptically transferred in microcalorimetric ampoules containing 3 mg of bioactive glass nanoparticles. This quantity was determined using the results of the yeast experiment assuming that human cell would be more sensitive to calcium concentration than yeast. There was also a control sample containing only chondrocyte cells and the culture medium. For making sure that the measured heat is only of the cell growth, the second control sample dedicated to bioceramic particles in the culture medium (without cells).

3. RESULTS

3.1. Elemental composition analysis

The result of EDX microanalysis of the glass particles showed the elements present in the composition. The peaks of O, Si, P and Ca indicate the consisting elements of prepared bioactive glass particles (Fig. 1).

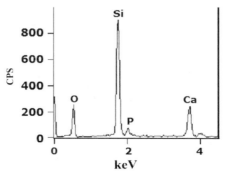

Fig. 1. Energy dispersive X-ray analysis (EDX) of the bioactive glass particles. The peaks of O, Si, P and Ca indicate the consisting elements of prepared bioactive glass.

The existent elements in prepared bioactive glass particles and estimated composition measured by X-ray fluorescence (XRF), is shown in Table 1. The molar percentage of oxides was expressed by computer according to the elemental analysis and considering the assumption that all the elements were in oxidized form.

Table 1- Bioactive glass estimated oxidic composition measured by X-ray fluorescence (XRF).

Oxide	Molar %
SiO_2	62.17
CaO	28.47
P_2O_5	9.25

3.2. X-ray diffraction analysis

The XRD pattern of the prepared glass after heating at 600 °C for 2 hours did not contain diffraction maxima, indicative of the internal disorder and the glassy nature of this material. The XRD patterns indicated that the obtained bioactive glass particles were amorphous (Fig. 2), proving the sol–gel method could prepare pure glasses.

3.4. Scanning electron microscope

Figure 3 shows SEM images of the bioactive glass particles. Heterogeneous surfaces consisting of random-sized particles can be seen. SEM confirmed particles are in the nanoscale range (<40 nm) but showed the particles were agglomerated.

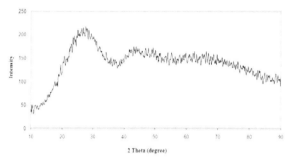

Fig. 2. XRD pattern of the prepared bioactive glass particles. Intensity of diffraction vs. angle of radiation (2θ, °). No peak of diffraction could be observed.

Fig. 3. SEM micrographs of the bioactive glass nanoparticles.

3.3. Cytotoxicity evaluation

3.3.1. Yeast experiment

Saccharomyces growth in YPD which was added with bioactive glass exhibited normal growth (i.e., comparable to the control (0 mg bioglass)) for bioactive glass concentration up to 3.3 mg/ml (10 mg/vial). For higher concentration, decrease in maximum activity and delays in growth were observed (Fig. 4). For concentration of 40 mg bioactive glass, in comparison with 30 mg bioactive glass, more activity of yeast was observed. However, the delay in yeast growth was evident for 40 mg bioactive glass. No yeast growth was detected for concentration of 50 mg bioactive glass.

3.3.2. Chondrocyte experiment

As can be perceived from Fig. 5, chondrocyte growth was not significantly affected by bioactive glass concentrations of 3 mg. Only a minimal effect was observed (slightly lower than maximal activity). All the cultures followed a similar growth pattern and decayed at a similar rate. The reason for this decease is the closed system of IMNC and consequently, inability of renewing the cultures. It can be concluded that for such concentrations of 63S bioactive glass, growth of human chondrocytes is not essentially affected (Fig. 5).

The detected heat from the ampoules which contain bioactive glass is comparable to the control sample and proves the metabolic activity of the cells. There was no detected heat for the second control sample. This shows that the detected heat was absolutely assigned to the cell activity. The results gained from the direct cytotoxicity evaluation bioactive glass with chondrocytes and performed through IMNC demonstrated the fact that bioactive glass had no deleterious effect on chondrocyte growth and are completely non-cytotoxic.

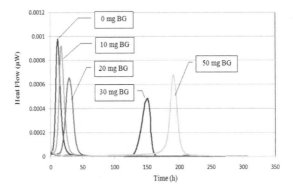

Fig. 4. Microcalorimetric curves showing the effect of bioactive glass particles (BG) on yeast growth.

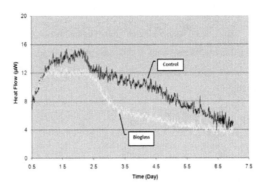

Fig. 5. Microcalorimetric curves showing the effect of adding bioactive glass nanoparticles to cultured human chondrocyte cells on chondrocyte growth.

4. DISCUSSION

The characterization results of bioactive glass particles showed that the composition of prepared bioactive glass was in good compatibility with documented composition for sol-gel 63S bioactive glass.

The obtained result from XRD showed that there was no crystalline phase in bioactive glass particles and this is completely in agreement with previous reports [9, 10].

The IMNC results showed that chondrocytes can live on, grow and proliferate in contact with 63S bioactive glass nanoparticles.

This may not be a novel result, but it would be valuable when firstly gained through IMNC (that is very simple and easy to handle) and secondly when the human chondrocyte cells were in direct contact with the bioceramic particles. Conventional methods for biocompatibility and cytotoxicity assay (e.g., MTT) do not suggest a way for direct evaluation of the particles' cytotoxicity. These methods usually evaluate cytotoxicity indirectly (using the extraction of materials) [11] or are designed for bulk structures (not for particles) [12].

The results also suggest that yeast model can be used as a preliminary step to determine the range of material amounts (bioceramic particles) needed for the experiments with human cells. In other words, the weight amount of needed bioceramic particles for the human cell experiment can be initially determined using yeast model assay.

These concentrations of particles could be obtained perfectly by coating implants. In these conditions, the biocompatibility of coated implants with chondrocyte cells can be assured. The obtained results from this study confirmed the growth, proliferation and viability of chondrocyte cells in contact with 63S bioactive glass.

Finally, this study suggests that isothermal microcalorimetry is a promising novel method for evaluating the cytotoxicity of materials to microorganisms such as human cells. This culture method detects heat, resulting from cellular metabolism and growth. IMNC can be used as a primary stage of biocompatibility assessments and yields complementary results when compared to the conventional in vitro methods.

5. CONCLUSION

The cytotoxicity of particulate materials such as bioactive glass nanoparticles can be evaluated using microcalorimetry method. This is a modern method for in vitro study of biomaterials biocompatibility and cytotoxicity which can be used alongside the old conventional assays. The obtained results showed that there was no negative, pernicious and toxic effect on chondrocyte growth and they could live on, grow and proliferate in contact with 63S bioactive glass particles. The results also suggested that yeast model (much easier to handle) could be used as a preliminary step to determine the ranges for human cell experiments.

REFERENCES

[1] Quteish D, Singrao S, Dolby AE. Light and electron microscopic evaluation of biocompatibility, resorption and penetration characteristics of human collagen graft material. J Clin Periodontol. 1991;18:305-11.

[2] Zhang H, Xiao-Jian Y, Jia-Shun L. Preparation and biocompatibility evaluation of apatite/wollastonite-derived porous bioactive glass ceramic scaffolds. J Biomed Mater Res. 2009;50:353–9.

[3] Xie Y, Depierre JW, Sberger LN. Biocompatibility of microplates for culturing epithelial renal cells evaluated by a microcalorimetric technique. J Mater Sci Mater Med. 2000;11:587-91.

[4] Beezer AE. Biological Microcalorimetry. Academic Press: London;1980.

[5] Braissant O, Wirz D, Gopfert B, Daniels AU. Use of isothermal microcalorimetry to monitor microbial activities. FEMS Microbiol Lett. 2009:1-8.

[6] Kemp RB, Guan YH. The application of heat flux measurements to improve the growth of mammalian cells in culture. Thermochimica Acta. 2000;349:23-30.

[7] Kemp RB. Hand book of thermal analysis and calorimetry. Elsevier;1999.

[8] Gough JE, Notingher I, Hench LL. Osteoblast attachment and mineralized nodule formation on rough and smooth 45S5 bioactive glass monoliths. J Biomed Mater Res A. 2004;68(4):640-50.

[9] Balamurugan A, Balossier G, Kannan S, Michel J, Rebelo AH, Ferreira JM. Development and in vitro characterization of sol–gel derived CaO–P_2O_5–SiO_2–ZnO bioglass. Acta Biomaterialia. 2007;3:255–62.

[10] Fathi MH, Doostmohammadi A. Preparation and characterization of sol–gel bioactive glass coating for improvement of biocompatibility of human body implant. Mater Sci Eng A. 2008;474:128–33.

[11] Yamamoto A, Honma R, Sumita M, Hanawa T. Cytotoxicity evaluation of ceramic particles of different sizes and shapes. J Biomed Mater Res A. 2004;68:244-56.

[12] Zhang K, Washburn NR, Simon CG. Cytotoxicity of three-dimensionally ordered macroporous sol–gel bioactive glass (3DOM-BG). Biomat. 2004;26: 4532-39.

BIOLOGICAL ASPECTS OF CHEMICALLY BONDED CA-ALUMINATE BASED
BIOMATERIALS

Leif Hermansson
Doxa AB, Axel Johanssons gata 4-6, SE 754 51 Uppsala, Sweden

ABSTRACT
    The paper describes biological features of the Ca-aluminate based biomaterials with regard to reaction chemistry, general chemistry and nanostructure including phases and porosity, biocompatibility and bioactivity, and similarity with and integration towards bone tissue. The nanostructure including nanosize porosity opens up for specific applications related to dentistry and orthopaedics. A field where nanosize porosity is essential is also within drug delivery systems for controlled release of medicaments.

INTRODUCTION
    Ca-aluminate materials belong to the chemical family inorganic cements. Materials in this group are mainly phosphates, aluminates, silicates and sulphates, and have found several applications as biomaterials, but have traditionally many application areas outside those of biomaterials, especially within the construction area [1,2].
    Ceramic biomaterials are often based on phosphate-containing solubable glasses and various calcium phosphate salts. These materials can be made to cure *in vivo* and are attractive as replacements for damaged bone or bone loss. Materials based on Ca-aluminate with chemistry similar to that of Ca-phosphates contribute to some additional features of interest within dental, orthopedic and drug delivery applications. This paper will summarize important aspects of the Ca-aluminate based materials with regard to biological aspects including reaction chemistry, biocompatibility and bioactivity, similarity with hard tissue, and integration with bone tissue. The almost impossibility of Ca-aluminate based materials to avoid nanostructures including nanosize porosity opens up for specific applications related to dentistry and orthopaedics. A field where nanosize porosity is essential is also within drug delivery systems for controlled release of medicaments. This paper will concentrate on the biological features of the Ca-aluminate based biomaterials.

MATERIALS DESCRIPTION
    Ca-aluminate materials belong to the chemically bonded ceramics, the materials which cure at low temperatures, at body temperature , in hydration reactions. In the $CaO-Al_2O_3$ system, several intermediate Ca-aluminate phases exist, and these are - using the cement chemistry abbreviation system - C_3A, $C_{12}A_7$, CA, CA_2 and CA_6, where $C=CaO$ and $A=Al_2O_3$ [3]. The C_3A phase is regarded as too reactive with too high reaction temperatures for biomaterials, and the CA_2 and CA_6 phases exhibit too low reaction rate. As biomaterials the greatest interest is related to $CaO-Al_2O_3$ molar ratio close to 1:1. The Ca-aluminate phases used in the testing with the main phase the mono-calcium aluminate (CA) were synthesized by the company Doxa AB. After crushing, the material was jet-milled to obtain particles with mean particle size below 5 μm and d(99) below 10 μm.
    In addition to the main binding phases (the Ca-aluminates and water), filler particles are included to contribute to some general properties of interest when used for different applications. The general contribution of added particles regards the microstructure (homogeneity aspects) and mechanical properties (especially hardness, Young's modulus and strength). For dental applications additives are mainly glass particles. For orthopedic application a high density oxide is selected. For many dental applications translucency is desired, why inert particles should have a refractive index

close to that of the hydrates formed [4]. A preferred high-density oxide for orthopedic application is zirconia, a material also used as a general implant material.

METHODS

Phase and microstructural studies

Studies were complemented by evaluating the chemical reactions and microstructure developed in the Ca-aluminate biomaterial and tissue at the highest level by transmission electron microscopy (HRTEM) in combination with focused ion beam microscopy (FIB) for site-specific high accuracy preparation, described in detail in [5]. Phase and elemental analyses were conducted using traditional XRD, HRTEM, XPS and STEM with EDX.

Biocompatibility including bioactivity

The Ca-aluminate materials have been evaluated comprehensively concerning their biocompatibility and toxicological endpoints as referred in the harmonized standard ISO 10993:2003, which comprises the following sections [6]:
Cytotoxicity (ISO10993-5), Sensitization (ISO10993-10), Irritation/Intracutaneous reactivity (ISO10993-10), Systemic toxicity (ISO10993-11), Sub-acute, sub-chronic and chronic toxicity (ISO10993-11), Genotoxicity (ISO10993-3), Implantation (ISO10993-6), Carcinogenicity (ISO10993-3) and Hemocompatibility (ISO10993-4).

In the cytotoxicity test primary cultures of human oral fibroblasts were used. A tissue culture insert retained test materials was assembled in a 12-well plate. The cytotoxicity was determined by MTT reduction assay at various time points. The MTT reduction assay was used to evaluate the cytotoxicity potential of the material. To stain the cells, the specimens were first removed from fibroblast culture. Then MTT solution (0.1 mg/ml) was added into each well and incubated between 1 to 3 h to obtain the maximal intracellular conversion of the dye. Subsequently, the medium was removed and the cultures were washed twice with PBS. 2-Propanol containing 0.04 M of HCl were added to the culture plate (100 µl/well) for 10 min to lyse the cells. The lysate was collected into an Eppendorf and centrifuged (10,000 g, 2 min) and the supernatant was then removed into a 96-well plate for spectrophotometric determination of absorbance at 590 nm. The results were calculated and expressed as a percentage of the corresponding control.

The in vitro bioactivity was evaluated with a method close to the recently published ISO standard 23317 [7].

Atomic absorption spectrometry

Determination of Ca and Al in the solution during the hydration process of the Ca-aluminate based material was performed using atomic absorption spectrometry. Standard solutions of different concentrations of Ca- and Al- ions were prepared according to the manual. The samples were prepared as plates with a surface area of 2,2 cm^2. The specimen weight was 0.41 g. For the test both distilled water (DW) and a saliva solution (SS) were used as storage medium. This inorganic saliva solution contained calcium chloride, magnesium chloride, sodium chloride, a phosphate buffer, hydrocarbonate and citric acid. The amount of liquid in each test was 10 ml. The temperature selected was 37 °C.

Acid erosion

The acid erosion test by impinging jet technique was conducted according to EI 29917:1994/ISO 9917:1991 [8], where removal of material is expressed as a height reduction using 0.1 M lactic acid as solution, pH 2.7. The test probe accuracy was 0.01 mm. Values below 0.05 mm per 24 h solution impinging are judged as acid resistant. The duration time of the test is 24 h. The test starts after 24 h hydration.

pH measurement
    Measurement of pH development during hydration of the material was conducted using a standard pH-meter. The samples were prepared according to the procedure for atomic absorption spectroscopy above. The analysis was made upon 1 ml solution.

Results and Discussion
    The results and the descussion will be presented under the following themes; Chemical and structural similarities with hard tissue, biocompatibility including bioactivity, hydrophilicity and chemical resistance and long-term stability, bacteriostatic and antibacterial aspects, as well as short description of dental and orthopedic application possibilities including carriers for drug delivery.

Chemical and structural similarities with hard tissue
    In water environment, the Ca-aluminate materials react in an acid-base reaction to form hydrates. The main reaction for the mono Ca-aluminate phase, using the cement chemistry abbreviation system, is shown below $(H=H_2O)$;

$$3CA + 12H \rightarrow C_3AH_6 + 2AH_3 \qquad \text{(Eq. 1)}$$

    The reactions are temperature dependent. At temperatures above 30 degree C and at body temperature the reaction is summarized as;
- dissolution of Ca-aluminate into the liquid
- formation of ions, and
- repeated precipitation of nanocrystals (hydrates)
    – katoite, $CaOAl_2O_36H_2O$ (CAH_6), and gibbsite $Al(OH)_3$ (AH_3).

    The reaction involves precipitation of nanocrystals on tissue walls and repeated precipitation until the Ca-aluminate is consumed, resulting in complete cavity/gap/void/ filling [9]. An example of the established contact zone between dentine and a Ca-aluminate based biomaterial is shown in Fig. 1.

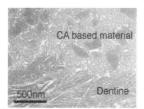

Fig. 1 Integration zone between dentine and an experimental Ca-aluminate based restorative material (bar 500 nm)

Biocompatibility including bioactivity
    Complementary reactions occur when the Ca-aluminate is in contact with tissue containing body liquid. Several mechanisms have been identified, which control how the Ca-aluminate material is integrated onto tissue. These mechanisms affect the integration differently depending on what type of

tissue the biomaterial is in contact with, and in what state (un-hydrated or hydrated) the Ca-aluminate is introduced. These mechanisms are summarized as follows and described in more details elsewhere [9];

Mechanism 1: Main reaction, the hydration step of Ca-aluminate (Eq. 1 above),
Mechanism 2 :   Apatite formation in presence of phosphate ions in the biomaterial,
Mechanism 3: Apatite formation in the contact zone in presence of body liquid,
Mechanism 4 :   Transformation of hydrated Ca-aluminate into apatite and gibbsite,
Mechanism 5: Biological induced integration and ingrowth, i.e. bone formation at the contact zone.
When phosphate ions or water soluble phosphate compounds are present in the biomaterial (powder or liquid) an apatite formation occurs according to the reaction

$$5Ca^{2+} + 3PO_4^{3-} + OH^- \rightarrow Ca_5(PO_4)_3OH$$

This complementary reaction to the main reaction occurs due to the presence of Ca-ions and a basic (OH^-) environment created by the Ca-aluminate material. The solubility product of apatite is very small [10], so apatite is easily precipitated. Body liquid contains among others the following ions HPO_4^{2-} and $H_2PO_4^-$. In contact with the Ca-aluminate system and water during setting and hydration, the presence of Ca-ions and hydroxyl ions, the hydrogen phosphate ions are neutralised according to

$$HPO_4^{2-} + H_2PO_4^- + OH^- \rightarrow PO_4^{3-} + H_2O,$$
whereafter the apatite-formation reaction occurs

$$5Ca^{2+} + 3PO_4^{3-} + OH^- \rightarrow Ca_5(PO_4)_3OH$$

The apatite is precipitated as nano-size crystals in contact zone [11]. See Fig. 2.

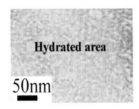 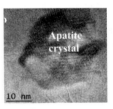

Fig. 2   a) Nano-size structure in the the contact zone to hard tissue,
b) apatite formation in the contact zone

Katoite is formed as a main phase according to the mechanism 1 above. The size of the hydrates is in the interval 15-30 nm, which is the same as the apatite crystals in dental tissue. The size of the hydrates formed is controlled by the solubility products of the phases formed, in the Ca-aluminate system katoite and in the Ca phosphate system apatite.

However, in long-time contact with body liquid containing phosphate ions, the katoite is transformed at the interface to body tissue into that at neutral pH even more stable apatite and gibbsite phases according to

$$Ca_3 \cdot (Al(OH)_4)_2 \cdot (OH)_4 + 2\ Ca^{2+} + HPO_4^{2-} + 2\ H_2PO_4^- \rightarrow$$
$$Ca_5 \cdot (PO_4)_3 \cdot (OH) + 2\ Al(OH)_3 + 5\ H_2O$$

When apatite is formed at the interface according to any of the reaction mechanisms 2-4 above, at the periphery of the bulk biomaterial, the biological integration may start. Bone ingrowth towards the apatite allows the new bone structure to come in integrated contact with the biomaterial. The transition from tissue to the biomaterial is smooth and intricate.

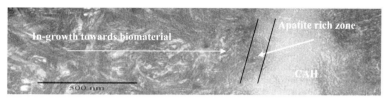

Fig. 3 Fine-tuned transition from tissue to biomaterial (bar = 500 nm)

The results of the in-vitro MTT reduction tests of the Ca-aluminate based material in human oral fibroblast cultures indicate no morphological changes in any of the test groups at different MTT reduction testing points. The cell culture was typically fibroblastic with a slender and elongated form in both control and the exposed groups. The average values of MTT reduction of the biomaterial were close to that of the controls. The variations found were within the 95% confidence interval to the control. Ɉeither the curing time nor exposure time le ngth of the tested material affected the cell survival. Some particles from the material were seen on the fibroblasts without any obvious change of cell morphology. These particles are a result of precipitation from the interaction between the material and surrounding solution.

The toxicological studies done on injectable calcium aluminate based bone filler according to ISO 10993-5 and 10993-10, did not reveal any toxicity or hypersensitivity.

Chemical resistance and long-term stability

The dissolution of Ca and Al is part of the hydration process and the continued formation of hydrates. Initially saturation in the solutions is reached within a short time and precipitation of hydrates occurs continuously, leaving an almost constant low level of free ions. As the hydration process continues, its intensity decreases with time. The precipitation rate is a function of pH, type of ions present and their solubility products. After the initial hydration time the ion concentration is determined by the solubility product of the phases formed. In the saliva solution, the original content of Ca ions in the saliva controls the Ca-concentration totally. Since the concentration of Ca in saliva is higher than what is obtained in distilled water it can be assumed that the filling material releases very limited amounts of Ca or Al once the material has hardened

During no part of the hydration in pure water did the ion concentration of Ca and Al exceed 30 and 5 ppm, respectively. In the buffer solution the ion content is predominately controlled by the saliva solution. Expressed as mass per surface area of material and day, the release of metal ions during the hardening process was below 3.3×10^{-5} mg/mm^2/day for aluminum and 1.96×10^{-4} mg/mm^2/day for calcium measured in water, and below 1.23×10^{-4} mg/mm^2/day for aluminum measured in artificial saliva. The same level of Ca as in the original saliva solution was found in the test solutions, approximately 60-70 ppm.

The pH is high during the early stage of the hydration and decreases with time and approaches neutrality – even in pure water. The reason for the high pH in the beginning is the general basic character of the material and the formation of OH⁻ during the hydration process as well as the absence of any buffering activity in the pure aqueous solution. In the saliva solution the pH is somewhat lower due to the buffering activity. This study was not designed to mimic the clinical situation where saliva is produced in a dynamic way creating an environment capable of buffering surrounding solution to neutrality. The pH in water and in saliva decreases from 11.5 and 10.3 initially down to 10.2 and 9.8 after 28 days in the constant storage solutions, respectively. After 1 hour in 10 ml recently exchanged storage solution the pH-range was always below 9 in material stored for over one week.

The total absence of material loss, measured as height reduction in the acid corrosion test, is related to the general basic nature of the material, with possibility of neutralization of the acid in the contact zone – especially in the earlier stage of the hydration process.

Hydrophilicity and water uptake

A confusing aspect sometimes mentioned related to hydration of biomaterials is the term hydrophilicity, why it seems appropriate to present actual definitions/descriptions.

Hydratisation: A chemical reaction where reactions with water forms new hydrate phases

Hydrophilicity: The wettability of phases formed, often measured as the contact angel of a water drop on the actual surface.

On-going hydration: The process for chemically bonded ceramics to fully convert the original oxide material to hydrates after initial curing

On-going water uptake: Long-term water uptake as free water or as hydrates, often in amounts less than 2 %. For especially dental materials in general this is regarded as a problem. However for chemically bonded ceramics such as Ca-aluminate based material this continuing water uptake just results in complete hydration. This small amount of reaction phases has only a miner effect on properties. For examples, Strength is increased somewhat, and dimensional changes are still very close to zero [12].

The possibility of some water uptake without changing of properties makes the Ca-aluminate based material very moisture tolerant. This is very important in general for most dental applications, since the oral environment is wet or humid. This makes the treatment easier with no extra need of water or humid protection.

Bacteriostatic and antibacterial aspects

The finding in studies recently performed [13] show that the bacteriostatic and antibacterial properties of the Ca-aluminate biomaterial may not primarily be related to pH or specific ions and ion concentration or reducing agents, but to the hydration procedure and the microstructure obtained. This also to some extent is an answer why highly biocompatible and even bioactive biomaterials can combine apparently contradictory features such as biocompatibility, bioactivity and apatite formation and environmental friendliness with bacteriostatic and antibacterial properties.

The bacteriostatic and antibacterial properties seem to be related to the development of the nanostructure and the nano-size porosity during hydration of the Ca-aluminate system. The initial low pH (< 8) of the system in the case of the presence of a polycarboxylic acid for cross-linking, is such not a hindrance for the antibacterial properties. The requirements of the microstructure of Ca-aluminate and also Ca-silicate based biomaterials to achieve antibacterial properties are related to the general nanostructure obtained; A nanoparticle/crystal size of hydrates in the interval 15-40 nm, a nanoporosity size of 1-4 nm and the number of pores per square micrometer of at least 500, preferably > 1000. Original particle size of the reacting chemically bonded ceramics should be less than 10 micrometer. The above mentioned requirements will guarantee that the nanostructure will be free of large pores meaning no escape of bacteria within the original liquid, paste or dental void, during the hydration.

The nanocrystals will participate on all walls, within the liquid, and on all inert particles and on bacteria within the original volume. The formation of nanocrystals will continue to all the void is filled. The bacteria will be totally encapsulated and will be chemically dissolved. Also the number of nanopores will be extremely which will have the possibility of catching and fasten bacteria to the hydrate surface – an analogue to how certain peptides may function as antibacterial material due to a structure with nanosize holes or nanosize topography of the structure.

Applications identified

The main hydration reaction, the general stability of the hydrates formed, and the contact zone reaction possibilities make the Ca-aluminate materials suitable as injectable biomaterials into tissue, within odontology as restoratives in enamel, dentine and augmentation in mandilla and maxilla, and within orthopedics as an augmentation material. The Ca-aluminate materials seem especially suitable for indications where stable materials are required. The first applications identified were as dental cement and restorative materials [14-18] and as stable materials within vertebroplasty, both for percutaneous vertebroplasty (PVP) and for kyphoplasty (KVP) [19-21]. The nanostructure of the Ca-aluminate based materials including nano-porosity, makes the material potential as carrier for drug delivery [22].

Below are presented in Figs. 4-5 some applications based on Ca-aluminate biomaterials.

 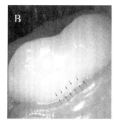

Fig. 4 A) High-resolution TEM of Ti-CA-paste interface, nano-mechanical integration,
B) Cemented ceramic crown.

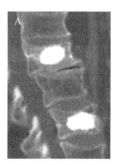 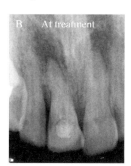 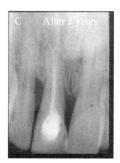

Fig. 5 A) Vertebral compression factures, restored by CA-material, and B) Bone loss at treatment,
and C) Bone recovery after endodontic retrograde treatment.

CONCLUSION

The Ca-aluminate based biomaterials exhibit some important features related to the biological response in contact with tissue. These are primarily related to

- chemical similarity with hard tissue
- nanostructural similarity with apatite crystasl in body tissue
- excellent biocompatibility including bioactivity
- chemical integration with body liquid forming apatite in the contact zone between the biomaterial and the hard tissue structures
- bacteriostatic and antibacterial properties
- favorable hydrophilicity and moisture-tolerance
- dimension stability and no shrinkage over time reduce stresses between the biomaterial and the biological surrounding tissue.

The Ca-aluminate technology provides a platform upon which Ca-aluminate based materials may work as a general biomaterial and as a complement to other chemically bonded ceramics based on phosphates, silicates or sulphates. Identified areas are in the first place within the dental and orthopedic areas, where injectable stable biomaterials are required.

ACKNOWLEDGEMENT

Results presented in this paper are based on two decades of research within Doxa AB and the Engineering Science Department, The Angstrom Laboratory, Uppsala University.

REFERENCES

[1] L. L. Hench, Biomaterials: a forecast for the future, *Biomaterials*, 19 (1998) 1419-1423.
[2] K. L. Scrivener and A. Capmas, *LEA'S Chemistry of Cement and Concrete*, P.C. Hewlett, Editor. 1988, Arnold. p. 709-771.
[3] A. Muan, EF. Osbourne, Phase equilibria among oxides. Adison-Wesley ,1965; New York
[4] H. Engqvist, J. Loof, S. Uppstrom, M. W. Phaneuf, J. C. Jonsson, L. Hermansson, N-O. Ahnfelt, Transmittance of a bioceramic calcium aluminate based dental restorative material *Journal of Biomedical Materials Research* Part B: Applied Biomaterials, 69 1 (2004) pp. 94-98
[5] H. Engqvist, G.A. Botton, M. Couillard, A novel tool for high resolution transmission electron microscopy of intact interfaces between bone and metallic implants, *J. Biomed. Mater. Res.* 2006 (78A 5) 20-24
[6] ISO 10993:2003
[7] ISO standard 23317 (2007)
[8] EN 29917:1994/ISO 9917:1991
[9] L. Hermansson, J. Lööf, T. Jarmar, Integration mechanisms towards hard tissue of the Ca-aluminate based biomaaterials, *Key Eng. Mater.* 2009;396-398:183-186.
[10] N. Axén, L-M. Bjursten, H. Engqvist, N-O. Ahnfelt, L. Hermansson, Zone formation at the interface between Ca-aluminate cement and bone tissue environment, *Ceramics, Cells and Tissue, 9th Annual Conference*, Faenza, Italy, 2004
[11] L. Hermansson, J. Lööf, T. Jarmar, Injectable ceramics as biomaterials: Today and tomorrow, in *Proc. ICC2*, Verona 2008
[12] L. Kraft, *Ph D Thesis,* Calcium aluminate based cement as dental restorative materials. Faculty of Science and technology, Uppsala University, Sweden. 2002.
[13] Patent pending, Doxa AB

[14] H. Engqvist, J-E. Schultz-Walz, J. Loof, G. A. Botton, D. Mayer, M. W. Phaneuf, N-O. Ahnfelt, L. Hermansson, Chemical and biological integration of a mouldable bioactive ceramic material capable of forming apatite in vivo in teeth, *Biomaterials* vol 25 (2004), 2781-2787

[15] L. Kraft, M. Saksi, L. Hermansson, C.H. Pameijer, A five-year retrospective clinical study of a calcium-aluminate in retrograde endodontics, J Dent Res. 2009;88(A):1333.

[16] L. Hermansson, A. Faris, G. Gomez-Ortega, J. Kuoppola and J. Lööf: Aspects of Dental Applications Based on Materials of the System, *Ceramic Engineering and Science Proceedings*, Volume 30, Issue 6, 71-80 (2009)

[17] S. Jefferies, C.H. Pameijer, D. Appleby, J Lööf and D. Boston: One year clinical performance and post-operative sensitivity of a bioactive dental luting cement – A prospective clinical study, Swed. Dent. J. 2009: 33, 193-199

[18] J. Lööf, Calcium-Aluminate as Biomaterial: Synthesis, Design and Evaluation, PhD Thesis Faculty of Science and Technology, Uppsala, University, Sweden (2008)

[19] H. Engqvist, M. Couillard, G. A. Botton, M. W. Phaneuf, N. Axén, N-O. Ahnfelt, L. Hermansson, In vivo bioactivity of a novel mineral based orthopaedic biocement, *Trends in Biomaterials and Artificial Organs*, 19 (2005), 27-32

[20] J. Lööf, A. Faris, L.Hermansson, H. Engqvist, In Vitro Biomechanical Testing of Two Injectable Materials for Vertebroplasty in Different Synthetic Bone, *Key Eng. Mater.* 361-363, (2008), 369-372

[21] T. Jarmar, T. Uhlin, U. Höglund, P. Thomsen, L. Hermansson, H. Engqvist, Injectable bone cemets for vertebroplasty studied in sheep vertebrae with electron microscopy, *Key Eng. Mater.*, 361-363 (2008), 373-37620)

[22] L Hermansson, Chemically bonded bioceramic carrier systems for drug delivery, *Ceram Eng. Sci. Proc.* 31, 77-88 (2010)

TITANIUM ALLOYS WITH CHANGEABLE YOUNG'S MODULUS FOR PREVENTING STRESS SHIELDING AND SPRINGBACK

Mitsuo Niinomi[1], Masaaki Nakai[1], Junko Hieda[1], Xiaoli Zhao[2], and Xfeng Zhao[2]

[1]Institute for Materials Research, Tohoku University, Sendai, Japan.
[2]Graduate Student of Tohoku University, Sendai, Japan
niinomi@imr.tohoku.ac.jp, nakai@imr.tohoku.ac.jp, hieda@imr.tohoku.ac.jp,
xiaolizhao@imr.tohoku.ac.jp, xfengzhao@imr.tohoku.ac.jp

ABSTRACT

While using low modulus titanium alloys, which can suppress stress shielding, several surgeons specializing in spinal diseases, such as scoliosis and spine fracture, pointed out that the amount of springback in the implant rods need to be smaller to better handle the implants during surgery; a high Young's modulus suppresses springback. Titanium alloys, which satisfy the requirements of both surgeons (a high Young's modulus, leading to suppressed springback) and patients (a low Young's modulus, leading to suppressed stress shielding) with regard to the Young's modulus of the implant rod, are currently being developed. We propose metastable Ti-12Cr as a candidate alloy for solving this problem. Ti-12Cr subjected to solution treatment (Ti-12Cr-ST) exhibits a low Young's modulus of <70 GPa. In contrast, the Young's modulus of Ti-12Cr subjected to cold rolling (Ti-12Cr-CR), is >80 GPa. This increase in the Young's modulus is due to the deformation-induced ω phase.

In order to develop titanium alloys with a changeable Young's modulus, we also propose the metastable Ti-30Zr-(Cr, Mo) alloy, which is developed based on the β-type titanium alloy Ti-30Zr-Mo with a low Young's modulus. Ti-30Zr-(Cr, Mo) and Ti-30Zr-3Cr-3Mo, which exhibit excellent tensile properties and changeable Young's modulus values, produce smaller springback than Ti-29Nb-13Ta-4.6Zr (TNTZ), which has a low Young's modulus. The latter is a typical β titanium alloy for biomedical applications. Thus, 3Cr3Mo is a potential candidate for biomedical implant applications.

INTRODUCTION

While using low-modulus titanium alloys, which suppress stress shielding, several surgeons specializing in spinal diseases, such as scoliosis and spine fracture, pointed out that the amount of springback in the implant rods needs to be smaller to offer a better handling ability during surgeries; a high Young's modulus suppresses springback. Self-adjustable titanium alloys, which satisfy the requirements of both surgeons (a high Young's modulus, leading to suppressed springback) and patients (a low Young's modulus, leading to suppressed stress shielding) with regard to the Young's modulus of the implant rod[1], are currently being developed. The target alloys are metastable β-type titanium alloys consisting of nontoxic and allergy-free elements with a low Young's modulus, where the deformation ω phase can be induced because it increases their Young's modulus values. In order to study the fundamental chemical composition of an alloy possessing the ability to self-adjust its Young's modulus, the increase in Young's modulus due to cold rolling was systematically investigated for the following alloys: binary Ti–(10 and 12) mass% Cr (Ti-10Cr and Ti-12Cr), Ti–(13 and 15) mass% Mo (Ti-13Mo and Ti-15Mo), Ti–(20 and 22) mass% V (Ti-20V and Ti-22V), and Ti– (6 and 8) mass% Fe (Ti-6Fe and Ti-8Fe) alloys. Moreover, Ti-12Cr has been reported to satisfy the abovementioned purpose.

Implants are often removed and then reimplanted. In this case, potential refracturing of the bones and calcium phosphate precipitation on the implants need to be prevented. It is reported that Zr has the ability to prevent the precipitation of calcium phosphate[2] and that Ti alloys with Zr content exceeding 25 mass% prevent the formation of calcium phosphate, which is the main component of human bones. Consequently, Ti-Zr and Ti-Zr-Mo-Cr metastable alloys with adjustable Young's modulus values are currently being developed.

In this paper, we describe the design concept for obtaining a changeable Young's modulus in β-type titanium alloys, β-type Ti-Cr alloys, β-type Ti-Zr alloys.

CONCEPT OF A CHANGEABLE YOUNG'S MODULUS

Deformation-induced transformation is the main factor that causes the Young's modulus to change from a low to a high value. Consequently, the chemical composition of titanium alloys in which the Young's modulus of only the deformed part changes from a low to a high need to be determined. Deformation-induced phases appear according to the stability of the metastable phase(s) and the type of alloying element used. In titanium alloys, the β phase is metastable. Consequently, the design target for titanium alloys with changeable Young's modulus values is to obtain metastable β-type titanium alloys (hereafter β-type titanium alloy). The β phase is stabilized by elements such as Nb, Ta, V, Mo, Cr, and Fe. As a result, the binary alloy systems of Ti-Nb, Ti-Ta, Ti–Cr, Ti–Mo, Ti–V, and Ti– Fe are good candidates for producing titanium alloys with changeable Young's modulus values. The deformation-induced phases that appear in β-type titanium alloys are the α″ and ω phases. The deformation-induced α″ phase decreases the Young's modulus of β-type titanium alloys because of lattice extension and the formation of deformation textures[2]. Furthermore, many researchers have reported on the appearance of the deformation-induced ω phase and its increased Young's modulus[3, 4-7]. Therefore, we used the results of previous studies to determine the chemical composition of β-type titanium alloys in which the deformation-induced ω phase transformation is likely to occur.

Ti-12Cr SYSTEM ALLOYS WITH CHANGEABLE YOUNG'S MODULUS

Ti-10Cr, Ti-12Cr, Ti-15Mo, Ti-20V, and Ti-22V were prepared and cold-rolled, regardless of the occurrence of the deformation-induced ω phase. Cold rolling, instead of bending, was employed to study the effect of the deformation-induced ω phase transformation on the mechanical properties including the change in the Young's modulus values. Mechanical properties were subsequently measured for cold-rolled specimens. Among the examined alloys, the Young's modulus values of Ti-10Cr, Ti-12Cr, Ti-15Mo, Ti-20V, and Ti-22V increased due to cold rolling, and that of Ti-12Cr showed the largest increase. Therefore, Ti-12Cr was judged as the most suitable β-type titanium alloy for biomedical implant applications.

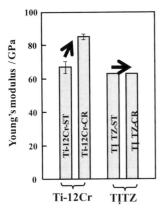

Fig.1 Young's moduli of Ti-12Cr and TNTZ subjected to solution treatment (ST) and cold rolling (CR).

Figure 1 compares the Young's moduli of Ti-12Cr, subjected to solution treatment (Ti-12Cr-ST) and cold rolling (Ti-12Cr-CR), with those of Ti-29Nb-13Ta-4.6Zr (TNTZ)[8], subjected to solution treatment (TNTZ-ST) and cold rolling (TNTZ-CR)[1] measured by resonance method. Ti-12Cr-ST exhibits a low Young's modulus of ~60 GPa; this value with that of TNTZ-ST, which is developed as a biomedical β-type titanium alloy with a low Young's modulus. TNTZ-CR also exhibits a low Young's modulus, which is almost equal to that of TNTZ-ST. Cold rolling leads to mealy no change in the Young's modulus of TNTZ. However, in the case of Ti-12Cr, the Young's modulus increases by cold rolling and that of Ti-12Cr-CR is >80 GPa.

Figure 2 shows the TEM diffraction patterns of the $[110]_\beta$ zone in Ti-12Cr-ST and Ti-12Cr-CR, respectively[1]. Only a circle-shaped streak is observed in combination with the spots derived from the β phase in Ti-12Cr-ST. However, in addition to the reflection from the β phase, extra spots, possibly derived from the ω phase, can be clearly observed for Ti-12Cr-CR. These results confirm the occurrence of the deformation-induced ω phase transformation in Ti-12Cr, which results from cold rolling.

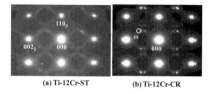

(a) Ti-12Cr-ST   (b) Ti-12Cr-CR

Fig. 2 TEM diffraction patterns of (a) Ti-12Cr-ST and (b) Ti-12Cr-Cr.

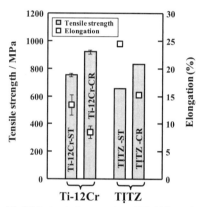

Fig.3 Mechanical properties of Ti-12Cr and TNTZ subjected to solution treatment (ST) and cold rolling (CR).

Figure 3 shows the mechanical properties of Ti-12Cr-ST, Ti-12Cr-CR, TNTZ-ST, and TNTZ-CR[1] measured by tensile tests using an Instron type testing machine. The tensile strengths of both Ti-12Cr-ST and TNTZ-ST increase and their elongations decrease. The tensile strengths of Ti-12Cr-ST and Ti-12Cr-CR are higher than those of TNTZ-ST and TNTZ-CR, respectively. The elongations of Ti-12Cr-ST and Ti-12Cr-CR are >10% and ~10%, respectively. High strength is an essential requirement from the viewpoint of practical applications, although such high strength leads to undesirable springback.

β-TYPE Ti-Zr SYSTEM ALLOYS WITH LOW YOUNG'S MODULUS FOR REMOVABLE IMPLANTS

It is essential to remove the internal fixator after surgery whenever there is a use of certain types of internal fixation devices (implanted into bone marrows such as femoral, tibial, and humeral marrows), screws for bone plate fixation, and implants for children. Otherwise, the fixator can grow into the bone. The assimilation of the removable internal fixators by the bone because of the precipitation of calcium phosphate may cause the refracturing of the bone during removal of the fixator. Therefore, it is essential to prevent the adhesion of alloys to the bone tissues, and to inhibit the precipitation of calcium phosphate.

It is reported that Zr has the ability to prevent the precipitation of calcium phosphate[9], and Ti alloys with Zr content exceeding 25 mass% prevent the formation of calcium phosphate, which is the main component of human bones. It is also reported that the tensile strength of these alloys is fairly high; however, the elongation decreases when the Zr content exceeds 56 mass%[10]. By taking the abovementioned facts into account, the microstructures and mechanical properties of

Ti-30Zr-xMo ($x = 0$, 2, 5, 6, 7, and 8 mass%, which are denoted as 0Mo, 2Mo, 5Mo, 6Mo, 7Mo, and 8Mo, respectively) were examined.

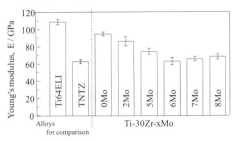

Fig. 4 Young's moduli of Ti-30Zr-xMo ($x = 0$, 2, 5, 6, 7, and 8 mass.%) alloys subjected to solution treatment and the alloys for comparison (Tti-6Al-4V ELI (Ti64ELI) and Ti-29Nb-13Ta-4.6Zr (TNTZ).

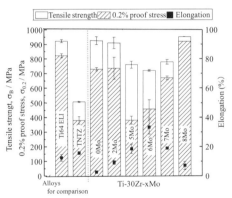

Fig. 5 Tensile properties of Ti-30Zr-xMo ($x = 0$, 2, 5, 6, 7, and 8 mass.%) subjected to solution treatment and the alloys considered for comparison (Ti-6Al-4V ELI(Ti64ELI) and Ti-29Nb-13Ta-4.6Zr (TNTZ).

elongation can be obtained for 6Mo and 7Mo.

The Young's moduli of Ti-30Zr-xMo alloys subjected to solution treatment and those of the alloys considered for comparison are shown in Fig. 4[11] by measured using a resonance method. The Young's moduli of Ti-30Zr-xMo alloys are lower than those of the alloys considered for comparison, except in the case of TNTZ. The Young's modulus of 0Mo is around 100 GPa, and those of Ti-30Zr-xMo alloys range from ~90 GPa (2Mo) to ~60 GPa (6Mo). The Young's modulus of 6Mo shows a minimum value of around 60 GPa, which is lower than that of 0Mo by 34%. 6Mo, 7Mo, and TNTZ exhibit a low Young's modulus.

Figure 5 shows the tensile properties of Ti-30Zr-xMo *alloys subjected to solution* treatment and those of the alloys considered for comparison[11] measured by tensile tests using an Istron type testing machine. The as-solutionized Ti-30Zr-xMo alloys exhibit relatively good tensile properties in comparison with other alloys. The tensile strengths of the as-solutionized Ti-30Zr-xMo alloys vary in a similar manner as in their Young's modulus values, as, shown in Fig. 6. A good balance between strength and

CHANGEABLE YOUNG'S MODULUS Ti-Zr SYSTEM ALLOYS FOR USE IN REMOVABLE IMPLANTS

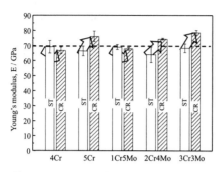

Fig. 6 Young's moduli of Ti-30Zr-4Cr (4Cr), -5Cr (5Cr), -1Cr-5Mo (1Cr5Mo), -2Cr-4Mo (2Cr4Mo), and -3Cr-3Mo (3Cr3Mo) subjected to solution treatment (ST) and cold rolling (CR).

Titanium alloys with a changeable Young's modulus that can take a low or a high value are required for use in removable implants. Based on the development of changeable Young's modulus Ti-30Zr-Mo system alloys for use in removable implants, Ti-30Zr-4Cr, Ti-30Zr-5Cr, Ti-30Zr-1Cr-5Mo, Ti-30Zr-2Cr-5Mo, Ti-30Zr-2Cr-4Mo, and Ti-30Zr-3Cr-3Mo respectively denoted as 4Cr, 5Cr, 1Cr5Mo, 2Cr4Mo, and 3Cr3Mo have been designed, and their mechanical properties such as springback and the change in the Young's modulus due to deformation were subsequently investigated.

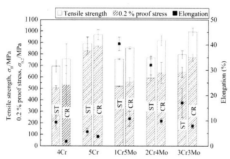

Fig. 7 Mechanical properties of Ti-30Zr-4Cr (4Cr), -5Cr (5Cr), -1Cr-5Mo (1Cr5Mo), -2Cr-4Mo (2Cr4Mo), and -3Cr-3Mo (3Cr3Mo) subjected to solution treatment (ST) and cold rolling (CR).

Figure 6 shows the Young's moduli of the designed alloys that are subjected to solution treatment (ST) and cold rolling (CR)[12]. All the novel designed alloys exhibit low Young's modulus values of <70 GPa; this value is much lower than the Young's modulus of Ti-6Al-4V ELI (Ti64 ELI), which is widely used for biomedical implant applications. After cold rolling, the Young's moduli of 4Cr and 1Cr5Mo decreased slightly, unlike in the case of those subjected to ST. However, the Young's moduli of 5Cr, 2Cr4Mo, and 3Cr3Mo increase by cold rolling. The Young's modulus of 3Cr3Mo after cold rolling increased to around 80 GPa, which an increase of 15%.

Figure 7 shows the tensile properties of each designed alloy that is subjected to solution treatment (ST) and cold rolling (CR)[12]. The tensile strengths of all the alloys subjected to solution treatment are >700 MPa. The elongations of the Ti-30Zr-Cr system alloys are <10% even under ST conditions. In contrast, the elongations of the Ti-30Zr-Cr-Mo system alloys are much larger than those of the Ti-30Zr-Cr system alloys. Apparently, Mo effectively suppresses the formation of deformation textures and improves the elongation of Ti-30Zr. After cold rolling, the tensile strength of each alloy increases more than that of the alloys subjected to solution treatment. The 0.2% proof stress behaves similarly to the tensile strength. However, the elongations from cold rolling tend to decrease. In Ti-30Zr-Cr-Mo alloys, the elongations after cold rolling are ~10%. High strength and a large elongation are essential requirements in practical applications. Therefore, in combination with its changeable Young's modulus, 3Cr3Mo is a preferable titanium alloy for use in biomedical implant applications.

$$R = \left[(\varepsilon_{i2} - \varepsilon_{i3}) / (\varepsilon_{i2} - \varepsilon_{i1}) \right] / (\sigma_i \times A)$$

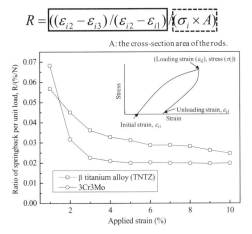

Fig. 8 Ratio of springback per unit load as function of applied strain for 3Cr3Mo and TNTZ and strains for calculation of springback ratio (inset).

Figure 8 shows the comparison between the ratio (R) of the springback and the loading strain per unit load as a function of the applied tensile strain for 3Cr3Mo and TNTZ[12]. The springback per unit load ratio for both 3Cr3Mo and TNTZ decreases with increasing applied strain. This decrease is considered to depend on the increase in stress that corresponds to the applied strain. The springback per unit load ratio of 3Cr3Mo is smaller than that of TNTZ, and it reaches a stable value after an applied strain of 3%. Thus, 3Cr3Mo has better potential for biomedical applications than TNTZ, from the viewpoint of springback.

SUMMARY

Ti-12Cr is a candidate β-type titanium alloy with a changeable Young's modulus. Ti-30Zr-7Mo is expected to be used in removable implants. Ti-30Zr-3Mo-3Cr is expected to be used in removable implants that have a changeable Young's modulus. The implants made using the three kinds of β-type titanium alloys discussed exhibit low Young's modulus values that can prevent stress shielding. Ti-30Zr-3Mo-3Cr exhibits less springback than Ti-29Nb-13Ta-4.6Zr.

REFERENCES

[1] M. Nakai, M. Niinomi, X. F. Zhao, X. Zhao and K. Narita, Self-Adjustment of Young's Modulus in Metallic Biomaterials during Orthopaedic Operation through Deformation-induced Phase Transformation, *Mater. Letters*, **65**, 688–690 (2011).

[2] S. Banerjee and P. Mukhopadhyay, *Phase Transformations: Examples from Titanium and Zirconium Alloys* (Elsevier, 2007).

[3] H. Matsumoto, S. Watanabe, N. Masahashi and S. Hanada, Composition Dependence of Young's Modulus in Ti-V, Ti-Nb, and Ti-V-Sn Alloys. *Metall. Mater. Trans. A*, **37**, 3239–3249 (2006).

[4] H. Matsumoto, S. Watanabe and S. Hanada, α' Martensite Ti–V–Sn Alloys with Low Young's Modulus and High Strength. *Mater. Sci. Eng. A*, **448**, 39–48 (2007).

[5] R. M. Wood, Martensitic Alpha and Omega Phases as Deformation Products in a Titanium-15% Molybdenum Alloy. *Acta Metall,.* **11**, 907–914 (1963).

[6] S. Hanada and O. Izumi, Transmission Electron Microscopic Observation of Mechanical Twinning in Metastable Beta Titanium Alloys. *Metall. Trans. A*, **17**, 1409–1420 (1986).

[7] S. Hanada and O. Izumi, Deformation Behavior of Retained β Phase in β-Eutectoid Ti-Cr Alloys. *J. Mater. Sci.*, **21**, 4131–4139 (1986).

[8] M. Niinomi, Fatigue Performance and Cyto-toxicity of Low Rigidity Titanium Alloy, Ti-29Nb-13Ta-4.6Zr, *Biomater.*, **24**, 2673–2683 (2003).

[9] E. Kobayashi, M. Ando, Y. Tsutsumi, H. Doi, T. Yoneyama, M. Kobayashi and T. Hanawa, Inhibition Effect of Zirconium Coating on Calcium Phosphate Precipitation of Titanium to Avoid Assimilation with Bone, *Mater. Trans,.* **48**, 301–306 (2007).

[10] E. Kobayashi, H. Doi, T. Yoneyama, H. Hamanaka, S. Matsumoto and K. Kudake, Evaluation of Mechanical Properties of Dental Casting Ti-Zr Based Alloys, *Jpn. Soc. Dent. Mater. Devices*, **14**, 321–328 (1995).

[11] X.L. Zhao, M. Niinomi, M. Nakai, T. Shimoto and T. Nakano, Development of High Zr-containing Ti-based Alloys with Low Young's Modulus for Use in Removable Implants, *Mater. Sci. Eng. C*, **31**, 1436–1444 (2011).

[12] X. L, Zhao, M. Niinomi, M. Nakai, G. Miyamoto and T. Furuhara, Microstructures and Mechanical Properties of Metastable Ti-30Zr-(Cr, Mo) Alloys with Changeable Young's Modulus for Spinal Fixation Applications, *Acta Biomater.*, **7**, 3230–3236 (2011).

BIOACTIVE GLASS IN BONE TISSUE ENGINEERING

Mohamed N. Rahaman[1], Xin Liu[1], B. Sonny Bal[2], Delbert E. Day[1], Lianxiang Bi[3], Lynda F. Bonewald[3]

[1]Missouri University of Science and Technology, Department of Materials Science and Engineering, and Center for Bone and Tissue Repair and Regeneration, Rolla, Missouri 65409
[2]University of Missouri, Department of Orthopaedic Surgery, Columbia, Missouri 65211
[3]University of Missouri-Kansas City, Department of Oral Biology, School of Dentistry, Kansas City, Missouri 64108

ABSTRACT
Bioactive glass has several appealing characteristics as a scaffold material for bone tissue engineering. While silicate bioactive glasses (such as 45S5 glass) have been widely investigated over the last few decades, borate bioactive glasses have been attracting considerable interest recently. Borate bioactive glasses convert faster and more completely to a hydroxyapatite-like material in a simulated body fluid than silicate bioactive glasses, and they have been shown to support faster bone formation and enhance angiogenesis. However, borate bioactive glasses show a more rapid degradation in strength than silicate bioactive glasses when immersed in a simulated body fluid or implanted in vivo. Advances in biomaterials processing have resulted in the creation of bioactive glass scaffolds with compressive strength and elastic modulus values in the range of trabecular to cortical bone. This article provides a review of our recent advances in bioactive glasses for the regeneration of loaded and non-loaded bone.

INTRODUCTION

Despite considerable work over many decades, the repair of large defects in load-bearing bone, such as segmental defects in the long bones, remains a challenging clinical problem. Focal bone defect repair is possible today using a variety of commercially-available synthetic bone substitutes, none of which can replace structural bone loss. Traditionally, bone allograft, metal spacers, and even bone cement have been used to restore segmental bone defects in the limbs. These methods are limited by high costs, limited availability, unpredictable long-term durability, and uncertain healing to host bone. The market for implants to repair segmental bone loss is estimated to be large; however, no product has successfully addressed the need for the replacement of structural bone loss in the limbs. New porous biocompatible implants that replicate the porosity, bioactivity, strength, and load-bearing ability of living bone are therefore needed.

Bioactive glass, while brittle, has several attractive characteristics as a scaffold material for bone repair. After *in vivo* implantation, bioactive glasses react with the body fluid and convert to hydroxyapatite (HA), the main mineral constituent of bone, which bonds firmly to host bone and soft tissue [1]. Calcium ions and soluble silicon released from the bioactive glass during the conversion to HA promote osteogenesis [2,3] and activate osteogenic gene expression [4,5]. The compositional flexibility of glass can be used so that it is a source of many of the trace elements known to favor bone growth, such as boron, copper, and zinc; as the glass reacts *in vivo* these elements are released at a biologically acceptable rate. Another inherent advantage of glass scaffolds is the flexibility of preparing three-dimensional (3D) scaffolds with a wide range of architectures that provide the requisite mechanical properties for load bearing, as well as an optimum physical and chemical environment for growing bone cells.

This article provides a review of our recent research in the development of bioactive glasses for the regeneration of loaded and non-loaded bone. Following a description of recently-developed bioactive glass compositions, the methods used to form bioactive glass into 3D scaffold architectures with mechanical strength in the range of trabecular to cortical bone are described. The performance of these bioactive glass scaffolds in vitro and in vivo are considered. Finally, our recent advances in the development of bioactive glass for loaded and non-loaded bone repair are summarized.

BIOACTIVE GLASS COMPOSITION

Silicate 45S5 glass in particulate, granular, or low-strength 3D architecture has been widely used over the last few decades to investigate bioactivity and bone attachment [1]. However, 45S5 glass has proved to be sub-optimal for bone repair because it crystallizes prior to appreciable viscous flow sintering, thereby resulting in a low-strength 3D structure [6]. Also 45S5 glass conversion to HA is incomplete in a simulated body fluid (SBF), suggesting that unconverted glass could remain in the body over an extended period [7]. A silicate bioactive glass, designated 13-93, has better processing characteristics by viscous flow sintering than 45S5 glass [8-11]. While it is based on the 45S5 composition (Table I), 13-93 glass has a higher SiO_2 content, plus additional network modifiers such as K_2O and MgO [12], which provides a larger window between the glass transition temperature and the onset temperature for crystallization. As a result the glass phase in porous constructs formed from 13-93 glass particles can be sintered to high density without crystallization, which is beneficial for producing scaffolds with enhanced strength. Furthermore, in vitro cell culture has shown no marked difference in the ability of 13-93 and 45S5 glass scaffolds to support the proliferation and function of osteoblastic cells [13]. However, 13-93 glass converts more slowly to HA than 45S5.

Table I. Bioactive glass compositions (in wt%)

Glass	SiO_2	B_2O_3	Na_2O	K_2O	MgO	SrO	CaO	P_2O_5	CuO
45S5	45.0		24.5				24.5	6.0	
13-93	53.0		6.0	12.0	5.0		20.0	4.0	
13-93B1	34.4	20.0	5.8	11.6	4.9		19.5	3.9	
13-93B2	16.5	38.2	5.7	11.5	4.9		18.8	4.3	
13-93B3		56.6	5.5	11.1	4.6		18.5	3.7	
13-93B2Sr	15.6	36.1	5.4	10.0	1.2	9.0	17.8	4.1	
13-93B3Cu		52.8	6.0	12.0	5.0		19.9	4.0	0.4

The last decade has seen the development of borate bioactive glass [14-16]. Because of their lower chemical durability, some borate bioactive glasses convert faster and more completely to HA than silicate 45S5 or 13-93 glass. Recent work has also resulted in the development of bioactive glass with controllable degradation rates [17, 18]. Using silicate 45S5 or 13-93 glass as the parent glass, the molar concentration of SiO_2 in these silicate glasses was partially or fully replaced by B_2O_3, resulting in borosilicate and borate bioactive glasses (Table I). For example, borate-based bioactive glasses designated 13-93B1, 13-93B2, and 13-93B3 were obtained by replacing 1/3, 2/3, and all of the molar concentration of SiO_2 in 13-93 glass with B_2O_3. The glass designated 13-93B2Sr, consists of a strontium-doped 13-93B2 glass [19], while the borate 13-93B3 glass doped with copper (0.4 wt% CuO) is designated 13-93B3Cu.

BIOACTIVE GLASS SCAFFOLDS

Scaffolds should be biocompatible, have a porous microstructure suitable for supporting tissue ingrowth, and also have mechanical properties comparable to those of the tissue to be replaced. An interconnected pore size (diameter or width of the openings between adjoining pores) of ~100 μm is considered to be the minimum requirement to permit tissue ingrowth and function [20]. Bioactive glass scaffolds with four different architectures have been created in this work (Fig. 1). The methods used to create the BG scaffolds and the characteristics of the scaffolds are summarized in Table II. In one method [13], a random packing of short fibers (100–200 μm in diameter × 3 mm in length) was thermally bonded in a graphite or mullite mold (Fig. 1a). The polymer foam replication technique [9] resulted in the creation of a scaffold architecture similar to that of dry human trabecular bone. Unidirectional freezing of an organic (camphene)-based suspensions [21] was used to create scaffolds with a columnar microstructure of oriented pores (Fig. 1c), while scaffolds with a grid-like microstructure (Fig. 1d) were created using a robocasting technique [22].

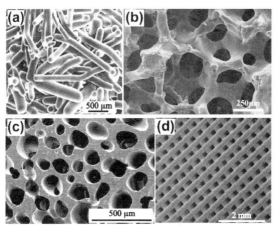

Figure1. Microstructures of bioactive glass (13-93) scaffolds created using different processing methods: (a) thermal bonding of short fibers; (b) polymer foam replication technique; (c) unidirectional freezing of suspensions; (d) robocasting.

Table II. Characteristics of bioactive glass scaffolds created using different methods

Method	Glass	Porosity (%)	Pore size (μm)	Compressive strength (MPa)	Reference
Thermal bonding of short fibers	13-93	45–50	>100	5	23
Polymer foam replication	13-93	75–85	100–500	11 ± 1	9
	13-93B1	80–85	100–500	7 ± 0.5	24
	13-93B3	80–85	100–500	5 ± 0.5	
Unidirectional freezing of suspensions	13-93	50	50–150	47 ± 5	25
Rapid prototyping	13-93	50	300	140 ± 70	11
	13-93B3	50	300	65±20	

IN VITRO EVALUATION OF BIOACTIVE GLASS SCAFFOLDS

Glass composition and microstructure have a strong effect on the conversion of the bioactive glass scaffold to HA and on the degradation of the strength of the scaffold as a function of immersion time in an SBF. Because of their lower chemical durability, borate 13-93B3 scaffolds convert much faster to HA than silicate 13-93 scaffolds (Fig. 2a). However, the as-fabricated 13-93B3 scaffold has a lower strength than the 13-93 scaffold, and its strength also degrades faster than 13-93 in a simulated body fluid (SBF) as a result of its faster conversion rate (Fig. 2b).

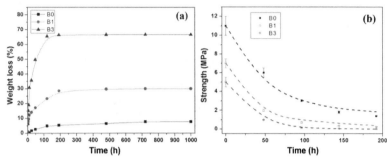

Figure 2. Effect of bioactive glass composition on (a) conversion to HA (as measured by the weight loss), and (b) compressive strength, as a function of immersion time in a simulated body fluid. The results are shown for silicate 13-93 (B0), borosilicate 13-93B1 (B1), and borate 13-93B3 (B3) bioactive glass scaffolds with the same trabecular microstructure [24].

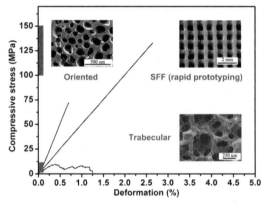

Figure 3. Compressive stress vs. deformation for silicate 13-93 bioactive glass with three different microstructures: "oriented" (prepared by unidirectional freezing of camphene-based suspensions), "grid-like" (prepared using a rapid prototyping technique), and "trabecular" (prepared using a polymer foam replication technique) [26].

Figure 3 shows the mechanical response in compression of as-fabricated 13-93 scaffolds with three different microstructures: (1) a trabecular microstructure (porosity = 85%; pore size = 100–500 μm) prepared using a polymer foam replication method [9]; (2) an oriented microstructure (porosity = 50%; pore diameter = 50–150 μm) prepared by unidirectional freezing of camphene-based suspensions [25]; and a grid-like microstructure (porosity = 50%; pore width = 300 μm) prepared by a rapid prototyping technique [11]. The scaffold with the oriented and grid-like microstructures show an elastic response followed by failure in a brittle manner, typical of dense ceramics and glass. The scaffold with the oriented microstructure has a compressive strength (47 ± 5 MPa) far higher than that of trabecular bone (2–12 MPa). In comparison, the scaffold with the grid-like microstructure has even more impressive mechanical properties, with a compressive strength (140 ± 70 MPa) in the range reported for cortical bone (100–150 MPa). The bioactive glass scaffold with the trabecular microstructure shows a brittle response with compaction of the scaffold at higher deformation, as evidenced by the peaks and valleys in the stress vs. strain curve. The compressive strength of the "trabecular" scaffold (11 ± 1 MPa) is comparable to the highest values reported for trabecular bone.

IN VIVO EVALUATION OF BIOACTIVE GLASS SCAFFOLDS

Glass composition and microstructure have also been observed recently to have a strong effect on the ability of bioactive glass scaffold to support bone ingrowth, mineralization, and angiogenesis and to heal defects in rat calvaria. In one study [27], scaffolds of 13-93, 13-93B1, and 13-93B3 glass (5 mm in diameter × 1–2 mm) with a "fibrous" microstructure (porosity ≈ 50%; pore size = 100–400 μm) (Fig. 1a) were implanted for up to 12 weeks in non-healing rat calvarial defects. Defects filled with 45S5 bioactive glass particles (150–300 μm) served as positive controls, while the empty defects served as negative controls. Sections of the implants were stained with hematoxylin and eosin (H&E), von Kossa, and periodic acid-Schiff (PAS) and then analyzed using histomorphometry to quantitate new bone, mineralization, and blood vessel invasion into the implants (Fig. 4). A greater amount of bone was formed within the borate 13-93B3 scaffold compared to 13-93 and 13-93B1. While osteointegration occurred with all scaffolds, greater mineralization was associated with the 13-93B3 implants (new bone and conversion of the scaffold to HA), as summarized in Table III. Scaffolds of borate 13-93B3 glass doped with copper (Cu) or seeded with bone marrow-derived MSCs also promoted angiogenesis when compared to silicate 13-93 scaffolds with the same microstructure (results not shown). Stimulation of angiogenesis by Cu was concentration dependent, with 0.4 wt% Cu providing the optimum enhancement.

Table III. Percentage new bone and mineralization matrix (von Kossa positive) as a function of bioactive glass composition after 12-week implantation in rat calvarial defects [27].

	Bioactive glass			
	13-93	13-93B1	13-93B3	45S5
% new bone	8.5	9.7	14.9	12.4
% von Kossa positive	34.1	39.2	45.7	30.7
% blood vessel area	2.0	2.4	2.2	3.8

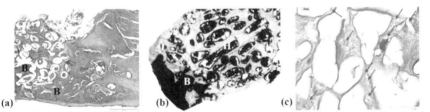

(a) (b) (c)

Figure 4. Histology of borate 13-93B3 scaffolds with a fibrous microstructure following 12-week rat calvarial implantation: (a) H&E stained section for new bone growth (B). The white circles show where hydroxyapatite (HA) was formed by conversion of the bioactive glass scaffold (magnification = 4×); (b) von Kossa section for mineralization showed that new bone (B) formed at the edge and within the scaffold, and the scaffold completely converted to HA (4×); (c) PAS stained section for blood vessel area (arrows) (10×). [27]

The effect of scaffold microstructure on bone regeneration, mineralization, and angiogenesis has also been investigated in a non-healing rat calvarial defect model using the methods described previously. Liu et al [28] implanted silicate 13-93 scaffolds with a trabecular microstructure (porosity = 75%, pore size = 100–400 μm) and an oriented microstructure (porosity = 50%; pore diameter = 50–150 μm) for 12 and 24 weeks and evaluated bone formation and conversion to HA using X-ray micro-computed tomography (microCT), histomorphometry, and SEM. Twelve weeks post implantation, new bone formed mainly on the dural side (bottom region) of the "oriented" scaffolds. In comparison, bone ingrowth occurred both on the dural side and into the circumferential region of the "trabecular" scaffolds (Fig. 5). For both groups of scaffolds, bone ingrowth increased with implantation time (24 weeks vs. 12 weeks) (results not shown).

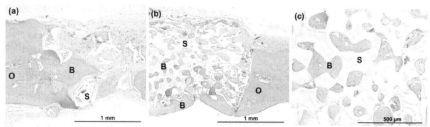

Figure 5. Comparison of bone formation in silicate 13-93 bioactive glass scaffolds with "trabecular" (a) and "oriented" (b, c) microstructures after implantation for 12 weeks in rat calvarial defects, showing new bone (B), host bone (O), and glass scaffold (S).

Bone regeneration, mineralization, and angiogenesis were evaluated in borate (13-93B3) and copper-doped borate (13-93B3Cu) scaffolds (Table I) with trabecular, oriented and fibrous microstructures (see Fig. 1). Scaffolds were implanted for 12 weeks in rat calvarial defects [27]. While bone regeneration was mainly osteoconductive, all scaffolds also showed osteoinductive bone growth (Fig. 6). The borate 13-93B3 scaffolds with a trabecular microstructure had a greater capacity for bone regeneration which make them promising for bone repair (Table IV). Copper appeared to increase osteoconduction in the fibrous 13-93B3 scaffolds.

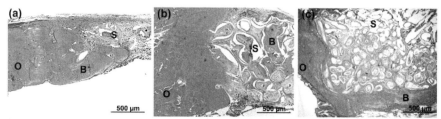

Figure 6. H&E stained sections of borate 13-93B3 scaffolds with (a) trabecular, (b) oriented, and (c) fibrous microstructure after implantation for 12 weeks in rat calvarial defects, showing new bone (B), host bone (O), and glass scaffold (S). The characteristics of the as-fabricated scaffolds are given in Table IV.

Table IV. Percentage new bone and mineralization matrix (von Kossa positive) as a function of scaffold microstructure in borate 13-93B3 (designated B3) and copper-doped 13-93B3 (B3Cu) bioactive glass scaffolds after 12-week implantation in rat calvarial defects [27].

Scaffold	Trabecular		Oriented		Fibrous		Control
Porosity (%)	77 ± 5		62 ± 4		51 ± 2		
Pore size (μm)	200 - 400		100 - 200		50-200		
Composition	B3	B3Cu	B3	B3Cu	B3	B3Cu	45S5
% new bone	33*	31	23	26	15	32*	19
% von Kossa positive	22	21	35	24	48	36	33

* denotes significantly higher bone regeneration than in fibrous 13-93B3 (B3) scaffolds; $p < 0.05$

IN VITRO VERSUS IN VIVO BEHAVIOR OF BIOACTIVE GLASS SCAFFOLDS

Bioactive glass scaffolds have shown markedly different behavior in vivo when compared to their behavior in vitro. As described previously, bioactive glasses convert to an HA-like material when immersed in an SBF in vitro. In comparison, when implanted in vivo, either in subcutaneous sites or calvarial defects in rats, the conversion to HA has been observed to occur at a faster rate. However, a remarkable difference between the in vitro and in vivo behavior is the response of bioactive glass scaffolds to cells. In vitro, the ability of bioactive glass scaffolds with the 13-93, 13-93B1 and 13-93B3 composition (Table I) to support proliferation and function of osteogenic MLO-A5 cells in conventional culture conditions decreased markedly with increasing B_2O_3 content of the glass (Fig. 7). In comparison, in vivo studies showed that borate 13-93B3 scaffolds enhanced tissue formation in rat subcutaneous sites [29] and calvarial defects in rats (Table III) [27].

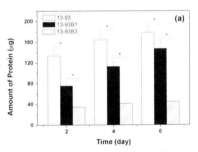 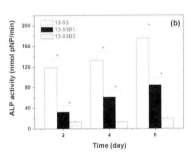

Figure 7. Ability of silicate 13-93, borosilicate 13-93B1, and borate 13-93B3 bioactive glass scaffolds with a trabecular microstructure to support (a) proliferation, and (b) differentiated function of osteogenic MLO-A5 cells. Mean ± sd; n = 4. *Significant difference for glass scaffolds with different compositions (p<0.01) [29].

As a result of the adverse response to cells in vitro, the toxicity of borate bioactive glass in vivo has been a concern. However, histological evaluation showed no adverse effects resulting from the implantation of borate 13-93B3 bioactive glass scaffolds subcutaneously in the dorsum of rats [23, 29] or in the calvaria of rats [27]. Further evaluation of borate 13-93B3 bioactive glass scaffolds in a rat subcutaneous implantation model showed that the bioactive borate glass caused no tissue damage to the kidney or the liver [23]. These results showed that borate bioactive glasses are non-toxic in the dynamic body environment of small animals, so they could be considered to be promising candidate materials for tissue regeneration in vivo.

The mechanical response of bioactive glass scaffolds in vivo has also been found to be markedly different from the response in vitro. As fabricated, bioactive glass scaffolds with oriented or grid-like microstructures show a brittle response typical of dense glass (Fig. 4); the highly porous trabecular scaffolds also shows a brittle response, but there is some compaction of the scaffolds with higher deformation. Our previous work had indicated that when implanted for 4 weeks or longer in rat subcutaneous sites [29], silicate 13-93 bioactive glass scaffolds became very compressible and no longer showed a brittle response. In a recent study [28], the in vitro mechanical response of 13-93 bioactive glass scaffolds was compared with the response of similar scaffolds after implantation in rat clavarial defects. Scaffolds with an oriented microstructure (porosity = 50%; pore diameter = 50–150 μm) and the geometry of discs (5 mm in diameter × 1.5 mm) were immersed in an SBF at 37°C for 12 and 24 weeks; similar scaffolds were implanted for the same time in rat calvarial defects. After immersion or implantation, the mechanical response of the scaffolds was evaluated using a diametral tensile test (deformation rate = 0.5 mm/min) [30].

Figure 8 shows the applied force vs. deformation of the scaffolds. In vitro, the scaffolds showed a brittle response, with a force to failure which decreased markedly with immersion time in the SBF. In comparison, after implantation for 12 weeks, the scaffolds showed an "elastic–plastic" response; following an initial elastic response, the scaffolds showed a large deformation with little increase in stress. The response of the scaffolds 24 weeks post-implantation showed a similar trend, but the force at any displacement was higher than that after the 12-week implantation, presumably because of the larger amount of bone ingrowth.

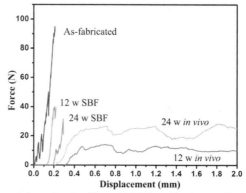

Figure 8. Force vs. deformation for silicate 13-93 bioactive glass scaffolds with an oriented microstructure after immersion in an SBF or implantation in rat calvarial defects for 12 and 24 weeks. For comparison, the response of the as-fabricated scaffold is also shown.

SUMMARY AND CONCLUSIONS

Bioactive glasses with controllable degradation rates have been developed recently by partially or fully replacing the SiO_2 content in silicate 13-93 bioactive glass with B_2O_3. Borate 13-93B3 bioactive glass converted more rapidly and more completely than silicate 13-93 glass. Processing techniques such as unidirectional freezing of suspensions and solid freeform fabrication were used to produce porous and strong bioactive glass scaffolds with compressive strengths in the range of trabecular bone to cortical bone. The ability to support tissue formation in vivo was shown to be dependent on the composition and the microstructure of the bioactive glass scaffolds. The conversion rate to hydroxyapatite, response to cells, and mechanical response of bioactive glass scaffolds in vivo is markedly different from the behavior in vitro. Consequently, the in vitro behavior may not be a reliable model for the in vivo response.

Borate bioactive glass scaffolds with a "trabecular" microstructure were found to have a greater capacity for bone regeneration and osteoinduction when implanted in rat calvarial defects, which make them promising for the repair of bone in low-load sites. Because of their far higher strength, bioactive glass scaffolds prepared by unidirectional freezing or solid freeform fabrication could potentially be useful for the repair and regeneration of load-bearing bones, such as the long bones of the limbs. The optimum combination of glass composition and scaffold microstructure of these porous and strong bioactive glass scaffolds for the repair of loaded bone is currently being evaluated.

ACKNOWLEDGEMENTS: This work was supported by the National Institutes of Health, National Institute of Arthritis and Musculoskeletal and Skin Diseases (NIH/NIAMS) Grant No. 1R15AR056119 (MNR; BSB), and the U.S. Army Medical Research Acquisition Activity, under Contract No. W81XWH-08-1-0765 (MNR, DED, LFB).

REFERENCES
[1] L.L. Hench, Bioceramics, *J. Am. Ceram. Soc.*, **81**, 1705-28 (1998).
[2] D.L. Wheeler, K.E. Stokes, H.E. Park and J.O. Hollinger, Evaluation of Particulate Bioglass® in a rabbit radius ostectomy model, *J. Biomed. Mater. Res. A*, **35**, 249-54 (1997).
[3] D.L. Wheeler, K.E. Stokes, R.G. Hoellrich, D.L. Chamberland and S.W. McLoughlin, Effect of Bioactive Glass Particle Size on Osseous Regeneration of Cancellous Defects, *J. Biomed. Mater. Res. A*, **41**, 527-33 (1998).
[4] I.D. Xynos, A.J. Edgar, L.D. Buttery, L.L. Hench and J.M. Polak, Ionic Products of Bioactive Glass Dissolution Increase Proliferation of Human Osteoblasts and Induce Insulin-like Growth Factor II mRNA Expression and Protein Synthesis, *Biochem. Biophys. Res. Commun.*, **276**, 461-5 (2000).
[5] I.D. Xynos, A.J. Edgar, L.D. Buttery, L.L. Hench and J.M. Polak, Gene-expression Profiling of Human Osteoblasts Following Treatment with the Ionic Products of Bioglass 45S5 Dissolution, *J. Biomed. Mater. Res.*, **55** 151-7 (2001).
[6] Q.Z. Chen, I.D. Thompson and A.R. Boccaccini, 45S5 Bioglass®-derived Glass-Ceramic Scaffolds for Bone Tissue Engineering, *Biomaterials*, **27**, 2414-25 (2006).
[7] W. Huang, D.E. Day, K. Kittiratanapiboon and M.N. Rahaman, Kinetics and Mechanisms of the Conversion of Silicate (45S5), Borate, and Borosilicate Glasses to Hydroxyapatite in Dilute Phosphate Solutions, *J Mater Sci Mater Med*, **17**, 583-96 (2006).
[8] Q. Fu, M.N. Rahaman, W. Huang, D.E. Day and B.S. Bal. Preparation and Bioactive Characteristics of a Porous 13-93 Glass, and its Fabrication into the Articulating Surface of a Proximal Tibia. *J. Biomed. Mater. Res. A*, **82**, 222-9 (2007).
[9] Q. Fu, M.N. Rahaman, B.S. Bal, R.F. Brown and D.E. Day. Mechanical and *In Vitro* Performance of 13-93 Bioactive Glass Scaffolds Prepared by a Polymer Foam Replication Technique, *Acta. Biomater.*, **4**, 1854-64 (2008).
[10] Q. Fu, M.N. Rahaman, R.F. Brown and B.S. Bal. Preparation and *In Vitro* Evaluation of Bioactive Glass (13-93) Scaffolds with Oriented Microstructures for Repair and Regeneration of Load-bearing Bones. *J. Biomed. Mater. Res. A*, **93**, 1380-90 (2010).
[11] T.S. Huang, N.D. Doiphode, M.N. Rahaman, M.C. Leu, B.S. Bal and D.E. Day, Porous and Strong Bioactive Glass (13-93) Scaffolds Prepared by Freeze Extrusion Fabrication, *Mater. Sci. Eng. C*, DOI: 10.1016/j.msec.2011.06.004 (2011).
[12] M. Brink, The Influence of Alkali and Alkaline Earths on the Working Range for Bioactive Glasses, *.J Biomed. Mater. Res.*, **36**, 109-17 (1997).

[13] R. F. Brown RF, D. E. Day, T.E. Day, S. Jung, M.N. Rahaman and Q. Fu, Growth and Differentiation of Osteoblastic Cells on 13-93 Bioactive Glass Fibers and Scaffolds. *Acta. Biomater.*, **4**, 387-96 (2008).

[14] D.E. Day, J.E. White, R.F. Brown and K.D. McMenamin, Transformation of Borate Glasses into Biologically Useful Materials, Glass Technology, **44**, 75-81 (2003)

[15] D.E. Day, E.M. Erbe, M. Richard and J.A. Wojcik, Bioactive Materials, US Patent No. 6709, 744 [March 23, 2004]

[16] X. Han, D.E. Day, Reaction of Sodium Calcium Borate Glasses to Form Hydroxyapatite, *J. Mater. Sci. Mater. Med.*, **18**, 1837-47 (2007).

[17] W. Huang, D.E. Day, K. Kittiratanapiboon and M.N. Rahaman, *J. Mater. Sci. Mater. Med.*, **17**, 583-96 (2006).

[18] A. Yao, D.P. Wang, W. Huang, Q. Fu, M.N. Rahaman and D.E. Day, *J. Am. Ceram. Soc.*, **90**, 303-6 (2007).

[19] H.B. Pan, X.L. Zhao, X. Zhang, K.B. Zhang et al, Strontium Borate Glass: Potential Biomaterial for Bone Regeneration, *J. R. Soc. Interface*, **7**, 1025-31 (2010).

[20] S.F. Hulbert, F.A. Young, R.S. Mathews, J.J. Klawitter, C.D. Talbert and F.H. Stelling, Potential of Ceramic Materials as Permanently Implantable Skeletal Prostheses, *J. Biomed. Mater. Res.*, **4**, 433–56 (1970).

[21] X. Liu, M.N. Rahaman and Q. Fu, Oriented Bioactive Glass (13-93) Scaffolds with Controllable Pore Size by Unidirectional Freezing of Camphene-based Suspensions: Microstructure and Mechanical Response, *Acta Biomater.*, **7**, 406–16 (2011).

[22] Q. Fu, E. Saiz and A.P. Tomsia, Direction Ink Writing of Highly Porous and Strong Glass Scaffolds for Load-bearing Bone Defects Repair and Regeneration, Acta Biomater., **7**, 3547-54 (2011).

[23] S. Jung, Borate Based Bioactive Glass Scaffolds for Hard and Soft Tissue Engineering, PhD Thesis, Missouri University of Science and Technology, 298, (2010).

[24] Q. Fu, M. N. Rahaman, H. Fu, and X. Liu, Silicate, Borosilicate, and Borate Bioactive Glass Scaffolds with Controllable Degradation Rates for Bone Tissue Engineering Applications, I: Preparation and In Vitro Degradation, *J. Biomed. Mater. Res. A*, **95**, 164-71 (2010).

[25] X. Liu, M.N. Rahaman, Q. Fu and A.P. Tomsia, Porous and Strong Bioactive Glass (13-93) Scaffolds Prepared by Unidirectional Freezing of Camphene-based Suspensions, *Acta Biomater.*, DOI:10.1016/j.actbio.2011.07.034 (2011).

[26] M.N. Rahaman, D.E. Day, B.S. Bal, Q. Fu, S.B. Jung, L.F. Bonewald and A.P. Tomsia, Bioactive Glass in Tissue Engineering, *Acta Biomater.*, **7**, 2355-73 (2011).

[27] L. Bi, S.B. Jung, D.E. Day, K. Neidig, V. Dusevich, D. Eick and L. Bonewald, Evaluation of Bone Regeneration, Angiogenesis, and Hydroxyapatite Conversion in Critical-Sized Rat Calvarial Defects Implanted with Bioactive Glass Scaffolds, submitted for publication (2011).

[28] X. Liu, M.N. Rahaman, R.F. Brown and Q. Fu, *In Vivo* Evaluation of Bioactive Glass (13-93) Scaffolds with Trabecular and Oriented Microstructures in a Rat Cavarial Model, submitted for publication (2011).

[29] Q. Fu, M. N. Rahaman, B. S. Bal, K. Kuroki, and R. F. Brown, Silicate, Borosilicate and Borate Bioactive Glass Scaffolds with Controllable Degradation Rates for Bone Tissue Engineering Applications, II: In Vitro and In Vivo Biological Evaluation, *J. Biomed. Mater. Res. A*, **95**, 172-9 (2010)

[30] M.B. Thomas, R.H. Doremus, M. Jarcho and R.L. Salsbury, Dense Hydroxylapatite: Fatigue and Fracture Strength after Various Treatments, from Diametral Tests, *J. Mater. Sci.*, **15**, 891-4 (1980).

SINTERING OF HYDROXYAPATITE

Monica Sawicki, Kyle Crosby, Ling Li, Leon Shaw

Department of Chemical, Materials, and Biomolecular Engineering
University of Connecticut, Storrs, CT 06269

ABSTRACT

Hydroxyapatite ($Ca_{10}(PO_4)_6(OH)_2$, HA) is the main inorganic mineral of bone. The use of HA in orthopedic implants supports bone ingrowth and osseointegration due to its bioactivity. The typical sintering temperature for dense HA bodies is 1100°C or higher. In this study, HA nano-rods were synthesized from a wet precipitation process which enables sintering of HA at temperatures lower than 1100°C. The relationships among the sintering temperature, grain size, and final density were established.

INTRODUCTION

Hydroxyapatite is the main inorganic component of human hard tissue. It is bioactive and supports bone ingrowth and osseointegration when used in orthopedic, dental and maxillofacial applications.[1] The inorganic mineral consists of nanoscale apatite crystals of calcium and phosphate. The Ca:P ratio is 1.67. These crystals form in the collagen fiber matrix and aid in lamellar sheet formation. The lamellar sheets and dense fibril formation give bone its structural stability. The preparation of synthetic HA has been thoroughly investigated via hydrothermal methods, sol-gel, microwave irradiation, and wet precipitation synthesis.[2-11]

Synthetic HA has the capability of forming a direct chemical bond with bone *in vivo*[12], but its use is restricted to non-load bearing applications, or as a coating on metallic implant surfaces due to its poor mechanical properties.[13] Grain size plays a pivotal role in the mechanical properties and stability of HA. With increased grain size, HA is prone to microcracking due to its hexagonal structure and anisotropic thermal expansion coefficient.[14] Sintering is used to produce dense HA bodies with fine microstructure. Particles coalesce through diffusion during sintering, and pores form between interstitial particles as the surface area is reduced. A smaller particle size has greater surface energy which increases the sintering rate, lowers sintering temperature, and enhances densification. The typical sintering temperature for dense hydroxyapatite is 1100°C or higher. Dense HA is characterized as having a porosity of less than 5 vol%, pore size of less than one micron, and a grain size greater than 0.2 micron.[15]

Sintering temperature is an important control factor in densification and sintering behavior of HA. Studies have shown that tricalcium phosphate (TCP) phases can form at temperatures of 900°C or higher during sintering. Elevated temperatures also lead to decomposition of HA to form TCP through dehydration.[16-17] Nonstoichiometry of synthesized powders leads to decomposition; pure HA is known to be stable up to ~1400°C.[18] Dehydration of HA leads to more porous HA bodies and results in decreased stability, mechanical properties, and density of HA bodies. Densification at a lower sintering temperature is essential in producing denser, stable, stronger, and pure HA bodies.

Nano-rod formation of HA via wet precipitation synthesis has reduced the sintering temperature of dense HA bodies (>99%) to 850°C and 900°C through morphology enhanced diffusion and driving force for densification. Grain sizes were reported at 67-83 nm.[19] In this study we conducted further studies to establish the relationships among the sintering temperature, grain size and final density. Wet precipitation was used to produce nano-rod HA. The as-synthesized precipitate was analyzed using transmission electron microscopy (TEM) and X-ray diffraction (XRD) to determine morphology and crystallite sizes using reflections that correlate to the long and short axis of the nano-rod. Samples were uniaxially pressed (300 MPa) and sintered to achieve low temperature sintering densification. The sintered HA bodies were analyzed to determine the density, microstructure, and grain size with respect to sintering temperature. The thermal stability of HA was assessed in terms of phases present after sintering.

EXPERIMENTAL PROCEDURE

HA Synthesis

HA nano-rods were synthesized using reagent grade calcium nitrate, $Ca(NO_3)_2 \cdot 4H_2O$ and diammonium phosphate, $(NH_4)_2HPO_4$, via a wet precipitation process.[19] HA was formed through the synthesis reaction

$$10Ca(NO_3)_2 \cdot 4H_2O + 6(NH_4)_2HPO_4 + 8NH_4OH \rightarrow Ca_{10}(PO_4)_6(OH)_2 + 20\ NH_4NO_3 + 46H_2O \quad (1)$$

The synthesis required separate calcium and phosphate solutions, with a molar ratio of 1.5:0.9, respectively. The calcium solution was prepared with 1.5 mol $Ca(NO_3)_2 \cdot 4H_2O$ in 900 mL of distilled water. The pH of the solution was raised to ~11-12 using concentrated NH_4OH, and then diluted to 1800 mL. The phosphate solution was prepared using 0.9 mol $(NH_4)_2HPO_4$ in 1500 mL of distilled water. The pH of the solution was raised to ~11-12 using NH_4OH, and diluted to 3200 mL. The pH of the dilute solutions was readjusted to ~11-12. The calcium solution was stirred continuously at room temperature with the drop wise addition of phosphate solution resulting in a milky, gelatinous precipitate which was removed through centrifugation after mixing for 24 hours. The slurry was washed with distilled water and left to sediment. The remaining distilled water was removed, and the sediment was washed ten times using distilled water. Centrifugation was used during the washing process to collect the washed, sediment precipitate. The precipitate was retrieved and dried at 90°C, then calcined for two hours at 300°C. The calcined precipitate was dry milled using a SPEX mill for about ten minutes to break down agglomerations.

Sintering and Characterization of HA

The SPEX milled HA powder was uniaxially pressed (300 MPa) into pellets using a ½" diameter steel die. The pressed pellets were sintered in air at temperatures of 900°C, 1100°C, and 1250°C. Sintering of samples at 900°C and 1100°C was conducted in a box furnace with heating and cooling rates of 5°C/min. Sintering of samples at 1250°C was conducted in an Al_2O_3 tube furnace under the same heating and cooling conditions. X-ray analysis (Bruker, D8 Advanced, Karlsruhe, Germany) was obtained for the as-synthesized and sintered HA. The morphology of the HA nano-rods was imaged using transmission electron microscopy (FEI Tecnai T-12). The green and sintered densities of all samples were determined using the mass and volume of HA pellets before and after sintering. The theoretical density of HA used to determine the percent density was 3.156 g/cm^3. The free surface and polished surface of sintered pellets were imaged using a field-emission scanning electron microscope (FESEM, JEOL JSM 6335F).

RESULTS AND DISCUSSION

As shown in Figure 1, the XRD scan of the as-synthesized HA precipitate powder matches well to the phase pure HA (JCPDF No. 74-0566). XRD scans of increasing temperature show improved /crystallinity with respect to HA peaks, but decomposition to tricalcium phosphate (TCP, JCPDF No. 09-0169) is evident for samples sintered at 1250°C (Figure 1). Many peaks within the scan are coupled due to combined HA and TCP reflections, such as (210) and (300), (002) and (211), (202) and (220), respectively.

The Scherrer formula was used to determine the crystallite size of the as-synthesized HA based on the (002) and (210) reflections of the XRD pattern. The (002) reflection corresponds to the long axis of the HA nano-rod, whereas the (210) reflection corresponds to the short axis of the nano-rod. The calculated sizes were ~77 nm and ~28 nm, respectively, leading to an average aspect ratio of ~2.8 for the HA nano-rods. TEM imaging (Figure 2) showed the morphology of the as-synthesized HA, but agglomerations of nano-rods led to inconclusive nano-rod measurement. A better dispersion technique must be utilized in order to determine the size and aspect ratio through TEM imaging to show agreement with crystallite size calculations.

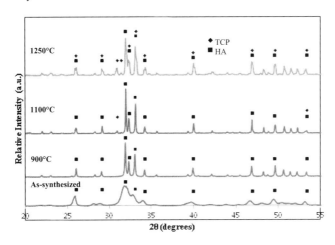

Figure 1. XRD data of the as-synthesized HA powder and sintered HA pellets.

Figure 2. TEM image of the as-synthesized precipitate HA nano-rods.

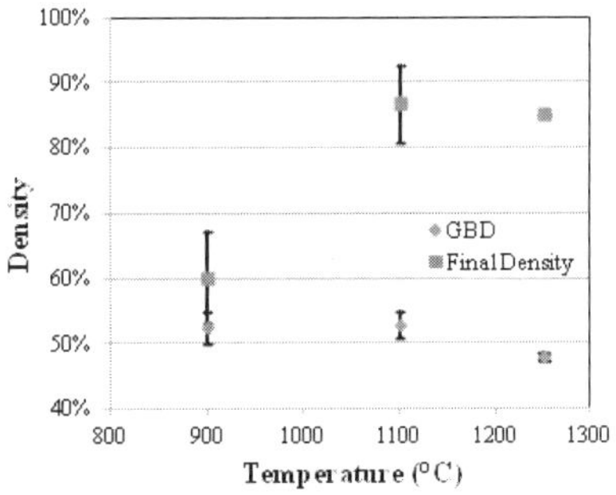

Figure 3. The green body density (GBD) and final density of uniaxially pressed (300 MPa) samples after sintering at 900°C, 1100°C, and 1250°C.

The green body density and sintered density of various pellets are presented in Figure 3. The density of pellets increases with temperature, but densification values are not satisfactory (only about

90% dense). Such relatively low sintered densities are likely due to the insufficient purity of the as-synthesized HA nano-rods, which leads to decomposition and phase change of the samples during sintering at higher temperatures. The phase change during sintering at 1250°C is confirmed by XRD, as shown in Figure 1. The low density of samples sintered at 1250°C is also partially related to lower green densities because lower green densities result in lower sintered densities. However, it is interesting to note that the highest percent densification, calculated from the change of the green density to the sintered density, is achieved at 1250°C (Figure 4). This observation is consistent with the well-established concept that the higher the sintering temperature, the denser the sintered body.

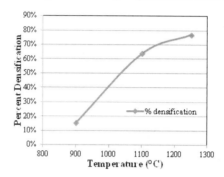

Figure 4. Percent densification of samples sintered at 900°C, 1100°C, and 1250°C.

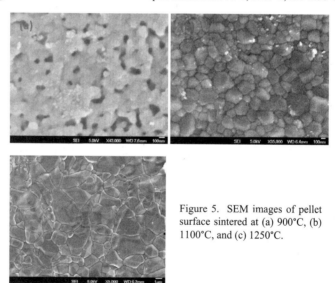

Figure 5. SEM images of pellet surface sintered at (a) 900°C, (b) 1100°C, and (c) 1250°C.

SEM observations (Figure 5) reveal the same trend as the density measurement shown in Figure 4, i.e., densification increases with temperature. In fact, SEM observations of the free surface of sintered bodies indicate that at some regions of the sample sintered at 1100°C is close to a fully dense condition (Figure 5b), while the sample sintered at 1250°C is almost 100% dense (Figure 5c). However, the densities measured using the sample weight and dimension for samples under both sintering conditions are less than 100%, as shown in Figure 3. To indentify the cause for the observed discrepancy, the sintered samples were also polished and examined for porosity. As shown in Figure 6, all samples contain porosities. However, the HA pellet sintered at 900°C has porosity everywhere, whereas the HA pellets sintered at 1100 and 1250°C only have isolated pores non-uniformly present inside the sample. Therefore, the low densities of sintered HA pellets can be attributed to the presence of pores. For the pellets sintered at 1250°C, another factor can contribute to the low density. This is the phase change of HA to TCP because TCP has a lower density (3.14 g/cm^3) than HA (3.156 g/cm^3). However, the contribution from the phase transformation can only account for a 0.5% reduction in the density if all HA is assumed to change to TCP. Therefore, the low sintered density is mainly due to the presence of pores.

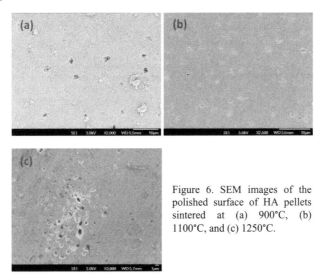

Figure 6. SEM images of the polished surface of HA pellets sintered at (a) 900°C, (b) 1100°C, and (c) 1250°C.

It is noted that the pores in samples sintered at 1250°C are distributed highly non-uniformly (Figure 6c), indicating that some regions can be sintered to 100% dense at 1250°C. A similar situation is present for 1100°C sintering (Figure 6b). We believe that porosity in these samples is mainly due to bubbling of H$_2$O resulting from decomposition of HA during sintering. The decomposition of HA to form TCP is defined by the following reaction[18]

$$Ca_{10}(PO_4)_6(OH)_2 \rightarrow 3Ca_3(PO_4)_2 + CaO + H_2O \qquad (2)$$

Such decomposition and thus porosity can be suppressed by making high purity HA nano-rods because pure HA is known to be stable up to ~1400°C.[18] This topic will be the focus of the future study.

The grain size for HA sintered at 900°C was estimated to be 40-80 nm, based on Figure 5(a). The grain sizes for HA sintered at 1100°C and 1250°C were determined via the line intercept method based on the images from Figures 5(b) and (c), and found to be ~200 nm and 1.89 μm, respectively. The observed trend is consistent with our expectation, i.e., the higher the sintering temperature, the larger the grain size. Furthermore, sintering has led to the change of HA particles from the nano-rod morphology to equiaxed grains, indicating that the nano-rod morphology is a metastable morphology. This result is in good accordance with the previous investigation.[19]

CONCLUSION

Nano-rod HA was synthesized using the wet precipitation method. TEM analysis shows nano-rod formation and agglomeration. XRD analysis of the as-synthesized HA matched well with the pure phase HA (JCPDF No. 74-0566). Uniaxially pressed samples were sintered at temperatures of 900°C, 1000°C and 1250°C. The resulting average density for sintered samples was 60.10±7.00%, 86.79±5.89%, and 84.95±0.92%, respectively. Decomposition of HA to TCP occurred at higher temperatures, leading to porosity formation and thus the reduced density. Grain growth took place during sintering, resulting in grain growth and morphology change from nano-rods (77 nm in length and 28 nm in diameter) to equiaxed grains with sizes of 40-80 nm, 200 nm and 1.89 m for sintering at 900°C, 1100°C and 1250°C, respectively.

ACKNOWLEDGMENTS

This research was sponsored by the U.S. National Science Foundation (NSF) under the contract number CBET-0930365. The support and vision of Ted A. Conway is greatly appreciated.

REFERENCES

[1] E. Shors, R. Holmes, in: L.L. Hench, J. Wilson (Eds.), An Introduction to Bioceramics, Singapore: World Scientific, 181, 1993

[2] J. Liu, X. Ye, H. Wang, M. Zhu, B. Wang, H. Yan, The Influence of pH and Temperature on the Morphology of Hydroxyapatite Synthesized by Hydrothermal Method, Ceram. Int., 29, 629-33, 2003

[3] Y. Fujishiro, H. Yabuki, K. Kawamura, Preparation of Needle-like Hydroxyapatite by Homogeneous Precipitation Under Hydrothermal Conditions, J. Chem. Tech. Biotechnol. 57, 349–53, 1993

[4] K. Ioku, S. Yamauchi, S. Fujimori, S. Goto, M. Yoshimura, Hydrothermal Preparation of Fibrous Apatite and Apatite Sheet, Solid State Ionics, 151, 147-50, 2002

[5] K. Kamiya, T. Yoko, K. Tanaka, Growth of Fibrous Hydroxyapatite in the Gel System. J. Mater. Sci., 24, 827-32, 1989

[6] A. Jillavenkatesa, R.A. Condrate, Sol–gel Processing of Hydroxyapatite, J. Mater. Sci., 33, 4111-9, 1998

[7] H.K. Varma, R. Sivakumar, Dense Hydroxyapatite Ceramics through Gel Casting Technique, Materials Letters, 29, 57-61, 1996

[8] J. Liu, K. Li, H. Wang , M. Zhu, H. Yan, Rapid Formation of Hydroxyapatite Nanostructures by Microwave Irradiation, Chemical Physics Letters, 396, 429-432, 2004

[9] S.Z.C Liou, S.Y. Chen, D.M. Liu, Synthesis and Characterization of Needlelike Apatitic Nanocomposite with Controlled Aspect Ratios, Biomaterials, 24, 3981–8, 2003

[10] A. J. Ruys, M. Wei, C. C. Sorrell, M. R. Dickson, A. Brandwood, B. K. Milthorpe, Sintering Effects on the Strength of Hydroxyapatite, Biomaterials, 16, 409-415, 1995

[11] L. M. Rodríguez-Lorenzo, M. Vallet-Regí, J. M. F. Ferreira, Fabrication of Hydroxyapatite Bodies by Uniaxial Pressing from a Precipitated Powder, Biomaterials, 22, 583-588, 2001

[12] J . B. Park and R. S . Lakes, Biomaterials: An Introduction, New York, NY: Plenum Press, 1992

[13]W. Weng, S. Zhang, K. Cheng, H. Qu, P. Du, G. Shen, J. Yuan, G. Han, Sol-gel Preparation of Bioactive Apatite Films, *Surf. Coat. Tech.*, 167, 292-296, 2003

[14]T.P. Hoepfner, E.D. Case, Physical Characteristics of Sintered Hydroxyapatite, *Bioceramics: Mat.and App. III*, 110, 53, 2000

[15] C.B. Carter, M.G. Norton, <u>Ceramic Biomaterials: Science and Engineering</u>, New York, NY: Science+Business Media, 2007

[16]L.L Hench, J. Wilson, <u>An Introduction to Bioceramics</u>, Singapore: World Scientific, 1993

[17]J. Zhou, X. Zhang, J. Chen, S. Zeng, K. De Groot, High-temperature Characteristic of Synthetic Hydroxyapatite, *J. Mater. Sci.: Mat. In Med.*, 4, 83-85, 1993

[18]G. Muralithran, S. Ramesh, The Effects of Sintering Temperature on the Properties of Hydroxyapatite, *Ceramics International*, 26, 221-230, 2000

[19]J. Wang, L.L. Shaw, Morphology-Enhanced Low-Temperature Sintering of Nanocrystalline Hydroxyapatite, *Adv. Mater.*, 19, 2364-2369, 2007

IN VIVO EVALUATION OF 13-93 BIOACTIVE GLASS SCAFFOLDS MADE BY SELECTIVE LASER SINTERING (SLS)

M. Velez, S. Jung
Mo-Sci Corporation,
Rolla, MO 65401-8277

K. C. R. Kolan, M. C. Leu
Department of Mechanical and Aerospace Engineering
Missouri University of Science and Technology
Rolla, MO 65409-0330

D.E. Day
Department of Materials Science and Engineering
Missouri University of Science and Technology
Rolla, MO 65409-0330

T-M.G. Chu
Department of Restorative Dentistry
Division of Dental Biomaterials
Indiana University
Indianapolis, IN 46202-2876

ABSTRACT

Bioactive glass is a synthetic material that reacts in vivo and forms an inorganic hydroxyapatite-like (HA) phase that mimics the HA found in human bone and can stimulate osteoconduction. Selective Laser Sintering (SLS) was used to fabricate "green" bone scaffolds from 13-93 bioactive glass particles (<75 μm) mixed with stearic acid. After heat treatment, the scaffolds had a hollow cylindrical shape that was intended to mimic the structure and mechanical properties of human trabecular bone. In vitro, the compressive strength of the SLS scaffolds was measured as a function of time for up to two months when immersed in Dulbecco's Modified Eagles Medium (DMEM) at 38°C. In vivo, the cylindrical scaffolds with and without bone morphogenic protein-2 (BMP-2), were implanted in rat femurs for up to three months. The data from the in vivo study is compared to similar biodegradable polymer scaffolds treated with BMP-2 in critical sized defects in rats.

INTRODUCTION

Selective Laser Sintering (SLS) is a layer-by-layer manufacturing process used to make complex geometrical parts which otherwise are difficult to manufacture by conventional methods [1,2]. The fabrication of scaffolds via SLS has the advantage of duplicating 3-D images attempting to mimic the external shape and internal structure of human bones. The use of bioactive glass has the potential advantage of rapid in vivo resorption, compared to bio-polymers, with high vascularization at short times after implantation [3,4]. Additionally, bioactive glasses in general, offer the opportunity for tailoring the chemical composition of the scaffolds and improving their properties as needed.

Cylindrical (4 mm in diameter x 5 mm length) prototype 13-93 bioactive glass bone scaffolds were made by SLS in the present work, see Fig. 1. The SLS fabrication process consisted of coating 13-93 bioactive glass particles with thermoplastic stearic acid as a binder, which melts when a computer controlled laser beam heats the coated glass particles. The stearic acid bonds the glass particles together to form a thin solid layer structure. After one layer is finished, the second layer begins with the final 3D component fabricated by continuous deposition of layers bonded together. After SLS processing, the stearic acid is removed by pyrolysis and the glass particles are sintered to high density. The compressive strength was measured for SLS scaffolds immersed in simulated body solutions at 38 °C for times up to 60 days.

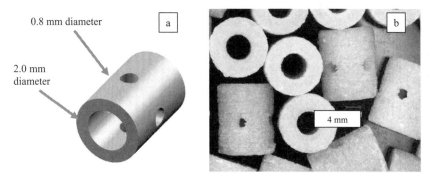

Figure 1. SLS scaffolds: (a) Schematic drawing of a SLS "cylindrical prototype scaffold"; and (b) sintered 13-93 scaffolds. Average cylinder dimensions 5 mm long, 4 mm outside diameter.

The SLS scaffolds were combined with metal inserts and implanted in rats (femur) after creating a bone defect. The implanted SLS scaffolds were monitored with X-Ray imaging. Twenty Long Evans Rats were used for implanting the 13-93 scaffolds. In ten scaffolds, bone morphogenetic protein 2 (BMP-2) at a dose of 10 μg was introduced into each scaffold the day before surgery. Ten rats received the scaffold with BMP-2 and ten rats received scaffolds without BMP-2 as control. From X-ray imaging, the bioactive glass SLS scaffolds were shown to induce bridging callus formation. The scaffolds/bone gap/interface was scanned under micro computer tomography (CT) to verify the 3D distribution of the bone structure formed.

SCAFFOLDS FABRICATION BY SELECTIVE LASER SINTERING
Bioactive glass 13-93 (nominal chemical composition by weight: 53% SiO_2, 4% P_2O_5, 20% CaO, 5% MgO, 6% Na_2O and 12% K_2O) was used to fabricate the SLS scaffolds. The glass was milled to a particle size below 75 μm, mixed in a V-blender with stearic acid and then dry ball-milled for 8 h with ZrO_2 grinding media to obtain powders with a 13-93 glass to stearic acid mixing ratio of 60:40 by volume. The SLS machine used was a DTM Sinterstation 2000, with a nominal 50W CO_2 laser, and a laser spot diameter of 450 μm. This machine has been used previously to investigate SLS processing of 13-93 bioactive glass with and without binder [5].
Figure 1a shows the CAD model and dimensions of the 3D cylindrical scaffolds fabricated by SLS. The cylinders were made by duplicating the bottom section sequentially – making a "hollow" cylinder for biological tests. Figure 1b shows several SLS scaffolds after sintering. The 13-93 scaffolds were fabricated using a scan speed of 304.8 mm/s and laser power of 3 W for melting the stearic acid at ~1 cal/cm^2 (energy density).

Sintering
Binder burnout and sintering schedules (Fig. 2) were developed based on the stearic acid melting and decomposition temperatures (74°C and 287°C, respectively) and the DSC curve of 13-93

bioactive glass. Sintering was done in air in alumina-castable cement substrates which were molded to the shape of the scaffolds to avoid collapsing during firing. The sintering shrinkage of fired cylindrical scaffolds was about 20%. Figure 3 shows an SEM image of the cross section of a scaffold wall showing low porosity (<5%) which is needed for high mechanical strength.

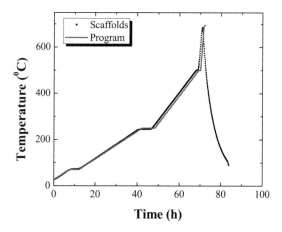

Figure 2. Schedule for sintering the SLS 13-93 scaffolds. Several experiments were conducted to optimize the maximum heating temperature, varying from 650 to 700 °C and for holding times between 15 and 60 minutes.

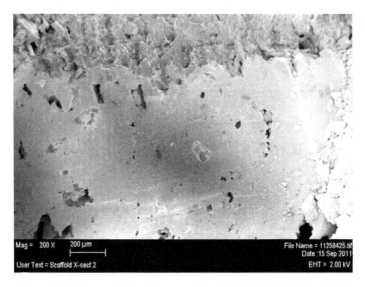

Figure 3. SEM image of the wall cross section of sintered 13-93 SLS scaffold.

COMPRESSION RESISTANCE OF SLS SCAFFOLDS

The compressive strength of sintered SLS scaffolds was measured according to ASTM-C773 using an Instron testing machine (Model 4302) at a crosshead speed of 1 mm/min. Ten scaffolds were tested at a time, and the average strength and standard deviation were determined. The SLS scaffolds were immersed in 500 ml of Dulbecco's Modified Eagle Medium (DMEM, invitrogen.com/GIBCO) solution at 38 °C. The compressive strength (Fig. 4) of the scaffolds was measured versus time in solution using graphite foils on top and bottom of the scaffolds (wet measurements). The compressive strength decreases with time in the DMEM solution, from 40 ± 10 MPa (dry, as sintered) to 26 ± 6 MPa after 60 days in DMEM. For comparison, tube-shaped BMP-2 carrier fabricated from poly(propylene fumarate)/tricalcium phosphate (PPF/TCP) as composite scaffolds have a compressive strength of 23 MPa and a reduced strength of 12 MPa after soaking for 12 weeks in SBF [6].

The presence of hydroxyapatite (HA) on the 13-93 SLS scaffolds immersed in "conventional" SBF solution [7] was proven previously [8], by SEM and XRD of the scaffolds surfaces.

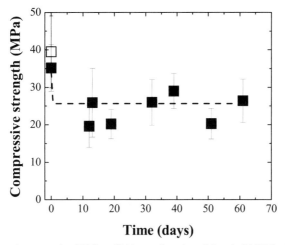

Time (days)

Figure 4. Compressive strength of SLS scaffolds as a function of time in DMEM solution at 38 °C. The load was applied at 1 mm/min. The scaffolds were tested wet with graphite foil on the top and bottom of the scaffold.

IN VIVO EVALUATION OF SLS SCAFFOLDS USING A RAT FEMORAL SEGMENTAL MODEL DEFECT

Twenty Long Evans Rats weighing about 450 g were used. The hollow SLS scaffolds were filled with dicalcium phosphate dehydrate cement. In ten scaffolds, bone morphogenetic protein 2 (BMP-2) at a dose of 10 μg was introduced into the dicalcium phosphate dehydrate cement region of the scaffold the day before surgery. With a lateral approach to the right femur, all muscles and periosteal tissue were stripped circumferentially from the diaphysis (mid-section of the bone). A 5 mm segmental defect was created in the region of the middle of the shaft 13 mm above the knee joint using a dental burr. After the creation of the critical size defect, incisions were made through knee joint capsule to expose the knee joint space.

A 1.60 mm K-wire with threaded tip (Synthes Inc, West Chester, PA) was drilled into the patella surface between the lateral and medial condyles into the femur marrow cavity (Fig. 5). The K-wire was allowed to pass through the scaffold and drill into the proximal end of the femur marrow cavity. The muscle layers were closed with 4-0 Vicryl sutures and the skin was closed with surgical staples. Ten rats received scaffolds with BMP-2 and ten rats received scaffolds without BMP-2 as a control.

The surgeries were performed following the animal surgical protocol NRD-667. Post-op X-rays show the scaffolds successfully placed in the defect. Further x-rays were taken at 3 and 6 week post-operation. After sample retrieval, the femurs will be analyzed with histology and micro CT to quantify the amount of residual scaffolds and the degree of bone formation in the samples.

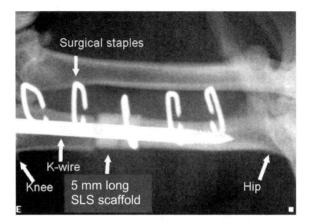

Figure 5. X-ray image of SLS 13-93 bioactive glass scaffold as was implanted in a rat femur.

The animals recovered from anesthesia without complications. In follow up X-rays at 3 and 6 week after surgery, no fracture or disintegration was noticed in any of the bioactive glass scaffolds. The results indicated that the scaffolds posses sufficient mechanical strength to sustain the gap in this load-bearing segmental defect model. Scaffolds with insufficient mechanical strength will not be able to sustain the gap and will collapse under the applied load [6].

From the X-ray results, the BMP-2 group seemed to close the gap as expected with 62.5% and 100% of the images showing cortical bone bridging for at least one side of the cortical bone at 3 and 6 weeks (Figs. 6 C and G). The significant finding is in the control group where only bioactive glass scaffolds without BMP-2 were used. None of the scaffolds showed significant bridging at 3 weeks (Figs. 6 A and B), but 40% showed cortical bridging in one side of the cortical bone and 20% showed bridging in both cortex at 6 weeks (Figs. 6 E and F).

This is an interesting finding when compared to our previous study using scaffolds made from poly(propylene fumarate)/tricalcium phosphate [9] (Figs. 6 D and H). In that study, no cortical bridging was noticed at 3 week, similar to the bioactive glass scaffold result in the present study. However, at 6 weeks, only 25% bridging in one cortex and no bridging of both cortex was noticed.

In other words, under X-ray, the bioactive glass scaffolds were shown to induce more bridging callus formation than the PPF/TCP scaffolds in this critical-sized defect model. A more significant difference between the bioactive glass group and the PPF/TCP scaffold [9] at longer time is expected.

This has significant implication since mechanical loosening is the most important cause to clinical failures in titanium segmental defect prosthesis [10]. The critical area is the junction between the bone end and the prosthesis. Future synthetic scaffolds most probably will be "composite scaffolds", formed by a combination of bioactive glass shapes, metal inserts, and polymer and organic components.

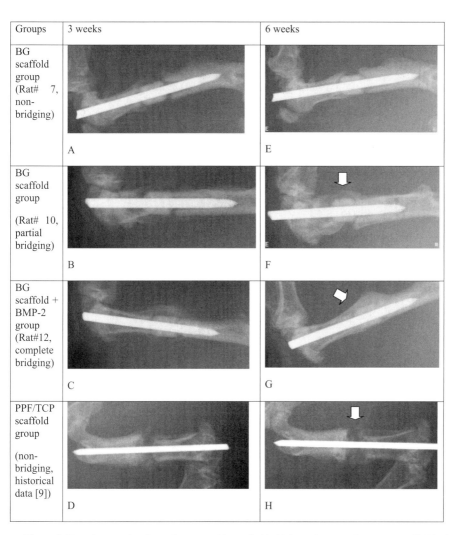

Groups	3 weeks	6 weeks
BG scaffold group (Rat# 7, non-bridging)	A	E
BG scaffold group (Rat# 10, partial bridging)	B	F
BG scaffold + BMP-2 group (Rat#12, complete bridging)	C	G
PPF/TCP scaffold group (non-bridging, historical data [9])	D	H

Figure 6. X-ray images showing at least one side cortical bridging. The 5 mm long SLS scaffold is the scale in images A to G. Arrows emphasize bridging or partial bridging in each case.

SUMMARY AND CONCLUSIONS

The SLS process is useful to manufacture relatively complex shapes using bioactive glass powders. Several 13-93 SLS scaffold prototypes were made, tailoring the dimensions needed for biological tests, using them as femur substitutes in segmented defects rats.

The compressive strength of the SLS scaffolds decreases with time when immersed in simulated body fluid (DMEM solution), and the strength is two times higher compared to that of poly(propylene fumarate)/tricalcium phosphate (PPF/TCP) composite scaffolds. The compressive strength of the scaffolds immersed in DMEM solution at 38 °C decreases from 40 ± 10 MPa (dry, as sintered) to 26 ± 6 MPa in 60 days.

From X-ray imaging, the scaffolds with BMP-2 seemed to close the defect gap as expected showing cortical bone bridging for at least one side of the cortical bone at 3 and 6 weeks. Forty percent of the scaffolds showed cortical bridging in one side of the cortical bone while twenty percent showed bridging in both cortexes at 6 weeks. The bioactive glass scaffolds were shown to induce more bridging callus formation than the PPF/TCP scaffolds in this critical-sized defect model.

ACKNOWLEDGEMENTS

This research work was funded in part by the Office of Naval Research, SBIR Phase I award number N00014-11-M-013 assigned to Mo-Sci Corp. The authors thank E. Bohannan and C. Wisner of Missouri University of Science and Technology and Y. He and J. Szabo of Mo-Sci Corp. for their assistance technical contributions.

REFERENCES

[1]M.C. Leu, E.B. Adamek, T. Huang, G.E. Hilmas, F. Dogan, Freeform Fabrication of Zirconium Diboride Parts Using Selective Laser Sintering, Proceedings of Solid Freeform Fabrication Symposium, pp. 194-205, Austin, TX, 2008.

[2]Z.H. Liu, J.J. Nolte, J.I. Packard, G. Hilmas, F. Dogan, M.C. Leu, Selective Laser Sintering of High-Density Alumina Ceramic Parts, Proceedings of International MATADOR Conference, pp. 351-354, Taipei, Taiwan, 2007.

[3]Q. Fu, M.N. Rahaman, B.S. Bal, R.F. Brown, D.E. Day, Mechanical and In Vitro Performance of 13-93 Bioactive Glass Scaffolds Prepared by a Polymer Foam Replication Technique, Acta Biomaterialia, **4**, 1854-64 (2008).

[4]Q. Fu, M.N. Rahaman, B.S. Bal, R.F. Brown, Preparation and In Vitro Evaluation of Bioactive Glass (13-93) Scaffolds with Oriented Microstructures for Repair and Regeneration of Load-bearing Bones, J. Biomedical Mat. Res., Part A, **93**[4] 1380-90 (2010).

[5]K.C.R. Kolan, M.C. Leu, G.E. Hilmas, M. Velez, Selective Laser Sintering of 13-93 Bioactive Glass, Proceedings of the 21st Annual International Solid Freeform Fabrication Symposium, pp. 504-512, Austin, TX, 2010.

[6]T.M. Chu, P. Sargent, S.J. Warden, C.H. Turner, R.L. Stewart RL, Preliminary Evaluation of a Load-bearing BMP-2 Carrier for Segmental Defect Regeneration. Biomed. Sci. Instrum., **42**, 42-47 (2006).

[7]C. Ohtsuki, T. Kokubo, M. Neo, S. Kotani, T. Yamamuro, T. Nakamura, Y. Bando, Bone-bonding Mechanism of Sintered b-3CaO-P2O5, Phosphorus Res. Bull., **1**, 191-196 (1991).

[8]K.C.R. Kolan, M.C. Leu, G.E. Hilmas, R.F. Brown, M. Velez, Fabrication of 13-93 Bioactive Glass Scaffolds for Bone Tissue Engineering using Indirect Selective Laser Sintering, Biofabrication, **3**[2] 2011, doi: 10.1088/1758-5082/3/2/025004.

[9]T.M. Chu, S.J. Warden, C.H. Turner, R.L. Stewart RL, Segmental Bone Regeneration using a Load-bearing Biodegradable Carrier of Bone Morphogenetic Protein-2, Biomaterials, **28**[3] 459-467 (2007).

[10]P. Ruggieri, A.F. Mavrogenis, G. Bianchi, V.I. Sakellariou, M. Mercuri, P.J. Papagelopoulos, Outcome of the Intramedullary Diaphyseal Segmental Defect Fixation System for Bone Tumors. J. Surg. Oncol., 2011, 103: n/a. doi: 10.1002/jso.21893.

[11]Stewart, R., Stannard, J., Volgas, D., Duke, J., Chu, Tien-Min, Have We Hit the Critical Defect Trifecta? Immediate Weight Bearing, Healing and a Biodegradable Carrier in a Canine Model, ORS 55th Annual Meeting, 2009.

EFFECT OF SINTERING TEMPERATURE ON MICROSTRUCTURAL PROPERTIES OF BIOCERAMIC BONE SCAFFOLDS

Juan Vivanco[1,2], Aldo Araneda[3], Heidi-Lynn Ploeg[1,2]

[1]Material Science Program, University of Wisconsin-Madison
Madison, Wisconsin, USA
[2]Department of Mechanical Engineering, University of Wisconsin-Madison
Madison, Wisconsin, USA
[3]Department of Industrial Engineering, Universidad Técnica Federico Santa María
Valparaiso, Chile

ABSTRACT

Bioactive calcium phosphate (CaP) scaffolds have emerged as synthetic alternatives to bone grafts, acting to target serious fracture healing and bone disease such as osteoporosis. Previous research has shown that these bioceramic materials due to their osteoinductive and biocompatible properties can potentially induce bone formation from the surrounding native tissue. Sintering temperature of CaP scaffolds has been shown to influence the microstructure and properties; however, a comprehensive study to determine the influence of sintering temperature on CaP scaffolds has yet to be performed. Thus, the objective of this work was to determine the microstructural properties of an injection molded CaP scaffold based on the crystallographic phases and grain arrangement for different sintering temperatures. The CaP scaffolds were fabricated using tricalcium phosphate and sintered at three different target temperatures. The microstructural and mechanical properties were characterized by different techniques such as: X-ray diffraction, scanning electron microscopy with associated energy dispersive X-ray spectroscopy, and micro-hardness. It was found that grain size, degree of density, crystallite size, and microhardness increased with increasing sintering temperature; whereas, the crystallographic and the Ca/P atomic ratio did not vary for a sintering temperature range of 950-1150°C. These results support the design and fabrication of bioceramic scaffolds with controlled microstructural properties to provide structural integrity and encourage bone ingrowth.

INTRODUCTION

A common strategy for bone tissue replacement and regeneration is to implant porous three-dimensional scaffolds. The mechanical environment requires the scaffold to provide appropriate strength and stiffness while providing adequate space for bone cells and their cell-cell communication. Furthermore, the biological environment requires nutrient transference and bone cells to stimulate cell migration and matrix deposition. One of the key micro-environmental aspects affecting cell differentiation is the base material of the scaffold and its interaction with cells. Biomaterials widely used as bone replacement grafts are those made of inorganic materials such as calcium phosphate (CaP) based bioceramics (natural and synthetic)[1] . Natural CaP occurs in the body through either normal or pathological biomineralization; whereas, synthetic CaP is usually prepared in-vitro by solution-based chemical reactions and a thermal sintering process[2]. Among CaP-based bioceramics, hydroxyapatite (HA, $Ca_{10}(PO_4)_6(OH)_2$) and tricalcium phosphate (TCP, $Ca_3(PO_4)_2$) are the most commonly used in clinical applications because of their biocompatibility, osteoconductivity, bioactivity, bioresorbability, and their chemical similarity to the mineral phase of bone[3-6]. Properties of CaP materials vary significantly with their crystallinity, grain size, porosity, and composition. High crystallinity, low porosity and small grain size tend to give higher stiffness, compressive strength and toughness. Some in vivo studies have shown that 95% of these calcium phosphates are resorbed in twenty-six to eighty-six weeks[7, 8]. In addition, their degradation depends on their phases, with crystalline TCP having a higher degradation rate than crystalline HA.[9, 10]

There is evidence that chemical composition and microporosity of a bioceramic scaffold can influence the capacity to induce bone formation[11]. Furthermore, microporosity has been observed to be affected by sintering temperature[12]. Although extensive research has been done concerning bioceramic scaffolds for use in bone tissue engineering applications, a comprehensive study to determine the influence of sintering temperature on CaP-based scaffolds is still warranted[13].

Thus, the objective of this work was to determine the microstructural properties of an injection molded CaP scaffold based on the crystallographic phases and grain arrangement for different sintering temperatures. The CaP scaffolds were fabricated using tricalcium phosphate, and sintered at three different target temperatures. The microstructural and mechanical properties of interest included crystalline phase, crystallite size, grain size, Calcium/Phosphorus atomic ratio (Ca/P), and micro hardness.

MATERIALS AND METHODS

Samples were fabricated by Phillips Plastic Corporation (Hudson, WI)[14] and sintered at target temperatures of 950°C, 1050°C, and 1150°C in air using the following heating scheme (Figure 1(a)): beginning at room temperature heated at a rate of 1°C/min to 600°C, soaked at 600°C for one hour, heated to either 950°C, 1050°C or 1150°C using a heating rate of 2°C/min and finally held at the target temperature for five hours. Samples were subsequently cooled to 600°C at a rate of 5°C/min and finally furnace cooled to room temperature. The scaffolds consisted of six approximately 500 μm^2 beams stacked upon one another in orthogonal directions to form a porous cubic structure (~5 mm^3). Figure 1(b) shows representative stereomicroscope images of a sintered β-TCP scaffold used in the current study.

The crystallographic phase and crystallite size were determined for each sintering temperature using X-ray diffraction (XRD). A Hi-Star 2D area detector with CuKα radiation ($\lambda = 0.1542$ nm) was used with settings placed at 40 kV and 40 mA. The XRD patterns were recorded in the 2θ range of 15°- 65°, with a step size of 0.02° and step duration of 0.5 seconds. The experimental XRD spectra were identified by computer matching with ones based on the structural data of similar apatite bioceramics available in the PDF cards from ICDD-database (International Centre for Diffraction Data, 2008). The crystallite sizes of the different scaffolds were estimated from the XRD spectra by means of the full width at half maximum intensity (FWHM) and the Scherrer equation: $t = (k*\lambda) / (B*\cos\theta)$; where t is the crystallite size (nm), k is the shape factor (0.9), λ is the wave length of the X-rays ($\lambda = 0.1541$ nm for CuKα radiation), θ is the Bragg diffraction angle (in radians), and B is the broadening of the diffraction peak measured at FWHM corrected for experimental broadening. The FWHM and data were processed by the software package PowderX[15].

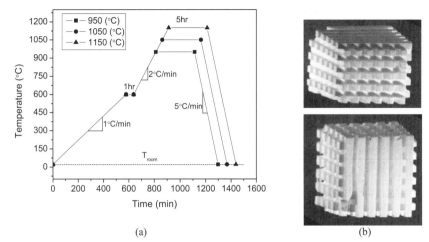

(a)   (b)

Figure 1: (a) Sintering scheme used in the β-TCP scaffold fabrication process for the target temperatures: 950, 1050, and 1150°C. (b) Stereomicroscope images of a representative bioceramic scaffold sintered at 950°C, bulk dimensions ~ 5 mm^3.

The grain size and Ca/P atomic ratio of the scaffold surfaces were investigated by scanning electron microscopy (SEM) with energy dispersive X-ray spectroscopy (EDS). Samples were mounted on aluminum stubs with double sided carbon tape and sputtered with gold for 30s at 45 mA and characterized using LEO DSM 1530 field emission SEM (SEM, Zeiss-LEO, Oberkochen, Germany) operating at 5kV. From SEM micrographs, average grain sizes were determined by using the linear intercept method (ASTM E 112-88). Furthermore, an EDS detector at 8 kV was used in tandem with the SEM for elemental analysis to determine the Ca/P atomic ratio in each sintering condition. For EDS analysis, samples were mounted on aluminum stubs and sputter coated with a thin layer (300 Å) of carbon.

Physical properties such as volume, density, and porosity were determined using Archimedes' principle, a fluid displacement method, using a 70% ethanol solution. Using known-volume bodies, in a previous study, a small difference between the measured and the true volume was obtained, with a mean percent difference of 0.975%[16]. Repeatability and consistency of physical properties have been observed in a related study with n=50 samples for each sintering temperature[16].

For mechanical characterization, surface samples were mechanically ground with silicon carbide paper of 600-800-1200 grit in ascending order and finally polished with a 0.25 μm diamond suspension. Mechanical properties of sintered bioceramics were evaluated in terms of their Knoop micro hardness (H_K) following the ASTM standards for advanced ceramics (ASTM C 1326-08). Micro hardness testing was conducted on the cross-section parallel to the surface of the scaffolds with a digitally controlled Buehler micro-indentation hardness tester (Micromet 5104, Lake Bluff, IL). Testing was conducted at room temperature using a test load of 5 N and a hold time of 15 seconds. At least ten measurements per scaffold were performed in accordance to the standard. The average and standard deviation of these readings was calculated and compared for further analysis.

RESULTS AND DISCUSSION
   SEM micrographs of TCP scaffolds sintered at 950, 1050, and 1150°C are shown in Figure 2. The micrographs show clear demarcation in the grain boundary, grain sizes and micro pores. Grain size has been considered a key factor to regulate degradability of bioceramic scaffolds [17], in this current study average grain was calculated using image software analysis (ImageJ 1.44, NIH, USA) with a total of six different measurements for each sintering condition. As sintering temperature increased so did the grain size with values of 0.74 ± 0.04, 3.59 ± 0.28, 8.07 ± 0.20 µm and material density with values of 2.27 ± 0.15, 2.84 ± 0.18, 3.22 ± 0.29 g/cm^3, for 950°C, 1050°C, and 1150°C, respectively (Figure 2). Furthermore, the porosity of the material decreased with increasing sintering temperature, with values of 24%, 6%, and 0% for the same levels of sintering temperature.

Mag. = 35K x   Mag. = 5K x

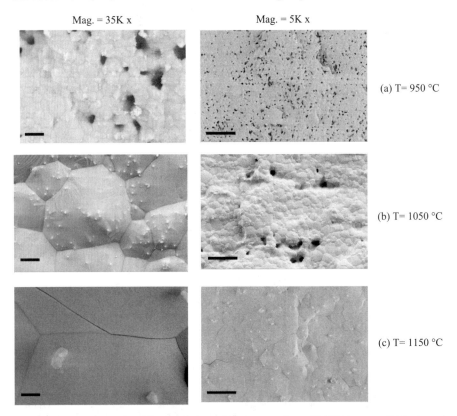

(a) T= 950 °C

(b) T= 1050 °C

(c) T= 1150 °C

Figure 2: SEM micrographs of the morphology of the cross sections of β-TCP bone scaffolds sintered at different temperatures: (a) 950°C, (b) 1050°C, and (c) 1150°C. Images are shown at different magnifications to distinguish microporosity and grain size. Scale bars are 1 µm (left images) and 10 µm (right images)

The microstructural evolution of the grain growth can be analyzed as an activation process following the Arrhenius-type temperature dependence for a constant sintering time as shown in Figure 3. The slope of the best fit line using the least square method was used to determine the apparent activation energy (Q) from the following equation: $D = A*exp[Q/RT]$; where D is the average grain size, T is the absolute sintering temperature, R is the universal gas constant (8.314 J/mol K), and A is a regression constant. From the Arrhenius plot, the apparent activation energy was calculated to be 41.57 kcal/mol; a value that lies in the levels published by other researchers for similar CaP-based materials with 42 kcal/mol[18] and 44 kcal/mol[19]. In addition, intra- and inter-granular cracks were observed at the higher sintering temperature as is shown in Figure 2(c). Intergranular cracks have been reported in other bioceramic materials and are a result of thermal expansion anisotropy during sintering[20]. In natural bone, micro-cracks are critical in remodeling and repair of damaged tissue; and, micro-cracking improves bone toughness by absorbing energy which prevents the propagation of a failure-inducing macro-crack[21]. It has been suggested that similarities between micro-cracking in bioceramic scaffolds and that in natural bone should be considered in the scaffold design in order to improve mechanical properties[20].

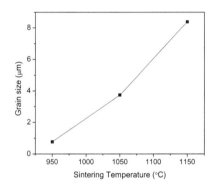
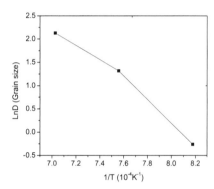

Figure 3: Microstructural evolution of bioceramic scaffolds. (a) Average grain size as a function of the sintering temperature. (b) Arrhenius plot of natural logarithm of the grain size versus the reciprocal of the sintering temperature. The apparent activation energy was 41.57 kcal/mol.

The crystallographic phase, which also plays a role in the solubility under physiological conditions, was evaluated using XRD technique. Depending on the sintering temperature, tricalcium phosphate (TCP) can exist in three polymorphs: the low temperature phase, β-TCP, is stable below 1180°C; the α-TCP phase is stable in the temperature range 1180-1400°C; and, the α'-TCP phase is observed above 1470°C[22]. In the current study, the XRD analysis for all sintered scaffolds matched with β-TCP of PDF card number 055-0898. For the three sintering temperatures examined, the phase purity of the β-TCP bioceramics was the same, as is observed in Figure 4(a). The mean crystallite size for each sintering temperature was estimated using the Sherrer's equation and integrating the XRD spectra with the software package PowderX. As sintering temperature increased so did the crystallite size, with values of 10.6, 12.0, 13.2 nm, for 950°C, 1050°C, and 1150°C, respectively.

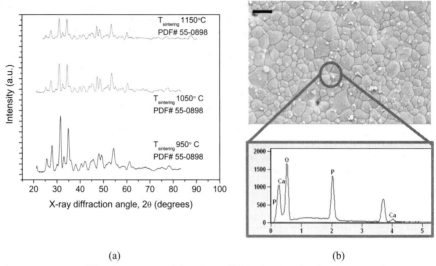

(a)   (b)

Figure 4: (a) X-ray diffraction patterns of the TCP scaffolds, showing that the phase, pure β-TCP, is retained after sintering at various temperatures and verified by PDF card as indicated. (b) SEM-EDS micrograph showing the microstructure (top image) and EDS spectra of elemental analysis profile (bottom image) of a representative β-TCP scaffold. Scale bar is 10 μm.

Previous research in bioceramic materials have shown that the biological properties, mechanical strength, thermal stability and degradability are strongly affected by the chemical composition of Ca/P[23]. In the current study, TCP sintered at different temperatures had a chemical similarity to that of the mineral phase of bone which is mainly hydroxyapatite with a Ca/P ratio of 1.67. EDS analysis found a stoichiometric value Ca/P atomic ratio of 1.55 (Figure 4(b)) a little higher than the theoretical Ca/P atomic ratio of pure β-TCP of 1.5. It is speculated that differences of Ca/P atomic ratio with respect to the theoretical value is due to impurities from the thermal process, mainly hydroxyapatite[24]. In addition, β-TCP crystallizes in the rhombohedral system (space groups R3c) with 21 formula units $Ca_3(PO_4)_2$ per hexagonal unit cell whose dimensions are: $a = b = 10.439$ Å, $c = 37.375$ Å, and $\alpha = \beta = 90°$, $\gamma = 120°$ [24]. Thus, the scaffolds sintered in the current study present similar crystallographic and stoichiometric properties to the mineral in bone tissue; and therefore, could be good candidates as a bone replacement material.

Mechanical properties of the TCP scaffolds at each sintering temperature were evaluated by micro indentation testing in terms of Knoop micro hardness (H_K) (ASTM C 1326-08). It was found that micro hardness increased with increasing sintering temperature with means values of 1.00±0.9, 2.53±0.06, and 2.85±0.09 GPa for 950, 1050, and 1150°C, respectively. The H_k for the 950°C scaffolds was significantly lower than 1050°C and 1150°C (p-value < 0.05), which were found to be equivalent as is shown in Figure 6. Furthermore, increasing hardness with sintering temperature is in consonance with other micro hardness testing reported in the literature for similar bioceramic materials and range of sintering temperatures[25-28]. A previous, related nanoindentation study found a similar trend of increasing hardness with increasing sintering temperature:1.66 GPa, 5.07 GPa and 5.52 GPa, for sintering temperatures of 950°C, 1050°C and 1150°C, respectively [29]. Although the studies found similar trends, the nanoindentation hardness was higher than the Knoop micro hardness because of an indentation size effect (ISE) common in ceramic materials, in which there is an increase of the mechanical properties with decreasing load.

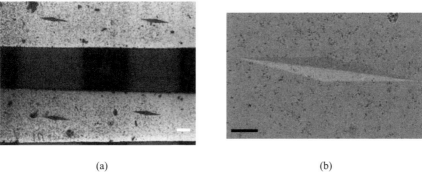

(a) (b)

Figure 5: Knoop micro hardness indentation (H_k) of β-TCP bone scaffolds sintered at different temperatures, (a) Microscope image of a micro indentation, (b) SEM micrograph showing a detailed micro indentation for a scaffold sintered at 950°C. Scale bar is 100 μm (left image) and 40 μm (right image).

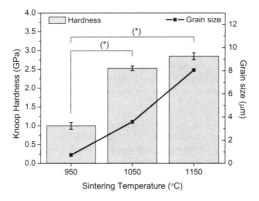

Figure 6: Knoop hardness (H_k) (left axis) for β-TCP scaffold sintered at different temperatures. Right axis represents grain size for the same specimens (*) P-value < 0.01.

CONCLUSION

We have investigated the effect of sintering temperature on microstructural properties of a bioceramic bone scaffold. The scaffolds were fabricated using tricalcium phosphate and sintered at 950°C, 1050°C, and 1150°C. The crystallographic phase, evaluated using XRD, corresponded to a β-TCP bioceramic, for all sintering temperatures. It was found that increasing sintering temperature increased the degree of densification of microstructure and increases the grain and crystallite sizes. The Ca/P atomic ratio did not vary for the range of temperatures investigated and was comparable to theoretical pure β-TCP and the mineral in bone tissue. Regarding mechanical properties, an increase in the sintering temperature resulted in an increase of micro hardness which at 950°C was significantly different than 1050°C and 1150C (p-value < 0.05).

Based on these findings, we conclude that by understanding the relationship between the processing parameters and the resulting microstructure, the sintering temperature can be controlled to optimize microstructural properties. The results of the current study support the use of this particular bioceramic for bone scaffold implants.

ACKNOWLEDGMENTS

The authors wish to thank Phillips Plastic Corporation for supplying the scaffold samples and the Chilean Government Graduate Research Fellowship.

REFERENCES

[1]R. Z. LeGeros: 'Calcium Phosphate-Based Osteoinductive Materials', *Chemical Reviews*, 2008, **108**(11), 4742-4753.

[2]S.-J. L. Kang: 'Sintering Densification, Grain Growth, and Microstructure', 265; 2005.

[3]M. Riminucci and P. Bianco: 'Building bone tissue: matrices and scaffolds in physiology and biotechnology', *Brazilian Journal of Medical and Biological Research*, 2003, **36**, 1027-1036.

[4]P. Kasten, I. Beyen, P. Niemeyer, R. Luginbühl, M. Bohner, and W. Richter: 'Porosity and pore size of [beta]-tricalcium phosphate scaffold can influence protein production and osteogenic differentiation of human mesenchymal stem cells: An in vitro and in vivo study', *Acta Biomaterialia*, 2008, **4**(6), 1904-1915.

[5]H. Yuan, Z. Yang, Y. Li, X. Zhang, J. D. De Bruijn, and K. De Groot: 'Osteoinduction by calcium phosphate biomaterials', *Journal of Materials Science: Materials in Medicine*, 1998, **9**(12), 723-726.

[6]Y. Kuboki, H. Takita, D. Kobayashi, E. Tsuruga, M. Inoue, M. Murata, N. Nagai, Y. Dohi, and H. Ohgushi: 'BMP-Induced osteogenesis on the surface of hydroxyapatite with geometrically feasible and nonfeasible structures: Topology of osteogenesis', *Journal of Biomedical Materials Research*, 1998, **39**(2), 190-199.

[7]K. David, M. E. P. Goad, A. Maria, R. Christian, T. Ali, C. Pramod, and D. D. Lee: 'Resorbable calcium phosphate bone substitute', *Journal of Biomedical Materials Research*, 1998, **43**(4), 399-409.

[8]J. Wiltfang, H. A. Merten, K. A. Schlegel, S. Schultze-Mosgau, F. R. Kloss, S. Rupprecht, and P. Kessler: 'Degradation characteristics of alpha and beta tri-calcium-phosphate (TCP) in minipigs', *Journal of Biomedical Materials Research*, 2002, **63**(2), 115-121.

[9]R. E.-G. Ahmed: 'Advanced bioceramic composite for bone tissue engineering: Design principles and structure-bioactivity relationship', *Journal of Biomedical Materials Research Part A*, 2004, **69A**(3), 490-501.

[10]V. Vicente, L. Meseguer, F. Martinez, A. Galian, J. Rodriguez, M. Alcaraz, and M. Clavel: 'Ultrastructural study of the osteointegration of bioceramics (whitlockite and composite beta-TCP + collagen) in rabbit bone', *Ultrastruct Pathol*, 1996, **20**(2), 179-188.

[11]P. Habibovic and K. de Groot: 'Osteoinductive biomaterials—properties and relevance in bone repair', *Journal of Tissue Engineering and Regenerative Medicine*, 2007, **1**(1), 25-32.

[12]R. Z. LeGeros and J. P. LeGeros: 'Calcium phosphate bioceramics: Past, present and future', Proceedings of the 15th International Symposium on Ceramics in Medicine; The Annual Meeting of the International Society for Ceramics in Medicine, Dec 4-8 2002, Sydney, NSW, Australia, 2003, Trans Tech Publications Ltd, 3-10.

[13]S. V. Dorozhkin: 'Bioceramics of calcium orthophosphates', *Biomaterials*, 2010, **31**(7), 1465-1485.

[14]M. U. Entezarian, R. U. Smasal, and J. Peskar, C.; (US). 'Methods, Tools, and Products for Molded Ordered Porous Structures', Patent PCT/US2008/010218, US, 2009.

[15]C. Dong: 'PowderX: Windows-95-based program for powder X-ray diffraction data processing', *Journal of Applied Crystallography*, 1999, **32**(4), 838.

[16]J. Vivanco, S. Garcia, E. L. Smith, and H. Ploeg: 'Material and Mechanical Properties of Tricalcium Phosphate-based (TCP) Scaffolds', American Society of Mechanical Engineering (ASME) Summer Bioengineering Conference, Lake Tahoe, CA, 2009.

[17]F. Zhang, J. Chang, K. Lin, and J. Lu: 'Preparation, mechanical properties and in vitro degradability of wollastonite/tricalcium phosphate macroporous scaffolds from nanocomposite powders', *Journal of Materials Science: Materials in Medicine*, 2008, **19**(1), 167-173.

[18]S. Li, H. Izui, and M. Okano: 'Densification, Microstructure, and Behavior of Hydroxyapatite Ceramics Sintered by Using Spark Plasma Sintering', *Journal of Engineering Materials and Technology*, 2008, **130**(3), 031012-031017.

[19]Y. W. Gu, N. H. Loh, K. A. Khor, S. B. Tor, and P. Cheang: 'Spark plasma sintering of hydroxyapatite powders', *Biomaterials*, 2002, **23**(1), 37-43.

[20]E. D. Case, I. O. Smith, and M. J. Baumann: 'Microcracking and porosity in calcium phosphates and the implications for bone tissue engineering', *Materials Science and Engineering A*, 2005, **390**(1-2), 246-254.

[21]G. C. Reilly, J. D. Currey, and A. E. Goodship: 'Exercise of young thoroughbred horses increases impact strength of the third metacarpal bone', *Journal of Orthopaedic Research*, 1997, **15**(6), 862-868.

[22]H.-S. Ryu, H.-J. Youn, B.-S. Hong, B.-S. Chang, C.-K. Lee, and S.-S. Chung: 'An improvement in sintering property of [beta]-tricalcium phosphate by addition of calcium pyrophosphate', *Biomaterials*, 2002, **23**(3), 909-914.

[23]K. Lin, L. Chen, J. Chang, and J. Lu: 'Influence of $Ca_2P_2O_7$ on sintering ability, mechanical strength and degradabilty of β–$Ca_3(PO_4)_2$ bioceramics', *Key Engineering Materials*, 2007, **361-363**, 23-26.

[24]C. Rey, C. Combres, C. Drouet, and S. Somrani: 'Tricalcium phosphate-based ceramics', in 'Bioceramics and their clinical applications', (ed. T. Kokubo), 326-366; 2008, England, Woodhead Publishing.

[25]K. A. Gross and L. M. Rodríguez-Lorenzo: 'Sintered hydroxyfluorapatites. Part II: Mechanical properties of solid solutions determined by microindentation', *Biomaterials*, 2004, **25**(7-8), 1385-1394.

[26]C. Y. Tang, P. S. Uskokovic, C. P. Tsui, D. Veljovic, R. Petrovic, and D. Janackovic: 'Influence of microstructure and phase composition on the nanoindentation characterization of bioceramic materials based on hydroxyapatite', *Ceramics International*, 2009, **35**(6), 2171-2178.

[27]C. X. Wang, X. Zhou, and M. Wang: 'Influence of sintering temperatures on hardness and Young's modulus of tricalcium phosphate bioceramic by nanoindentation technique', *Materials Characterization*, 2004, **52**(4-5), 301-307.

[28]A. Slosarczyk and J. BiaŁOskorski: 'Hardness and fracture toughness of dense calcium–phosphate-based materials', *Journal of Materials Science: Materials in Medicine*, 1998, **9**(2), 103-108.

[29]J. Vivanco, J. Slane, R. Nay, A. Simpson, and H.-L. Ploeg: 'The effect of sintering temperature on the microstructure and mechanical properties of a bioceramic bone scaffold', *Journal of the Mechanical Behavior of Biomedical Materials*, 2011,

APPLICATION OF POLYMER-BASED MICROFLUIDIC DEVICES FOR THE SELECTION AND MANIPULATION OF LOW-ABUNDANT BIOLOGICAL CELLS

Małgorzata A. Witek[1,2,3], Udara Dharmasiri[1,2], Samuel K. Njoroge[1,2], Morayo G. Adebiyi[2], Joyce W. Kamande[1,2,3], Mateusz L. Hupert [1,2,3], Francis Barany[4], and Steven A. Soper [1,2,3,5]

(1) Center for Bio-Modular Multi-Scale Systems, Louisiana State University, 8000 G.S.R.I. Rd., Blg. 3100, Baton Rouge, LA 70820
(2) Department of Chemistry, Louisiana State University, 232 Choppin Hall, Baton Rouge, LA 70803
(3) Current Address: Department of Biomedical Engineering, University of North Carolina, Chapel Hill, NC, 27599
(4) Department of Molecular Biology, Weill Cornell Medical College, New York, NY 10065
(5) School of Nano-Bioscience and Chemical Engineering, Ulsan National Institute of Science and Technology, Ulsan, South Korea

ABSTRACT

Microfluidics are providing new opportunities for diagnostics. Innovative technologies developed in our group allow for the selection, enumeration, and molecular profiling of low-abundant (<100 cells/mL) circulating tumor cells (CTCs) directly from blood. Cancer cells overexpressing the integral membrane protein, EpCAM, were immunospecifically isolated using anti-EpCAM antibodies immobilized onto the walls of a selection bed poised on a microfluidic device. The cells were enzymatically released from the surface and hydrodynamically carried for enumeration through pair of platinum (Pt) electrodes. Following counting, CTCs were withdrawn electrophoretically from the bulk flow, and preconcentrated into an anodic reservoir. The molecular processing invoked a strategy consisting of a primary PCR for amplification of gene fragment, and an allele-specific ligation reaction to detect mutations within specific gene. The strategy developed offers the ability to profile point mutations harbored in rare CTCs in whole blood without the interferences from highly abundant leukocytes and erythrocytes. The entire CTC's enumeration and preconcentration steps can be implemented in under 40 min for 1 mL samples.

INTRODUCTION

The ability to enumerate low-abundant cells and perform subsequent analyses can be applied in cancer research [1], forensic science, homeland security, and environmental testing [2]. Some impactful applications involve the isolation and characterization of intact fetal cells in maternal blood for non-invasive prenatal diagnosis of genetic disorders [3], stem cells for cell-based therapies [4], detection of Circulating Tumor Cells (CTCs) in blood for diagnosis and prognosis of various cancers [5-10], T lymphocytes for determining the progression rate in human immunodeficiency virus (HIV) infections [11, 12], or autoimmune disease [13].

The major challenge of any analytical strategy involving low-abundant cells begins with the selection of the target cells from a heterogeneous population in which the target is a minority. For example, in the case of CTCs, in 1 mL of whole blood there are $>10^9$ red blood cells (RBCs), $>10^6$ white blood cells (WBCs) and only as little as 1-5 CTCs, requiring an enrichment factor $>10^9$ [14, 15]. Four metrics must be considered when isolating low-abundant cells. They are (i) throughput; the number of cell identification or sorting steps per unit time; (ii) recovery; an indicator of the fraction of target cells collected from the input sample; (iii) purity; which depends on the number of "interfering" cells excluded from the analysis; and (iv) viability, whether the isolated cells retain their biological function after the selection process [16, 17]. In addition, highly efficient quantification of the number of enriched low-abundant cells must be provided in most cases [18].

It is established that most cancer-related mortalities result from metastatic disease [19], which can be spawned by the release and deposition of tumor cells into certain organs using peripheral blood and/or bone marrow as the carrier(s). CTCs are detected in blood of metastatic cancer patients using different enrichment methods. There are two types of formats for selecting CTCs from clinical samples, macro-scale and micro-scale techniques. The macro-scale formats for enriching CTCs include: (i) magnetic capture using micrometer-size ferromagnetic beads coated with molecular recognition elements, typically called immunomagnetic-assisted cell sorting, IMACS [20-22]; (ii) size-based separations using nuclear tracked membranes or microfabricated filters [23, 24]; (iii) fluorescence-assisted cell sorting, [25, 26]; and (iv) reverse-transcription of mRNAs with the mRNAs used as surrogates for cell identification through the PCR amplification of cDNAs unique to the target [27]. Many of these macro-scale techniques have provided either low recoveries with high purity, or relatively poor purity but with high recoveries [22, 28-32]. Immuno-magnetic approaches for CTC enrichment typically yields recoveries ~70% but extremely favorable purity [29]. Size-based selection employing filter membranes with varying pore sizes (8 – 14 μm) can filter large volumes (up to 18 mL) of blood and recover nearly 85% of the CTCs, however significant numbers of leukocytes are also retained causing low purity and complicating the enumeration process [33]. Investigations utilizing reverse-transcription PCR to report CTC levels, have the ability to detect one CTC in an excess of 10^6 mononucleated cells [34], however, these assays are prone to high inter-laboratory variability and require extensive sample handling and manipulation.

Affinity-based selection of low-abundant cells is often employed in microsystems and exploits the non-covalent interactions between a ligand, such as an Ab, aptamers, or peptide, and a target specific receptor (e.g., membrane protein) [35]. Microchip technology has the potential to select with high recoveries, enumerate, and characterize rare CTCs in cancer patients. For example, Nagrath *et al.* developed a microfluidic device for selecting CTCs directly from peripheral blood.[36] The potential utility of this CTC chip in monitoring response to anti-cancer therapy was also investigated.[36] Adams *et al.* and Dharmasiri *et al.* generated a polymer–based fluidic device for the high recovery and efficient enumeration of CTCs directly from peripheral blood using either antibodies or aptamers as the recognition elements [18, 37].

In addition to the enumeration of CTCs, mutations in certain genes carried by CTCs can be used to guide therapy and provide opportunities for personalized treatment. For example, colorectal cancer (CRC) patients with mutated Kras oncogene do not benefit from anti-EGFR mAB therapy, whereas patients with

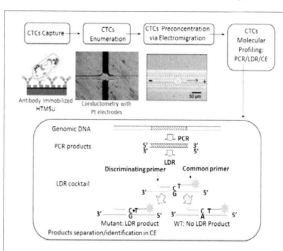

Scheme 1. Overview for the cell selection, enumeration, electrokinetic enrichment, and molecular profiling strategy adopted for the analysis of low-abundance CTCs resident in peripheral blood.

wild-type Kras genotypes do benefit from the chemotherapy treatment.[38] The significant challenge in genotyping genomic DNA from mass-limited samples, such as CTCs isolated from cancer patients, is the small copy number of the assay's input material.

Herein we present system capable of selecting and enumerating CTCs; a polymer–based fluidic device for the high recovery of CTCs directly from peripheral blood using anti-EpCAM antibodies. High-throughput micro-sampling unit was coined with an electrokinetic manipulation unit that was interfaced to a fluidic chip. The electrokinetic manipulation unit used hydrodynamic flow and an electric field to direct CTCs released from the selection surface into a reservoir (2 μL) to allow for the molecular interrogation of low numbers of CTCs. In the present study we used Kras mutations in the CRC cell lines, SW620, and HT29, as a model.

These cells express EpCAM which has first been identified as a tumor-specific antigen on several carcinomas and epithelial malignancies, including colon, breast, and prostate [39]. Because most Kras mutations are localized to codon 12 and to a lesser extent, codons 13 and 61, a PCR coupled to a ligase detection reaction (LDR) assay was performed on codon 12 of the Kras oncogene of the CTCs.[40, 41] Following PCR/LDR, the resulting products were analyzed via capillary gel electrophoresis (CGE). The attractive nature of this assay format for the mass-limited samples, is the use of two amplification steps; PCR provides an exponential amplification of the interrogated genes, while the LDR gives a linear amplification of products formed.

EXPERIMENTAL SECTION

Reagents and cells. SW620 and HT29 colorectal cancer cell lines, growth media, HEPES buffer, phosphate buffered saline (PBS) and trypsin were purchased from American Type Culture Collection (ATCC, Manassas, VA). Citrated rabbit blood was secured from Colorado Serum Company (Denver, CO). Poly(methylmethacrylate), PMMA, substrates and cover plates (0.5 mm thickness) were purchased from Good Fellow (Berwyn, PA). Pt wires were purchased from Alfa Aesar (Boston, MA). Polyimide-coated fused silica capillaries were obtained from Polymicro Technologies (Phoenix, AZ).

Antibody immobilization to the HTMSU. Chemicals used for PMMA surface cleaning and modification included reagent grade isopropyl alcohol, 1-ethyl-3-[3-dimethylaminopropyl] carbodiimide hydrochloride (EDC), N-hydroxysuccinimide (NHS), fetal bovine serum and 2-(4-morpholino)-ethane sulfonic acid (MES) with all acquired from Sigma-Aldrich (St. Louis, MO). Monoclonal anti-EpCAM antibody was obtained from R & D Systems (Minneapolis, MN). Tris-glycine buffer was obtained from Bio-Rad Laboratories (Hercules, CA).

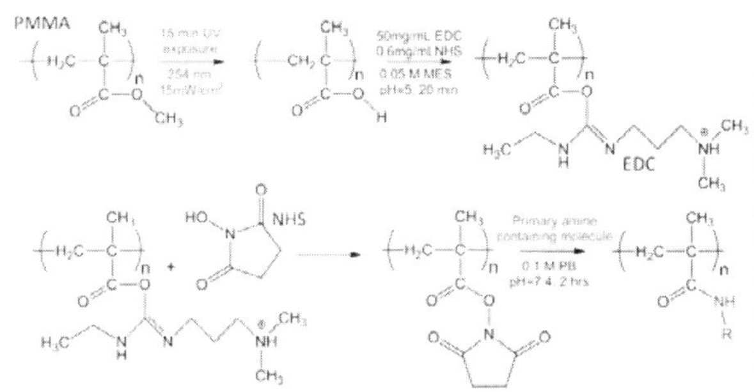

Scheme 2. Modification chemistry of PMMA induced by UV-irradiation. The pendant carboxylic acids formed can be used to attach – NH_2 containing groups via EDC/NHS coupling chemistry (R = Ab).

All solutions were prepared in nuclease-free water (Invitrogen, Carlsbad, CA). Nuclease-free microfuge tubes were purchased from Ambion (Foster City, CA) and used for preparation and storage of all samples and reagents. A fluorescein cell membrane labeling derivative, PKH67, was secured from Sigma-Aldrich. All oligonucleotide probes and primers were obtained from

Integrated DNA Technologies (IDT, Coralville, IA). Antibody immobilization was carried out in a two-step process (see Scheme 2). The UV-modified thermally assembled HTMSU device was loaded with a solution containing 50 mg mL^{-1} EDC, 0.6 mg mL^{-1} NHS in 150 mM MES (pH = 6) for 1 h at room temperature to obtain a succinimidyl ester intermediate. After this incubation, the EDC/NHS solution was removed by flushing nuclease-free water through the device. Then, an aliquot of 1.0 mg mL^{-1} of the monoclonal anti-EpCAM antibody solution contained in 50 mM PBS (pH = 7.4) was introduced into the HTMSU and allowed to react for 4 h. The device was then rinsed with a solution of PBS (pH = 7.4) to remove any non-specifically bound anti-EpCAM antibodies.[18]

Fabrication of the HTMSU. The HTMSU and electro-manipulation unit were hot embossed into PMMA substrates via micro-replication from a metal mold master. Fabrication of the HTMSU followed steps previously reported [18]. The electro-manipulation unit possessed one entry, through reservoir (a) (see Figure 1) and two exits, via reservoirs (b) and (c). Reservoir (a) was connected to a "T" intersection labeled (e) in Figure 1. All of the channels were rectangular in shape with dimensions of 50 μm × 100 μm, width and depth, respectively. Pt wires of 125 μm diameter served as external electrodes in the (b) and (c) reservoirs of the electro-manipulation unit.

Cell visualization. A Zeiss Axiovert 200M (Carl Zeiss, Thornwood, NY) inverted microscope was utilized in cases where the cells required optical visualization to assist in device operational optimization. The PMMA devices were fixed onto a programmable motorized stage of an Axiovert 200M inverted microscope and video images were collected during each experiment at 30 frames s^{-1} using a monochrome CCD (JAI CV252, San Jose, CA). A Xe arc lamp was

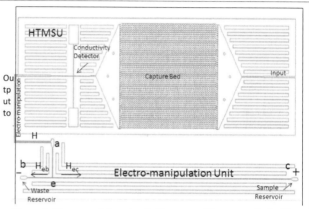

Figure 1. Diagrams of the microfluidic system: The capture bed with a series of 51 curvilinear channels . The electro-manipulation unit contained 80 μm wide, 100 μm deep and 5 cm long linear channels. The total volume of the receiving reservoir was 2 μl.

used to excite fluorescent dyes incorporated into the cells' membrane.

Cell capture/release and enumeration using the HTMSU. A pre-capture rinse was performed of the HTMSU prior to sample processing using 0.2 mL of 150 mM PBS at 50 mm s^{-1} linear velocity to maintain isotonic conditions. Then, the appropriate volume of a cell suspension was introduced at 27.5 μl min^{-1} volumetric flow rate (linear velocity through the selection bed was 2 mm s^{-1}).[18] Next, a post-capture rinse was performed with 0.2 mL of 150 mM PBS at 50 mm s^{-1} to remove any non-specifically adsorbed cells. Following a post capture rinse, a 0.25% trypsin solution in 0.2 mM Tris/19.2 mM glycine buffer (pH = 8.3) was infused into the HTMSU.

Electrokinetic cell manipulation. The released cells were transported into the electro-manipulation unit at 1 μL min^{-1} (3.3 mm s^{-1}) from reservoir (a) (see Figure 1). When cells were sensed by the conductivity detector an electric field was applied between reservoirs (b) and (c) within the electro-manipulation unit using a home-bulit, computer controlled high-voltage power supply) Cell migration velocities in the present study were calculated by measuring consecutive cell events observed microscopically and imaged at video frame rates (30 frames s^{-1}). The average velocity was based on values obtained from at least 5 different cell events.

Colorectal cancer cell lysis. Genomic DNA was released from CTCs using a Lyse-and-Go PCR reagent kit (Pierce Biotechnology, IL, USA). Following manufacturer's recommendations, a 5 µL of Lyse-and-Go PCR reagent was added to the selected CTCs, and heated using the following temperature program: 65 °C for 30 s; 8 °C for 30 s; 65 °C for 90 s; 97 °C for 180 s; 8 °C for 60 s; 65 °C for 180 s; 97 °C for 60s; 65 °C for 60 s. Prior to the addition of the PCR cocktail, the samples were placed at 80 °C.

PCRs, LDRs, gel electrophoresis and capillary electrophoresis. PCRs were carried out on selected CTCs using the following gene-specific primer sequences: exon 1 forward –5' TTA AAA GGT ACT GGT GGA GTA TTT GAT A 3', (T_m = 55.4 °C) and exon 1 reverse – 5' AAA ATG GTC AGA GAA ACC TTT ATC TGT 3' (T_m = 56.3 °C). Forty-five µL of PCR cocktail consisting of 10 mM Tris–HCl buffer (pH = 8.3) containing 50 mM KCl, 1.5 mM $MgCl_2$, 200 µM dNTPs, and 0.4 µM of each forward and reverse primers were added to the lysate previously held at 80 °C in the thermocycler. After a 2-min initial denaturation, 1.5 U of AmpliTaq DNA polymerase (Applied Biosystems, Foster City, CA, USA) was added under hot-start conditions and amplification was achieved by thermally cycling (30 cycles) at 95 °C for 30 s, 60 °C for 2 min, and a final extension at 72 °C for 3 min.

To test the fidelity and yield of the PCR, slab gel electrophoresis was performed. CTC solutions were mixed with 1 µL of a loading dye and 1x TBE buffer. The mixture was then loaded into a well of an ethidium bromide pre-stained 3% agarose gel (Bio-Rad Laboratories, Hercules, CA). The slab gel electrophoresis was run at 5 V cm^{-1} for 30 min with images acquired using a Gel Logic 200 Visualizer (Carestream Molecular imaging, New Haven, CT).

LDRs were executed in a total volume of 20 µL using a commercial thermal cycling machine (Eppendorf Thermal Cycler, Brinkmann Instrument, Westbury, NY, USA). The reaction cocktail consisted of 10 mM Tris–HCl (pH = 8.3), 25 mM KCl, 10 mM $MgCl_2$, 0.5 mM NAD^+ (nicotinic adenine dinucleotide), 0.01% Triton X-100, 10 nM of the discriminating primer (5' AAACTTGTGGTAGTTGGAGCTGT 3', T_m = 71.3 °C), a fluorescently labeled phosphorylated common primer (5' Phos/TGGCGTAGGCAAGAGTGCCT/Cy5.5Sp 3', T_m = 63.5 °C) and 2 µL of the PCR product. Forty U of Taq DNA ligase (New England Biolabs) was added to the cocktail under hot-start conditions and the reactions thermally cycled 20 times for 30 s at 94 °C and 2 min at 65 °C. The LDR products were stored at 4 °C. The CGE analysis used a CEQ 8000 Genetic Analysis System (Beckman Coulter, Fullerton, CA, USA).

RESULTS AND DISSCUSSIONS

CTC selection. Both cell lines, SW620 and HT29 overexpress the EpCAM, with SW620 cells harboring Kras c12.2V mutations and HT29 possessing wild-type DNA. SW620 and HT29 cells are typically between 15 and 30 µm in diameter (average ~23 µm)[42] and EpCAM occurs at a frequency of 1×10^6 and 2.5×10^5 molecules per cell, respectively.[43] The specific selection of SW620 or HT29 CTCs was based on the recognition capabilities of anti-EpCAM antibodies that were tethered to the walls of the HTMSU selection bed. The number of interactions between the CTC membrane antigens and the channel wall containing the recognition elements are important in determining the recovery of rare CTCs from blood. Based on our previous results, curvilinear-shaped channels were employed to provide a high number of cell/wall interactions to improve recovery. The cell-free marginal zone apparent in straight channels was not observed in these curvilinear channels.[18] Chang's model of cell adhesion in flowing systems has been applied in previous reports to describe the encounter rate between the solution-borne cells and the surface-tethered selection elements.[44] According to the Chang/Hammer model for cell adhesion, efficient antigen/antibody adhesion is determined primarily by two factors; the encounter rate (k_o) and the probability (P) of interaction between the membrane bound antigen and the channel wall's tethered antibody.[44] The optimum linear velocity for selection of CTCs, determined from our previous studies, was 2 mm s⁻¹). The recovery of SW620 cells was found to be 96 ±4% at an optimized translational velocity of 2.0 mm s⁻¹.s.

The selected cells were observed using fluorescence and brightfield microscopy as shown in Figure 2. The contact area between the selection surface and the CTC was determined to be $448 \pm 18 \ \mu m^2$. The anti-EpCAM antibody density on the selection bed wall has been reported to be 2.3×10^{11} molecules cm^{-2},[18] therefore, at the maximum 2.4×10^5 EpCAM/anti-EpCAM antibody interactions were involved in one CTC binding event.

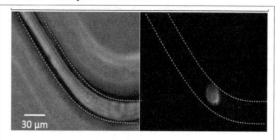

Figure 2. Brightfield (left) and fluorescence (right) micrographs (40x) of selected SW620 CTCs .

Cell detachment from the selection surface. Release of the cells intact is critical because the enumeration process depends on counting via conductivity whole cells and the molecular profiling is performed on rare cells selected from peripheral blood. As such, the genetic material should not be diluted extensively or mixed with potential interfering materials, which could occur if the cells were damaged prior to the enrichment phase of the assay. Enzymatic digestion of the extracellular domain of EpCAM and/or anti-EpCAM antibodies using 0.25% trypsin revealed that the average time for release of the captured SW620 CTCs was ~16 min.

Conductivity enumeration of the CTCs. The selected and released CTCs were subsequently enumerated using a conductivity sensor described previously [18, 37]. A typical data stream is shown in Figure 3. The transducer measured changes in the conductivity of the release buffer with respect to the CTCs present in the buffer as they passed between the two Pt electrodes. Tris-glycine buffer was selected as the major component in the release buffer due to its low conductance, which improved the SNR for the conductivity detection of single CTCs. There were 10 peaks observed in the conductance response shown in Figure 3 that could be assigned to SW620 cells based on a SNR threshold of 3 (99.7% confidence level) giving a recovery of 100% (seed level = 10 CTCs mL⁻¹) for this data set.

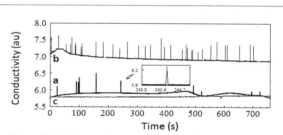

Figure 3. Conductometric responses generated for 1 mL of whole blood seeded with (a) 10 SW620 CTCs, (b) 32 HT29 CTCs and (c) 0 CTCs processed using the HTMSU at a linear flow velocity of 2.0 mm s⁻¹ .

A similar set of experiments was undertaken for HT29 cells and showed similar recoveries as these seen for the SW620 cells. One mL of a blank, which consisted of whole blood without any CTCs seeded into it, was analyzed by the HTMSU and enumeration via conductivity under the same conditions as described for peripheral blood seeded with CTCs. The resultant trace "c" is shown in Figure 3. In this case, no peaks (S/N≥3) were observed.

Electrokinetic enrichment of CTCs. In the electro-manipulation unit, the total hydrodynamic flow (H) arriving from the inlet (a) was divided at the T junction into a major flow (H_{eb}) and a minor flow (H_{ec}, see Figure 1). The ratio between H_{eb} and H_{ec} was equal to 9:1, which was set by the pressure drop ratio ($\Delta P_{eb}/\Delta P_{ec} = 0.1$) between the two channels. As such, the device was designed to have 10% of the hydrodynamic driven input entering the CTC collection reservoir, (c) in Figure 1, and 90% of the hydrodynamic driven flow entering reservoir (b) as seen in Figures 4 A,B.

To provide efficient collection and pre-concentration of the selected CTCs, an electric field was applied to direct the solution-borne CTCs from the hydrodynamic flow and divert them into the collection reservoir (c). As shown in Figure 4 C,D CTCs were electrophoretically diverted from the hydrodynamic flow into (c) due to their intrinsic electrophoretic mobility and the applied electric field, which overcame the force exerted on the cell by the pressure-driven flow. The volume of reservoir (c) was 2 µL providing a 500x enrichment factor for the CTCs (input volume = 1 mL). In addition, reservoir (c) was free from leukocytes and erythrocytes that were the majority cells in the peripheral blood sample input into the HTMSU.

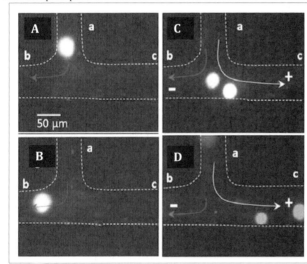

Figure 4. Micrographs (40x) showing the electro-manipulation of SW620 CTCs. The SW620 CTCs were introduced into the system at (a) (see Figure 1). (A, B) The cells were travelling at a hydrodynamic flow rate of 1 µl min^{-1} in the absence of an electric field. The hydrodynamic flow direction is given by the red arrow. (C, D) SW620 CTC movement in the presence of an electric field (100 V cm^{-1}). The CTC main travel direction is given by the white arrow when in the presence of both the hydrodynamic and electrokinetic flows.

According to the Smoluchowski equation, the electrophoretic mobility of cells is determined by surface charged groups.[45] CTCs possess a net negative surface charge (Q) at pH = 8.3 that can be calculated from $Q = 4 \rho \pi r_p^2$, where ρ is the charge density (-5.8 × 10^{-14} C µm^{-2}) [46] and r_p is the cell radius (11.5 µm for CTCs). Therefore, the average net surface charge for a typical CTC is approximately -9.6×10^{-11} C. Because Q is partially set by protonation/deprotonation of membrane-bound proteins and other groups, the CTCs' charge can be controlled by altering the properties of the buffer solution (*i.e.,* pH, ionic strength, salt composition). This, along with the size of the CTC and its zeta potential will determine its electrophoretic mobility. The electrophoretic force (F_{EP}) acting on a particle with a net charge Q under the influence of an electric field is given by $F_{EP} = QV/d$, [47] where V is the applied voltage and d is the distance between the cathode and anode. The electrophoretic force acting on the CTC in the present case was calculated to be 9.6×10^{-9} N.

It was observed that SW620 cells resident in 50 mM PBS underwent extensive lysis at an electric field strength of 100 V cm^{-1}. However, utilizing a Tris-glycine buffer (0.2 mM Tris/19.2 mM glycine) improved the CTC manipulation because this buffer has been found to increase the target cells' zeta potential and thus, mobility.[48] In addition, the buffer capacity must be such to negate possible changes in pH caused by electrolysis at the electrodes and thus, maintain a constant charge on the cells. The SW620 cells were found not to lyse at fields up to 100 V cm^{-1} using the Tris-glycine buffer and the migration velocity was determined to be 130 ±15 μm s^{-1}. This electric field strength may cause membrane permeation, but did not result in cell lysis.[49] The cells were found to migrate from the T junction to (c) in ~5 min. The use of higher electric fields accelerated the cell migration rate, however, at fields ≥200 V cm^{-1} extensive cell lysis was observed. We also found that applying an electric field of 100 V cm^{-1} along with a hydrodynamic flow rate of 1 μl min^{-1} (linear velocity = 3.3 mm s^{-1}) was sufficient to direct the CTCs into (c) with nearly 100% efficiency. Figure 4C,D illustrates two cells moving toward the anode. Above a hydrodynamic flow rate of 1 μl min^{-1}, the CTCs preferentially followed the hydrodynamic flow towards the waste reservoir (reservoir (b) in Figure 1).

For an incompressible Newtonian fluid of low Reynolds number, the fluid motion can be determined through the use of the Stoke's equation. The Stoke's frictional force on a charged particle is due to the relative motion of the particle with regard to the EOF and is given by $F_{Stoke's}$ = -6πηr_p(U_h+U_{EOF}), where η is the solution viscosity and is equal to 7.98×10^{-4} N s m^{-2} in the current experiments,[50] r_p is the radius of the cell, and U_h is the linear velocity of the hydrodynamic flow, H_{eb} = 3.0 × 10^{-3} m s^{-1}. The EOF value for 0.25% w/v trypsin/0.2 mM Tris/19.2 mM glycine buffer was determined to be 3.1 ±0.23 × 10^{-4} cm^2 V^{-1} s^{-1} at pH = 8.3. Therefore, the linear velocity (U_{EOF}) for the corresponding EOF was 3.1 × 10^{-4} m s^{-1}. At pH = 8.3, the PMMA surface consists of negatively charged functional groups producing a cathodic EOF,[51] which is in the same direction as the hydrodynamic flow of U_h in the e-b section of the channel. The total Stoke's force on the SW620 cells at the T junction was therefore calculated to be 5.0 × 10^{-10} N.

At the T junction, the electrophoretic force (9.6 × 10^{-9} N) was larger than the Stoke's force (5.0 × 10^{-10} N) acting on the cell. Therefore, CTCs were directed into the collection reservoir (c) consistent with that observed in our experiments. In addition, the linear velocity due to the hydrodynamic flow in channel e-c of the electro-manipulation unit, U_{ec}, was 3.0×10^{-4} m s^{-1}, which was 10% of the total hydrodynamic velocity (U_h = 3.0 × 10^{-3} m s^{-1}). This flow is in the opposite direction to that of the EOF in the e-c channel. As such, the hydrodynamic flow was cancelled by the EOF. The apparent electrophoretic mobility of the SW620 cells was calculated to be 1.23 x 10^{-4} cm^2 V^{-1} s^{-1}. This value was close to the experimentally determined of 1.30 ±0.15 cm^2 V^{-1} s^{-1}. Chip-to-chip reproducibility of the cell velocities varied slightly, however, the direction of the cell transport was reproducible from chip-to-chip. For example, in 3 different PMMA devices, the RSD for the velocities were 10–15%.

PCR/LDR/CGE analysis of CTCs. A method that can detect single point mutations in gDNA is the ligase detection reaction (LDR) coupled to a PCR (see Scheme 1). Following PCR amplification of the appropriate gene fragments containing the desired point mutations, the amplicon was mixed with two LDR primers; a common and discriminating primer that flanked the point mutation of interest. The discriminating primer contained a base at its 3'-end coinciding with the single base mutation site. If these bases were not complementary, ligation of the two primers did not occur. A perfect match resulted in a successful ligation event of the two primers and a product length that was the sum of the two oligonucleotide primers.[41]

Molecular profiling required the analysis of point mutations in gDNA from only a few CTCs, and as such, highly efficient reactions were required on this mass-limited sample. Unfortunately, the presence of trypsin used for cell release could inhibit the enzymatic action of both

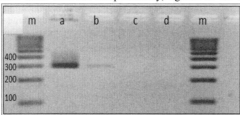

Figure 5. Gel electropherogram for PCR performed on SW620 cells obtained from the HTMSU selection followed by electrokinetic enrichment for; (a) gDNA template from SW620 (no selection or enrichment – positive control); (b) 10 SW620 cells selected from whole blood using HTMSU and enrichment; (c) whole blood with no SW620 cells (d) no gDNA template.

the *Taq* polymerase used for PCR and/or the ligase enzyme used for the LDR, thereby lowering the limit-of-detection of the molecular assay. The cell releasing agent from the selection surface consisted of 0.25% trypsin, which possessed a net positive charge at the pH used for CTC release and electrokinetic manipulation (pI = 10.3).[52] Therefore, trypsin should migrate almost exclusively toward the cathode, facilitating transfer of trypsin into the waste reservoir (b) (see Figure 1). We noticed no degradation in the efficiency of the PCRs and LDRs following selection from blood samples using the HTMSU and subsequent trypsin release indicating efficient removal of trypsin from the enriched CTCs.

Enriched CTCs were lysed and the PCR/LDR assay was performed to search for point mutations in codon 12 of the Kras oncogene (c12.2V). The PCR phase of the assay was first evaluated using different numbers of SW620 cells ranging from ~10 to ~5,000. The cells were added to a PCR tube, lysed and the PCR reagents added to the tube. Cells were successfully amplified to yield a 300 bp amplicons, as confirmed by gel electrophoresis. To demonstrate the capability of our HTMSU with electrokinetic cell enrichment to permit subsequent molecular profiling, 1 mL of whole blood containing low-abundance CTCs was processed. Ten SW620 cells were selected and enriched from 1 mL of whole blood and subjected to PCR, yielding the results shown in Figure 5. Successful PCRs on 10 CTCs was confirmed by both the positive (CTCs present) and negative (no CTCs present) control experiments.

Three samples, one containing whole blood without CTCs, a second one containing 32 HT29 CTCs, and a third sample containing 10 SW620 CTCs were processed using the

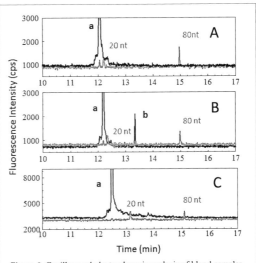

Figure 6. Capillary gel electrophoresis analysis of blood samples after processing using the HTMSU and the electro-manipulation unit seeded with (A) 0 CTCs; (B) 10 SW620 CTCs; and (C) 50 HT29 CTCs. Peak 'a' represents the primer peak and peak 'b' is the 43 nt LDR product. DNA size markers of 20 and 80 nt were co-electrophoresed with the LDR products.

HTMSU and enriched with the electro-manipulation unit. The processed samples were then subjected to PCR/LDR/CGE and yielded the results shown in Figures 6A-C. As expected, our results confirmed the presence of no mutations in whole blood that contained no CTCs (Figure 6A). The sample containing 32 HT29 cells did not show any LDR product as well (Figure 6B) because these cells do not harbor the point mutation being interrogated. Conversely, the results of whole blood that contained 10 SW620 cells produced the desired 43 nt LDR product indicative of a c12.2V point mutation as shown in Figure 6C.

CONCLUSIONS

We demonstrated the integration of the rare cell selection device with an electro-manipulation unit to allow the ability to search for point mutations in the gDNA of selected CTCs even for cases when the CTC number was low and found to be a significant minority in a mixed population (*i.e.*, whole blood). A series of analytical processes was carried out on this mass-limited sample including affinity selection of the CTCs, quantification of the selected CTCs via conductivity sensing and electrophoretic enrichment of the selected CTCs for the subsequent PCR/LDR/CGE interrogation of potential point mutations within their gDNA. We showed the recovery of CTCs with ~96 ±4% efficiency from whole blood and ~100% electrokinetic enrichment of the selected CTCs. The strategy developed offers the ability to profile point mutations harbored in rare CTCs in whole blood without the interferences from highly abundant leukocytes and erythrocytes. The entire series of processing steps can be implemented in under 40 min for 1 mL samples.

REFERENCES

[1]   B. Mostert, *et al.*, "Circulating tumor cells (CTCs): Detection methods and their clinical relevance in breast cancer," *Cancer Treatment Rev.,* vol. 35, pp. 463-474, 2009.

[2]   J. Utikal, *et al.*, "Immortalization eliminates a roadblock during cellular reprogramming into iPS cells," *Nature,* vol. 460, pp. 1145-8, 2009.

[3]   Elisavet K., *et al.*, "Evaluation of Non-Invasive Prenatal Diagnosis from Fetal Nucleated Red Blood Cells (NRBCs) isolated from Maternal Circulation," *Chromosome Res.,* vol. 17, pp. 230-231, 2009.

[4]   E. de Wynter and R. E. Ploemacher, "Assays for the assessment of human hematopoietic stem cells," *J.Biol.Regul.Homeost.Agents,* vol. 15, pp. 23-7, 2001.

[5]   W. J. Allard, *et al.*, "Tumor cells circulate in the peripheral blood of all major carcinomas but not in healthy subjects or patients with nonmalignant diseases," *Clinical Cancer Research,* vol. 10, pp. 6897-904, 2004.

[6]   L. Bertazza, *et al.*, "Circulating tumor cells in solid cancer: tumor marker of clinical relevance?," *Curr.Oncol.Rep.,* vol. 10, pp. 137-46, 2008.

[7]   S. Mocellin, *et al.*, "The Prognostic Value of Circulating Tumor Cells in Patients with Melanoma: A Systematic Review and Meta-analysis," *Clinical Cancer Research,* vol. 12, pp. 4605-4613, 2006.

[8]   S. Nagrath, *et al.*, "Isolation of rare circulating tumour cells in cancer patients by microchip technology," *Nature,* vol. 450, pp. 1235-1239, 2007.

[9]   C. Siewert, *et al.*, "Rapid enrichment and detection of melanoma cells from peripheral blood mononuclear cells by a new assay combining immunomagnetic cell sorting and immunocytochemical staining," *Recent results cancer res.,* vol. 158, pp. 51-60, 2001.

[10]   P. Wuelfing, *et al.*, "HER2-Positive Circulating Tumor Cells Indicate Poor Clinical Outcome in Stage I to III Breast Cancer Patients," *Clinical Cancer Research,* vol. 12, pp. 1715-1720, 2006.

[11]   C. A. Jansen, *et al.*, "Prognostic value of HIV-1 gag-specific CD4+ T-cell responses for progression to AIDS analyzed in a prospective cohort study," *Blood,* vol. 107, pp. 1427-1433, 2006.

[12]   E. Ramirez de Arellano, *et al.*, "Genetic analysis of the long terminal repeat (LTR) promoter region in HIV-1-infected individuals with different rates of disease progression," *Virus Genes,* vol. 34, pp. 111-116, 2007.

[13]   D. P. Collins, *et al.*, "T-lymphocyte functionality assessed by analysis of cytokine receptor expression, intracellular cytokine expression, and femtomolar detection of cytokine secretion by quantitative flow cytometry," *Cytometry,* vol. 33, pp. 249-255, 1998.

[14]   M. Cristofanilli, *et al.*, "Circulating tumor cells, disease progression, and survival in metastatic breast cancer," *New England Journal of Medicine,* vol. 351, pp. 781-791, Aug 2004.

[15]   M. Cristofanilli, *et al.*, "Circulating tumor cells: A novel prognostic factor for newly diagnosed metastatic breast cancer," *Journal of Clinical Oncology,* vol. 23, pp. 1420-1430, Mar 2005.

[16]   J. El-Ali, *et al.*, "Cells on chips," *Nature,* vol. 442, pp. 403-411, 2006.

[17]   A. P. Kodituwakku, *et al.*, "Isolation of antigen-specific B cells," *Immunol.Cell Biol.,* vol. 81, pp. 163-70, 2003.

[18]   A. A. Adams, *et al.*, "Highly Efficient Circulating Tumor Cell Isolation from Whole Blood and Label-Free Enumeration Using Polymer-Based Microfluidics with an Integrated Conductivity Sensor," *J. Am. Chem. Soc.*, vol. 130, pp. 8633-8641, 2008.

[19]   C. Leaf, "Why we're losing the war on cancer [and how to win it]." *Fortune*, pp. 76-94, 2004.

[20]   H. Mohamed, *et al.*, "Circulating tumor cells: capture with a micromachined device," *NSTI Nanotech 2005, NSTI Nanotechnology Conference and Trade Show, Anaheim, CA, United States, May 8-12, 2005*, vol. 1, pp. 1-4, 2005.

[21]   V. I. Furdui and D. J. Harrison, "Immunomagnetic separation of rare cells on chip for DNA assay sample preparation," *Micro Total Analysis Systems 2001, Proceedings micro TAS 2001 Symposium, 5th, Monterey, CA, United States, Oct. 21-25, 2001*, pp. 289-290, 2001.

[22]   P. Grodzinski, *et al.*, "A Modular Microfluidic System for Cell Pre-concentration and Genetic Sample Preparation," *Biomedical Microdevices*, vol. 5, pp. 303-310, 2003.

[23]   P. Wilding, *et al.*, "Manipulation and flow of biological fluids in straight channels micromachined in silicon," *Clinical chemistry*, vol. 40, pp. 43-7, 1994.

[24]   G. Vona, *et al.*, "Isolation by size of epithelial tumor cells : a new method for the immunomorphological and molecular characterization of circulatingtumor cells," *The American journal of pathology*, vol. 156, pp. 57-63, 2000.

[25]   M. Balic, *et al.*, "Cancer metastasis: advances in the detection and characterization of disseminated tumour cells facilitate clinical translation," *The National medical journal of India*, vol. 18, pp. 250-5, 2005.

[26]   R. A. Ghossein, *et al.*, "Detection of prostatic specific membrane antigen messenger RNA using immunobead reverse transcriptase polymerase chain reaction," *Diagnostic molecular pathology the American journal of surgical pathology, part B*, vol. 8, pp. 59-65, 1999.

[27]   A. C. Lambrechts, *et al.*, "Comparison of immunocytochemistry, reverse transcriptase polymerase chain reaction, and nucleic acid sequence-based amplification for the detection of circulating breast cancer cells," *Breast cancer research and treatment*, vol. 56, pp. 219-31, 1999.

[28]   V. I. Furdui and D. J. Harrison, "Immunomagnetic T cell capture from blood for PCR analysis using microfluidic systems," *Lab on a Chip*, vol. 4, pp. 614-618, 2004.

[29]   M. Balic, *et al.*, "Comparison of two methods for enumerating circulating tumor cells in carcinoma patients," *Cytometry B Clin Cytom*, vol. 68, pp. 25-30, 2005.

[30]   H. J. Gross, *et al.*, "Model study detecting breast cancer cells in peripheral blood mononuclear cells at frequencies as low as 10^{-7}," *Proceedings of the National Academy of Sciences of the United States of America*, vol. 92, pp. 537-541, 1995.

[31]   G. Vona, *et al.*, "Isolation by size of epithelial tumor cells - A new method for the immunomorphological and molecular characterization of circulating tumor cells," *American Journal of Pathology*, vol. 156, pp. 57-63, Jan 2000.

[32]   J. Yang, *et al.*, "Cell separation on microfabricated electrodes using dielectrophoretic/gravitational field flow fractionation," *Analytical Chemistry*, vol. 71, pp. 911-918, Mar 1 1999.

[33]   L. Zabaglo, *et al.*, "Cell filtration-laser scanning cytometry for the characterisation of circulating breast cancer cells," *Cytometry Part A*, vol. 55A, pp. 102-108, Oct 2003.

[34]   R. A. Ghossein, *et al.*, "Detection of prostatic specific membrane antigen messenger RNA using immunobead reverse transcriptase polymerase chain reaction," *Diagnostic Molecular Pathology*, vol. 8, pp. 59-65, 1999.

[35]   O. K. Koo, *et al.*, "Targeted Capture of Pathogenic Bacteria Using a Mammalian Cell Receptor Coupled with Dielectrophoresis on a Biochip," *Anal.Chem.*, vol. 81, pp. 3094-3101, 2009.

[36]   S. Nagrath, *et al.*, "Isolation of rare circulating tumour cells in cancer patients by microchip technology," *Nature (London, U. K.)*, vol. 450, pp. 1235-1239, 2007.

[37]   U. Dharmasiri, *et al.*, "Highly efficient capture and enumeration of low abundance prostate cancer cells using prostate-specific membrane antigen aptamers immobilized to a polymeric microfluidic device," *Electrophoresis,* vol. 30, pp. 3289-3300, 2009.

[38]   O. Bouche, *et al.*, "The role of anti-epidermal growth factor receptor monoclonal antibody monotherapy in the treatment of metastatic colorectal cancer," *Cancer Treat. Rev.,* vol. 36, pp. S1-S10, 2010.

[39]   D. B. Seligson, *et al.*, "Epithelial Cell Adhesion Molecule (KSA) Expression," *Clinical Cancer Research,* vol. 10, pp. 2659-2669, April 15, 2004 2004.

[40]   M. Hashimoto, *et al.*, "Polymerase chain reaction/ligase detection reaction/hybridization assays using flow-through microfluidic devices for the detection of low-abundant DNA point mutations," *Biosens. Bioelectron.,* vol. 21, pp. 1915-1923, 2006.

[41]   M. Khanna, *et al.*, "Ligase detection reaction for identification of low abundance mutations," *Clinical Biochemistry,* vol. 32, pp. 287-290, Jun 1999.

[42]   M. Hosokawa, *et al.*, "Size-Selective Microcavity Array for Rapid and Efficient Detection of Circulating Tumor Cells," *Anal. Chem. (Washington, DC, U. S.),* vol. 82, pp. 6629-6635, 2010.

[43]   P. Stephan Jean, *et al.*, "Development of a frozen cell array as a high-throughput approach for cell-based analysis," *Am J Pathol,* vol. 161, pp. 787-97, 2002.

[44]   K.-C. Chang and D. A. Hammer, "The forward rate of binding of surface-tethered reactants: effect of relative motion between two surfaces," *Biophys. J.,* vol. 76, pp. 1280-1292, 1999.

[45]   J. N. Mehrishi and J. Bauer, "Electrophoresis of cells and the biological relevance of surface charge," *Electrophoresis,* vol. 23, pp. 1984-1994, 2002.

[46]   K. M. Lipman, *et al.*, "The surface charge of isolated toad bladder epithelial cells. Mobility effect of pH and bivalent ions," *J. Gen. Physiol.,* vol. 49, pp. 501-16, 1966.

[47]   Y. Kang and D. Li, "Electrokinetic motion of particles and cells in microchannels," *Microfluid. Nanofluid.,* vol. 6, pp. 431-460, 2009.

[48]   P. G. Righetti, "Isoelectric focusing," *Electrokinet. Sep. Methods,* pp. 389-441, 1979.

[49]   P. C. H. Li and D. J. Harrison, "Transport, Manipulation, and Reaction of Biological Cells On-Chip Using Electrokinetic Effects," *Anal. Chem.,* vol. 69, pp. 1564-1568, 1997.

[50]   L. Korson, *et al.*, "Viscosity of water at various temperatures," *J. Phys. Chem.,* vol. 73, pp. 34-9, 1969.

[51]   M. A. Witek, *et al.*, "Cell transport via electromigration in polymer-based microfluidic devices," *Lab Chip,* vol. 4, pp. 464-472, 2004.

[52]   C. Novillo, *et al.*, "Isolation and characterization of two digestive trypsin-like proteinases from larvae of the stalk corn borer, Sesamia nonagrioides," *Insect Biochem. Mol. Biol.,* vol. 29, pp. 177-184, 1999.

LASER PROCESSED TANTALUM FOR IMPLANTS

Amit Bandyopadhyay, Solaiman Tarafder, Vamsi Krishna Balla and Susmita Bose

W. M. Keck Biomedical Materials Research Lab
School of Mechanical and Materials Engineering
Washington State University
Pullman, WA 99164-2920
E-mail: amitband@wsu.edu

ABSTRACT

Metallic biomaterials currently in use for load-bearing orthopedic applications are bioinert. Recent *in vitro* and *in vivo* studies demonstrated that Ta is a promising metal that is bioactive, but its applications have been limited by processing challenges. We report how to process Ta to create net shape porous and dense structures as well as coatings using Laser Engineered Net Shaping (LENS). Porous Ta samples with relative porosities between 60 to 90% have been successfully fabricated and characterized. *In vitro* cell materials interactions, using human osteoblast cell line showed six times better biocompatibity than Ti. *In vivo* results in a rat distal femur defect model shows that bone tissue integrated within 8 weeks of implantation and increasing porosity helped tissue integration. Our results demonstrate that Ta can be processed using LENS™ based agile manufacturing technology both in coating and porous metal form.

INTRODUCTION

Metallic biomaterials currently in use for load bearing orthopedic applications are bioinert and lack in sufficient osseointegration required for implant's longevity. One consideration to improve the healing process is focused on the use of bioactive calcium phosphate ceramics on metal implants. Plasma-sprayed HA coatings on metallic implants have been used in dentistry and orthopedics since the mid 1980s [1, 2] due to (i) rapid fixation and stronger bonding between the bone and the implant, and (ii) faster bone in-growth and/or on-growth at the bone-implant interface [3-5]. Clinical results also shown promising short and long-term continued fixation for up to 10 years [6-10]. However, the performance and long-term clinical use of HA coated implants has been doubted in some cases due to many concerns, especially with regard to long-term mechanical and biological stability. First important issue is the uncontrollable resorption and degradation of HA coatings *in vivo*, which could result in disintegration of the coating leading to the loss of both the coating–substrate bond strength and the implant fixation. Possible coating delamination and disintegration could form particulate debris [11, 12], which increases the wear of polyethylene liner or third-body wear leading to an increased incidence of osteolysis [13-15]. Although HA is a bioactive ceramics, several factors including purity (phase composition), crystallinity, Ca/P ratio, microstructure, porosity, surface roughness, thickness, processing route and implant type and surface texture, have been found to influence the HA coating quality on metallic implant surfaces [16-18].

For enhanced osseointegration and faster healing coupled with long-term stability of metallic implants, the coatings should be biologically active, of high purity, fully dense, mechanically strong and tough, and strongly bonded to the substrate. However, none of the bioactive ceramics and bioinert metals/alloy satisfies all of those requirements. Also, as discussed, the plasma sprayed HA and porous metal coated implants have shown some serious concerns in terms of long-term *in vivo* stability. Recently, Ta is gaining more attention as a new metallic biomaterial. Tantalum has been shown to be corrosion resistant [19] and bioactive *in vivo* [20]. In several *in vitro* and animal studies, the porous Ta metal has provided a scaffold for bone in-growth and mechanical attachment with bone [21-24]. These porous Ta components offer a low modulus of elasticity, high surface frictional characteristics, and excellent osseointegration properties (i.e., bioactivity, biocompatibility, and in-growth properties) [25, 26]. Although Trabecular Metal™ is commercially available, relatively high cost of manufacture and inability to produce a modular all tantalum implant has limited its widespread acceptance. Besides its high affinity towards oxygen, extremely high melting temperature of Ta (3017°C) makes it difficult to process Ta implant structures via conventional processing routes. Nevertheless, in our recent work, we have successfully demonstrated direct processing of Ta coatings and porous structures using high power laser in Laser Engineering Net Shaping (LENS™) [27-29]. *In vitro* biocompatibility study showed significantly better human fetal OB (hFOB) cell-materials interactions with six times higher living cell density on laser processed Ta coating surface than on Ti surface indicating its potential for enhanced/early biological fixation. The enhanced cell attachment and proliferation on Ta surface were found to be a direct consequence of its high wettability and surface energy [27]. Similarly, laser processed porous Ta structures have shown significantly better hFOB cell adhesion, growth, motality, and differentiation than porous Ti indicating its enhanced/early bioactivity [29]. In another study, interface microstructural features and *in vitro* cell-materials interactions of laser deposited Ta coatings and RF induction plasma sprayed HA coatings on Ti substrate have been compared [28]. The results showed equally excellent cellular adherence and growth on Ta and HA coatings. Ta coating surface showed ~ 22% higher surface energy of 55 mN m^{-1} than the surface energy of HA coating, which was 45 mN m^{-1}. However, the polar component of the surface energy for Ta coating (19.5 mN m^{-1}) was comparable to that of HA coating (15.5 mN m^{-1}). In the light of the above, the proposed Ta coatings on Ti6Al4V and Co-Cr-Mo alloys implants are of significant importance for enhanced osseointegration with uniform bone in-growth at the bone-implant interface. In the present work we have compared laser processed Ta towards its *in vivo* tissue integration abilities using a rat distal femur model.

MATERIALS AND METHODS

Ta metal powder (Grandview Materials Inc., Columbus, OH) with 99.5% purity and particles size between 45 and 75 μm was used. Porous Ta samples with 3 mm diameter and 6 mm height were deposited on a substrate of 3 mm thick rolled, commercially pure (CP) Ti plates using LENS™-750 (Optomec Inc. Albuquerque, NM) equipped with a 500W continuous wave Nd:YAG laser. LENS™ process uses a Nd:YAG laser, up to 2 kW power, focused onto a metal substrate to create a molten metal pool on the substrate. Metal powder is then injected into the metal pool which melts and solidifies. The substrate is then scanned relative to the deposition head to write metal line with a finite width and thickness. Rastering of the part back and forth to create a pattern and fill material in desired area allows a layer of material to be deposited. Finally, this procedure is repeated many times along the Z-direction, i.e., height, until the entire object represented in the three-dimensional CAD model is produced on the substrate, which is a solid or tailored porosity object. Porous Ta samples were fabricated in a glove box containing argon atmosphere with O_2 content less than 10 ppm to limit oxidation of Ta during processing. Laser power of 450W was chosen to partially melt metal powders during deposition process to create the porous structures. Scan speeds between 10 and 20 mm/s were used to fabricate structures with varying porosity. Similarly, powder feed rates of 126 and 141 g/min were used to vary the porosity in the samples. Also, the distance between the two successive metal roads or laser scans was varied between 0.381 and 1.27 mm to tailor pore size. Porous Ta samples with 60 and 70% densities have been produced using those process parameters. To compare performance, porous CP-Ti was also produced in LENS™ keeping the part density at 90%. The average hardness of laser deposited Ta coatings was 392 ± 37 Hv, which is higher than the average hardness of Ti substrate 189 ± 4 Hv.

RESULTS

In vitro Biocompatibility and Cell-materials Interactions

We have compared the *in vitro* biocompatibility and cell-materials interactions of Ta coatings with CP Ti [27] using established human osteoblast cell line hFOB 1.19 for culture durations between 3 and 14 days. The number of living cells determining the hFOB cell proliferation on laser processed Ta coating and Ti control surfaces was determined by MTT assay. Figure 1 shows the morphology of hFOB cells on Ti control and Ta after 3 and 14 days of culture. Ti surfaces showed less cell distribution even after 14 days of culture. Figure 1a and c shows that the cells on Ti surface exhibit few filopodia extensions indicating that they largely failed to attach and spread on the control surface. Cells on Ta coatings appeared cuboidal and three-dimensional morphology with more filopodia extensions, as shown in Figure 1b, suggesting significantly better cell attachment and spreading. After 14 days, cells on Ta surface became elongated, and formed a confluent cellular layer. The cells on Ta surface generated abundant amount of the extracellular matrix (ECM), a sign for early stage of osteoblast differentiation. However, Ti surface showed no confluent cellular layer even after 14 days.

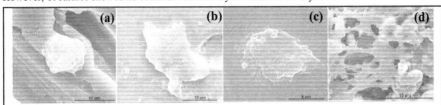

Figure 1. FE-SEM micrographs illustrating the hFOB cell morphologies, (a) Ti surface, after 3 days, (b) Ta surface, after 3 days, (c) Ti surface, after 14 days, and (d) Ta surface, after 14 days [27].

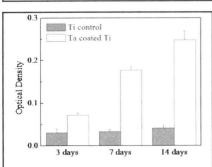

Figure 2. MTT assay of hFOB cells observed on Ta and Ti surfaces as a function of culture time. A higher optical density represents a higher concentration of living cells [27].

The amount of cells on the samples as determined by MTT assay is shown in Figure 2. High cell density was observed on Ta surface than on Ti control for all culture durations. With the increase in culture time, differences in cell density increased. Bout six times higher cell density was observed on Ta compared to Ti control surface after 14 days of culture. This is an important finding as none of the currently used metallic biomaterial showed such a high number of living cell density during *in vitro* biocompatibility evaluation. More filopodia extensions of cell morphology on Ta indicate its favorable surface characteristics for bone cell attachment in the early stages of cell culture. Better cell material interactions of Ta surface have also been confirmed by the large number of living cells observed during the MTT assay.

In vivo osteogenesis

For *in vivo* study, 60% and 70% dense tantalum implants of 3 mm diameter and 5 mm height were designed and fabricated by LENS[TM]. A 90% Dense Ti was used as control. All implants for *in vivo* study were sterilized by autoclaving at 121°C for 30 min prior to implantation.

Surgery and implantation procedure

Sprague-Dawley rats between 280 to 320 grams (Charles Rivers Laboratories International, Inc., Wilmington, MA, USA) were used. Surgeries were performed according to the protocol approved by the Institutional Animal Care and Use Committee (IACUC), Washington State University. Prior to surgery, rats were housed in individual cages with alternating 12 h cycles of light and dark in temperature and humidity controlled rooms. Following acclimatization, all animals underwent surgery to create femoral defect in the distal femur. Rats were anesthetized using IsoFlo® (isoflurane, USP, Abbott Laboratories, North Chicago, IL, USA) coupled with an oxygen (Oxygen USP, A-L Compressed Gases Inc., Spokane, WA, USA) regulator. The defect was created in the distal femur by means of a drill bit. The cavity was rinsed with physiological saline to wash away remaining bone fragments. Following implantation, undyed braided-coated polyglycolic acid synthetic absorbable surgical suture (Surgical Specialties Corporation, Reading, PA, USA) was used for stitching. Disinfectant was applied to the wound site to prevent infection. At 4 and 8 weeks post-surgery, rats were euthanized by overdosing with halothane in a bell jar, followed by administration of a lethal injection of potassium chloride (70%) into the heart.

Histomorphology

The bone-implant specimens were fixed in 10% buffered formalin solution, and dehydrated in graduated ethanol (70%, 95%, and 100%), ethanol-acetone (1:1), and 100% acetone series. After embedding samples in Spurr's resin, each undecalcified implant block was sectioned perpendicular to implant axis using a slow speed diamond wheel. After polishing, the sections were gold coated, and observed by back scattered electron microscope (BSEM) using field-emission scanning electron microscope (FESEM) (FEI Inc., Hillsboro, OR, USA) installed with a back scattered electron detector.

BSEM micrographs after 4 weeks implantation, as shown in Figure 3, show new bone formation around porous Ta implants with pores getting infiltrated with new bone, while Ti control implant shows the presence of a fibrous interzone (FIZ) not yet filled with new bone. The presence of this FIZ was still there even after 8 weeks of implantation. Both the porous Ta implants i.e., 60% and 70% dense showed excellent bone tissue ingrowth inside the pores after 8 weeks. However, 60% dense Ta implant showed better bone tissue ingrowth inside the pores than the 70% dense Ta implant for both 4 and 8 weeks samples. New bone formation was also observed into the central pores of the 60% dense Ta scaffolds.

CONCLUSIONS

In this work, we have processed porous Ta structures using laser engineered net shaping (LENS™). Porous structures were compared for their biocompatibility both *in vitro* and *in vivo*. Porous Ta samples with relative porosities between 60 to 90% have been successfully fabricated and characterized. *In vitro* cell materials interactions, using human osteoblast cell line showed six times better cell proliferation than laser processed Ti. *In vivo* results in a rat distal femur defect model shows the presence of a fibrous interzone (FIZ) even after 8 weeks for 90% dense Ti samples. However, both the porous Ta implants i.e., 60% and 70% dense showed excellent bone tissue ingrowth inside the pores after 8 weeks. Our results demonstrate that Ta can be processed using LENS™ based agile manufacturing technology maintaining excellent biocompatibility towards load bearing implant applications.

ACKNOWLEDGEMENTS

Authors would like to acknowledge the National Science Foundation (Grant No. CMMI 0728348) and the National Institutes of Health (Grant No. NIH-R01-EB-007351) for the financial support. Financial support from the W. M. Keck Foundation to establish a Biomedical Materials Research Lab at WSU is also acknowledged.

Figure 3. BSEM micrographs of Ti control, 60% dense Ta, and 70% dense Ta implants showing the development of new bone formation after 4 and 8 weeks implantation in rat distal femur. FIZ: fibrous interzone, NB: new bone.

REFERENCES
1. www.docstoc.com
2. R. J. Furlong, J. F. Osborn, Fixation of hip prostheses by hydroxyapatite ceramic coatings, J Bone Joint Surg 73B (1991) 741-745.
3. R. G. T. Geesink, Experimental and clinical experience with hydroxyapatite-coated hip implants, Orthopedics 12 (1989) 1239-1242.
4. R. G. T. Geesink, K. de Groot, C. P. Klein, Bonding of bone to apatite-coated implants, J Bone Joint Surg 70B (1988) 17-22.
5. J. Pavón, E. Jiménez-Piqué, M. Anglada, E. Saiz, A. P. Tomsia, Monotonic and cyclic Hertzian fracture of a glass coating on titanium-based implants, *Acta Materialia*, **54**, (2006), pp. 3593-3603.
6. P. K. Stephenson, M. A. Freeman, P. A. Revell, J. Germain, M. Tuke, C. J. Pirie, The effect of hydroxyapatite coating on ingrowth of bone into cavities in an implant, J Arthroplasty 6 (1991) 51-58.
7. A. K. Roynesdal, E. Ambjornsen, S. Stovne, H. R. Haanas, A comparative clinical study of three different endosseous implants in edentulous mandibles, Int J Oral Maxillofac Implants 13 (1998) 500-505.
8. W. J. Donnelly, A. Kobayashi, T. W. Chin, M. A. R. Freeman, H. Yeo, M. West, G. Scott, Radiological and survival comparison of four fixation of a proximal femoral stem, J Bone Joint Surg 79B (1997) 351-360.

9. W. D. Capello, J. A. D'Antonio, J. R. Feinberg, M. T. Manley, Hydroxyapatite-coated total hip femoral components in patients less than fifty years old. Clinical and radiographic results after five to eight years of follow-up, J Bone Joint Surg 79A (1997) 1023-1029.

10. R. G. H. H. Nelissen, E. R. Valstar, P. M. Rozing, The effect of hydroxyapatite on the micromotion of total knee prostheses. A prospective, randomized, double-blind study, J Bone Joint Surg 80A (1998) 1665-1672.

11. T. W. Bauer, R.C.T. Geesink, R. Zimmerman, J.T. McMahon, Hydroxyapatite-coated femoral stems. Histological analysis of components retrieved at autopsy, J Bone Joint Surg 73A (1991) 1439-1452.

12. J. P. Collier, V.A. Surprenant, M.B. Mayor, M. Wrona, R.E. Jensen, H.P. Surprenant, Loss of hydroxyapatite coating on retrieved, total hip components, J Arthroplasty 8 (1993) 389-393.

13. V. G. Varanasi, E. Saiz, P. M. Loomer, B. Ancheta, N. Uritani, S.P. Ho, A. P. Tomsia, S. J. Marshall, G. W. Marshall, Enhanced osteocalcin expression by osteoblast-like cells (MC3T3-E1) exposed to bioactive coating glass (SiO_2–CaO–P_2O_5–MgO–K_2O–Na_2O system) ions, *Acta Biomaterialia*, **5** (2009), pp. 3536-3547.

14. E. W. Morscher, A. Hefti, U. Aebi, Severe osteolysis after thirdbody wear due to hydroxyapatite particles from acetabular cup coating, J. Bone Joint Surg (Br) 80B (1998) 267-272.

15. R. D. Bloebaum, J. A. Dupont, Osteolysis from a press-fit hydroxyapatite-coated implant. A case study, J Arthroplasty 8 (1995) 195-202.

16. L. Sun, C.C. Berndt, K.A. Gross, A. Kucuk, Material Fundamentals and Clinical Performance of Plasma-Sprayed Hydroxyapatite Coatings: A Review, J Biomed Mater Res B: Appl Biomater 58 (2001) 570-592.

17. R. H. Rothman, W. J. Hozack, A. Ranawat, L. Moriarty, Hydroxyapatite-coated femoral stem. A matched-pair analysis of coated and uncoated implants, J Bone Joint Surg 78A (1996) 319-324.

18. J. E. Dalton, S. D. Cook, In vivo mechanical and histological characteristics of HA-coated implants vary with coating vendor, J Biomed Mater Res 29 (1995) 239-245.

19. I. H. Oh, N. Nomura, N. Masahashi, S. Hanada, Mechanical properties of porous titanium compacts prepared by powder sintering, Scripta Mater 49 (2003) 1197-1202

20. H. Kato, T. Nakamura, S. Nishiguchi, Y. Matsusue, M. Kobayashi, T. Miyazaki, H. Kim, T. Kokubo, Bonding of alkali- and heat-treated tantalum implants to bone, J Biomed Mater Res 53 (2000) 28-35.

21. J. D. Bobyn, G. J. Stackpool, S. A. Hacking, M. Tanzer, J. J. Krygier, Characteristics of bone ingrowth and interface mechanics of a new porous tantalum biomaterial, J Bone Joint Surg Br 81-B (1999) 907-914.

22. S. A. Hacking, J. D. Bobyn, K. Toh, M. Tanzer, J. J. Krygier, Fibrous tissue ingrowth and attachment to porous tantalum, J Biomed Mater Res 52 (2000) 631-638.

23. J. D. Bobyn, K. K. Toh, S. A. Hacking, M. Tanzer, J. J. Krygier, Tissue response to porous tantalum acetabular cups: a canine model, J Arthroplasty 14 (1999) 347-354.

24. H. Matsuno, A. Yokoyama, F. Watari, U. Motohiro, T. Kawasaki, Biocompatibility and osteogenesis of refractory metal implants, titanium, hafnium, niobium, tantalum and rhenium, Biomaterials 22 (2001) 1253-1262.

25. T. Miyazaki, H. M. Kim, T. Kokubo, C. Ohtsuki, H. Kato, T. Nakamura, Mechanism of bonelike apatite formation on bioactive tantalum metal in a simulated body fluid, Biomaterials 23 (2002) 827-832.

26. T. Kokubo, H. M. Kim, M. Kawashita, Novel bioactive materials with different mechanical properties, Biomaterials 24 (2003) 2161-2175.

27. Vamsi Krishna Balla, Shashwat Banerjee, Susmita Bose and Amit Bandyopadhyay "Direct Laser Processing of Tantalum Coating on Titanium for Bone Replacement Structures," *Acta Biomaterialia*, **6** [6], pp. 2329-2334 (2010).

28. Mangal Roy, Vamsi Krishna Balla, Susmita Bose and Amit Bandyopadhyay, "Comparison of Tantalum and Hydroxyapatite Coatings on Titanium for Applications in Load Bearing Implants" *Advanced Engineering Materials (Advanced Biomaterials)*, **12** (11), pp. B637-B641 (2010).

29. Vamsi Krishna Balla, Subhadip Bodhak, Susmita Bose and Amit Bandyopadhyay "Porous Tantalum Structures for Bone Implants: Fabrication, Mechanical and In vitro Biological Properties," *Acta Biomaterialia*, **6** [8], pp. 3349-3359 (2010).

THE ROLE OF BACTERIAL ATTACHMENT TO METAL SUBSTRATE AND ITS EFFECTS ON MICROBIOLOGICALLY INFLUENCED CORROSION (MIC) IN TRANSPORTING HYDROCARBON PIPELINES

Faisal M. AlAbbas[1], Anthony Kakpovbia[1], David L Olson[2], Brajendra Mishra[2] and John R. Spear[3]

[1]Inspection Department, Saudi Aramco, Dhahran, Saudi Arabia, 31311
[2]Department of Metallurgical and Materials Engineering
[3]Department of Civil and Environmental Engineering
Colorado School of Mines, Golden, Colorado, USA, 80401

ABSTRACT
    During oil and gas operations, pipeline networks are subjected to different corrosion deterioration mechanisms, one of which is microbiologically influenced corrosion (MIC) that results from accelerated deterioration caused by different microbial activities present in hydrocarbon systems. Bacterial adhesion is a detrimental step in the MIC process. The MIC process starts with the attachment of planktonic microorganisms to a metal surface that leads to the formation of a biofilm with subsequent metal deterioration. The tendency of a bacterium to adhere to the metal surface can be evaluated using thermodynamics approaches via interaction energies. In this paper, thermodynamic and surface energy approaches to model bacterial adhesion are reviewed; the factors that affect bacterial adhesion to a metal surface are presented; and the subsequent physical-chemical interaction between the biofilm and substratum and its implication for MIC in pipeline systems are discussed.

INTRODUCTION
    Microbiologically influenced corrosion (MIC) is of considerable concerns to the oil and gas industry. MIC has been reported in oil and gas treating facilities such as refineries and gas fractionating plants, pipeline systems and exporting terminals. MIC can be responsible for an increase in corrosion rate due the presence of microbial metabolic activities that accelerate the rate of anodic and/or cathodic reactions [1]. MIC does not produce a defined type of damage; however, it mostly results in a localized type of corrosion that manifests in pitting, crevice corrosion, under deposit corrosion, cracking, enhanced erosion corrosion and dealloying [2-4]. It is believed that MIC is one of the most damaging mechanisms to pipeline steel materials. Microbial activities are thought to be responsible for greater than 75% of the corrosion in productive oil wells and for greater than 50% of the failure of pipeline system [5,6]. MIC has been estimated to account for 20 to 30 percent of all internal pipeline corrosion costs. In 2006, MIC was suspected as one of the two major factors that shutdown the major Alaska Prudhoe Bay oil field pipeline. This leak caused turmoil in the global oil market [7].
    Different microorganisms thrive in oil and gas transporting systems. The reason is that all of the essential elements for life are present in these environments. Microbial life needs four basic things to thrive in an environment; a carbon source, water, an electron donor and an electron acceptor [8]. Hydrocarbon acts as an excellent food source for a wide variety of microorganisms. Water also exists in mixed solution with hydrocarbon. Other elements including sulfur, nitrogen, carbon and phosphorus that are needed to support microbial life are also present in the process feed. The main type of bacteria associated with metals in pipeline systems are sulfate reducing bacteria (SRB), iron and CO_2 reducing bacteria and iron and manganese oxidizing bacteria [9]. Among these, SRB has been recognized to be the major MIC causative agent in pipeline systems. Sulfate reductive activity, is thought to be responsible for more than 75% of the corrosion in productive oil wells and for more than 50% of the failures of buried pipelines and cables [10]. Practically, MIC is really the result of synergistic interactions of different microbes, consortia, that coexist in the environment and are able to affect the electrochemical processes through co-operative metabolisms [11].

131

The MIC process starts by the attachment of planktonic microorganisms to a metal surface that then leads to the formation of a complex biofilm. During the growth of the biofilm and through their metabolic activities, bacteria catalyze numerous invisible slow electrochemical reactions at the cell / metallic surface interface. There, metabolic reactions may be corrosive in nature or may dissolve a protective surface –oxide films, or both [12].

The literature concerning bacterial attachment and biofilm development is significant for MIC investigations. This paper will provide concise reviews that address the following:

1. Biofilm developmental stages.
2. Factors that affect bacterial adhesion to the metal surface.
3. Thermodynamic and surface energies model approaches of bacterial adhesion.
4. Subsequent physical-chemical interaction between the biofilm and substratum in pipeline systems with a focus on SRB.

BIOFILM DEVELOPMENTAL STAGES

The MIC process starts with a biofilm formation on a metal substrate. Immobile cells attach to the steel substrate, grow, reproduce and produce an extracellular polymeric substance (EPS) that results in a complex biofilm formation [2,3]. The biofilm formation encompasses three different stages. The first stage starts with the absorption of macromolecules, such as protein, lipids, polysaccharides and humic acids that work as a conditioner of the steel surface. These macromolecules change the physical chemsitry of the interface including the hydrophobicity and electrical charge. During this stage, microorganisms, surface and aqueous medium characteristics play a significant role in the extent of bacterial transfer rate, adhesion and resultant biofilm size. The microbial characteristics include surface charge, cell size and hydrophobicity. Surface properties include chemical compositions, roughness, inclusions, crevice, oxides or coating and zeta potential, whereas the aqueous medium properties include flow regime of the system and ionic strength [2].

The second stage involves the microorganisms movement from the bulk phase to the surface. The bacterial transportation process is affected by kinetic mechanisms. The initial bacteria attachment is formed through a reversible adsorption process, which is governed by electrostatic attraction, physical forces and hydrophobic interactions [13,14]. This initial attachment is a crucial step in the process of biofilm development. Whether the transporting cell will adhere or not to the surface depends on the surface properties, hydrodynamics and physiological state of the microbe. The adhesion force is affected by the physicochemical property of the substrate and the surface property of the microbial cell. The attached bacteria are called sessile bacteria and they are more important to the MIC process than the planktonic bacteria [13]. When sessile cells reside on a steel surface, their metabolic products introduce multiple cathodic reactions and thus promote corrosion.

The third stage includes extracellular polymeric substance (EPS) production. The adhered microorganisms produce a slime adhesive organic substance known as EPS. It has heterogeneous composition that includes exo-polysaccharides, nucleic acids, proteins, glycoproteins, and phospholipids [15-17]. It has been reported that exo-polysaccharides account for 40-95% of the macromolecules in microbial EPS [18]. EPS promotes the colonization process on the surface as it makes it possible for negatively charged bacteria such as SRB to attach to either negatively or positively charged surfaces. The further growth of the biofilm depends on the microorganism's colonization rate. The microbial transport to the interface is mediated by: (1) diffusion by Brownian motion (2) convection by system flow and (3) motile movement [2]. The biofilm development on the surface is an autocatalytic process whereby the initial microbial migration increases surface irregulaties and promotes the formation of dense biofilm. Figure 1 shows a developed dense biofilm by SRB, *Desulfovibrio africanus sp.*, on a surface of carbon linepipe steel [19].

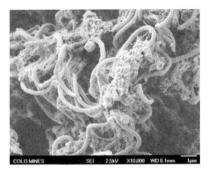

Figure 1. FESEM Image for a dense biofilm developed by SRB, *Desulfovibrio africanus sp.*, on a surface of carbon linepipe steel [19].

FACTORS AFFECTING BIOFILM DEVELOPMENT

Surface, bacteria and medium characteristics play significant role during biofilm development.

Surface Properties
The surface properties that have significant impact on bacterial attachment and biofilm development include surface roughness, polarizations, oxides coverage and chemical compositions. The initial roughness is known by the pattern or texture of surface irregulaties that are introduced by the manufacturing process. There is conflicting literature on the influence of the surface roughness on the bacterial attachment process. Some literature reports higher bacterial colonization and adhesion on high roughness surface while others found the opposite. Korber et. al. (1997) [20] postulated that the roughest surface increases surface area at the microorganism-materials interface that may then lead to more film attachment by providing more contact points. Sreekumari et al. (2001) [21] tested the bacterial attachment to 304L stainless steel welds and base metal. They reported more attachment to the weld metal than the base metal, which was correlated to the average grain size. A larger area of attachment was associated with smaller grain size as weld joints have smaller grains and grain boundaries. Little et al. (1988) [22] confirmed that porous welds provide more sites for bacterial colonization than base metal. Medilanski at al. (2002) [23] demonstrated that smoother and rougher surfaces enhance the bacterial attachment. They tested four different bacterial strains on the surface of SS 304 that had five different surface finishes with roughness values (Ra) that ranged from 0.03 to 0.89 μm . Minimal adhesion was observed at Ra= 0.16 μm while both smoother and rougher surfaces showed more adhesion.

Surface Polarization is another surface characteristic that affects microorganism adhesion to the surface. Armon et al. (2001) [24] investigated the polarization affects on the adhesion of *P. fluorescens* to stainless steel and carbon steel surfaces. Maximum absorption was reported in a potential range of -0.5 V to 0.5 V / SCE (standard calomel electrode). Deviation outside that range caused a gradual decrease in bacterial adsorption. de Romero et al. (2006) [25] evaluated the cathodic protection influence on the attachment of the SRB, *Desulfovibrio desulfuricans*, to a pure carbon steel surface. It was found that an applied cathodic polarization of -1000 mV / SCE was not sufficient to counteract the bacterial growth and attachment.

Surface coverage such as oxides and corrosion products has detrimental influence on the microorganism attachment. The effect of metal oxides on adhesion is one of the research interests for bacterial adhesion. Different oxides can be developed over a surface during the corrosion process. Examples include iron oxides (i.e. Fe_2O_3), chromium oxides (i.e. Cr_2O_3) and titanium oxides (i.e. TiO_2). Most of the research work has focused on iron oxides [26] that are known to increase bacterial adhesion. Iron hydroxides and other form of oxides on the metal surface provide firm attachment sites to bacteria.

The metal oxides provide a positively-charged surface that can significantly increase the bacterial deposition to the surface [2,22,26]. Baikun Li *et al.* (2004) [26] reported that metal oxides can increase the adhesion of negatively-charged bacteria to surfaces primarily due to their positive charge and hydrophobicity. They found significant increase in bacterial adhesion to glass surfaces covered with different metal-oxide coating compared to uncoated glass. The attachment increase was attributed to the increase in surface roughness, surface charge and surface hydrophobicity due to the metal oxides. It has also been shown that unstable or deteriorated corrosion products or oxides can detach biofilm associated with them [2].

The chemical properties of the surface have been known to directly influence the microorganism's adhesion and distribution in a biofilm [27]. Metals are the most common and economical material that have been used in oil and gas pipeline systems. Bacterial attachment and subsequent biofilm formation can occur on wide variety of metals including carbon steel, aluminium, stainless steel and copper alloys with different extent. Some metals such as aluminium or copper are considered toxic to bacteria [28]. On the other hand, copper has been reported to enhance the growth rate of some bacteria, whilst decrease the growth in other microbial populations [30,31]. Microbes have enormous physiological range of tolerance and use of metals, and this is an example of that. When compared to low alloy carbon steel and stainless steel surfaces, copper displays the most inhibitory effects on various microorganisms [32]. Gerchakov *et al.* (1977) [33] reported that a stainless steel has more initial bacterial attachment compared to 60/40 copper-zinc brass and copper-nickel surfaces. Stainless steel is generally known for its high corrosion resistance due to the formation of thin passive chromium –oxide film. However, it is vulnerable to bacterial attachment especially to the metal-depositing organisms (MOB) that has been known for MIC on stainless steel. Low alloy carbon steel is the most common steel used in pipeline systems and has been known for their high propensity to MIC. Addition of alloying elements such as silca and sulfur has been reported to increase the low alloy steel susceptibility to MIC. Reports show sulfide inclusion sites were the most favorable sites for bacterial colonization [2-4]. Figure 2 shows the biofilm formed on the surface of a low alloy carbon steel and stainless steel coupons respectively.

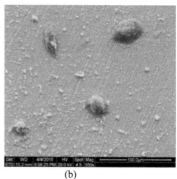

(a)   (b)

Figure 2. Biofilm formed by SRB, *Desulfovibrio capillatus,* on a surface of (a) low alloy carbon steel (API 5L X80); and (b) stainless steel (SS 316) coupons.

Medium characteristics

Medium concentration, pH, total organic and inorganic ionic strength can influence the microbial settlement potential [1-4]. The change in electrolyte pH influences the microbial cell surface charge. Commonly, at neutral pH bacteria are negatively charged, but a few strains have been reported that exhibit a net positive charge [34]. Increasing the cell negative charge will increase the repulsion against a negatively charged surface, subsequently decreasing the bacterial attachment. X.X. Sheng *et al.* (2008) [35]

examined the effect of solution pH on the attachment of three different bacteria, *Desulfovibrio desulfuricans, Desulfovibrio singaporenus* and a *Pseudomonas sp.* to a stainless steel surface. They found that for all bacterial strains tested, the adhesion force reached its highest value when the pH of the solution was near the isoelectric point of the bacteria at the zero point charge. The adhesion forces at pH 9 were higher than at pH 7 due to the increase in the attraction between iron ions (Fe^{2+}) and negative carboxylate groups ($COO-$). The carboxylate groups are highly ionized at pH 9. These negatively charged $COO-$ groups, in turn, bind with positive Fe^{2+} by electrostatic interactions on the stainless steel surface, and induce the large adhesion force in the solution with a high pH.

The effect of electrolyte ionic strength (I) has been investigated extensively inside and outside the laboratory. Some studies have shown an increase of bacterial adhesion in electrolyte concentrations that range from 0 to about 0.1–0.2 M, above this concentration, an increase in I either increased or decreased adhesion. Similarly, organic material adsorption, such as protein, showed an increase with increased I in an interval from 0 to 50 mM KCl. However there are several studies that concluded no correlation between bacterial species attachment to hydrophobic surfaces and changes in electrolyte concentrations [34]. Again, this could be due to the multitude of microbial metabolisms possible, and what happens to be at one place at one time, or what kind of microbe is used for the research study. Fletcher *et al.* (1988) [36] found that increasing the concentration of several cations in an electrolyte solution such as sodium, calcium and ferric ions affect the attachment of *P. fluorescens* to a glass surface by reducing the repulsion forces between the negatively charged bacterial cells and a glass surface.

In general, increasing the total organic carbon (TOC) will provide more nutrients to the bacteria and hence increase the bacterial colonization. Cowan *et al.* evaluated the effect of nutrients on bacterial colonization on a glass surface. They related the bacterial colonization of a surface to their ability to grow toward turbidity in the water column, and the deposition onto the surface increased with the density of suspended cells. Moreover, carbon limitations were shown to influence the adhesive strength of attached bacteria. Phosphorus and nitrogen are also important nutrients for microorganisms [8]. Limitation on these elements adversely impacts the growth of most microorganism. It has been reported that an electrolyte with a carbon –nitrogen ratio greater than 7:10 is considered nitrogen limited for microbial growth. The nitrogen depletion in the medium results in lower amounts of produced EPS and a thinner biofilm [2].

Microorganism Properties

Microbial cell characteristics have a significant role in the adhesion process. The cell surface protects the microbe and provides structural support. A microbial cell can be classified based on surface charge into two major groups; Gram negative and Gram positive microbes [8]. The difference between them is related to the cell wall configuration, and the great majority of microbial cells in the environment tend to be Gram negative. During the adhesion process, a Gram negative bacteria will be more attracted to a positively charged surface and vice versa. It has been shown that proteinaceous appendages including pili and flagella initiates the bacterial adhesion by establishing bridges between surface and cells [39].

The interaction between the microbial cells themselves plays an important role in biofilm formation. Research has shown that chemical signaling plays an important role in the formation of microbial biofilm. A class of diffusible molecules known as N-acylated homoserine lactones (AHLs) which are released by the bacteria into the local environment can interact with neighboring cells in a form of chemical signaling or communication [39]. Consequently, with this communication EPS is generally considered to be important in cementing bacterial cells together in the biofilm structure, making for a stronger, protected and communicative community. X. Sheng *et al.* (2008) [35] measured the cell–cell interaction forces of three different bacteria, *Desulfovibrio desulfuricans, Desulfovibrio singaporenus* and *Pseudomonas sp.* The reported force curves indicate the long-range of repulsive force for the cell–cell interactions. They reported that surface charges for both bacterial cell and substratum greatly influenced the adhesion force by controlling the electrostatic interactions. The electrostatic interaction resulted in stronger repulsive forces in the cell–cell interaction as compared to the cell–metal surface interaction. The surface energies, charges, interaction forces and other properties for bacterial cell, surface and environment should be considered to compute the free energy of the adhesion process.

THERMODYNAMIC AND SURFACE ENERGIES APPROACHES OF BACTERIAL ADHESION

Bacterial adhesion to a substrate is complex and involves several different factors [2]. Different approaches have been used to describe, understand and model the adhesion process. There are at least three approaches: Thermodynamics, DLVO [Derjaguin, Landau, Verwey, Overbeekand] and Extended DLVO that was introduced by Van Oss *et al.* These approaches are based on the fundamental interaction forces between a microbial cell and a surface, and in order to have an adequate description of this interaction, both long range and short range forces should be considered [40].

Thermodynamics approach

The thermodynamic approach assumes the system is in equilibrium and the bacterial attachment is a reversible process. The interfacial free energies between the interacting surfaces are compared and calculated as schematically illustrated in Figure 3. This comparison is expressed in the so-called free energy of adhesion. Based on that, the work of adhesion (W_{adh}) and free energy of adhesion(ΔG_{adh}) is obtained. The work of adhesion can be calculated by the Dupré Equation as follows:

$$W_{adh} = (-\gamma_{sm} + \gamma_{ml} + \gamma_{sl})dA \qquad (1)$$

The terms γ_{sm}, γ_{ml} and γ_{sl} are the solid-microorganism, solid-liquid, and microorganism-liquid interfacial free energies, respectively. The free energy of adhesion (ΔG_{adh}) is calculated by the following:

$$\Delta G_{adh} = -W_{adh} = (\gamma_{sm} - \gamma_{ml} - \gamma_{sl})dA \qquad (2)$$

The microbial adhesion will be favorable when the ΔG_{adh} is negative (< 0) and will not be energetically favorable if ΔG_{adh} is positive. Different theories are deployed to compute the interfacial energies and are based on the measurement of contact angles on solid surface and bacteria lawn. In those theories, the contact angle is related to the interfacial energy by Young's equation:

$$\gamma_{lv} \cos \theta = \gamma_{sv} - \gamma_{sl} \qquad (3)$$

The subscripts denote the respective surface free energy between the liquid (l), solid (s), or vapor (v). When contact angles on microbial lawns are measured, the subscript (s) should be replaced by (m) for microbial. Different approaches have been used to calculate the interfacial energies [40]:

1. Equation of state [41]: It requires one polar liquid (i.e water) for calculation and uses the following equation:

$$\gamma_{sl} = \frac{\left(\sqrt{\gamma_{sv}} - \sqrt{\gamma_{lv}}\right)^2}{1 - 0.015\sqrt{\gamma_{sv}\gamma_{lv}}} \qquad (4)$$

2. In the second approach, the surface free energies are separated in a polar or Lifshitz-van der Waals (γ^{LW}) and a polar or acid-base (γ^{AB}) component. So one polar (i.e water) and non-polar (i.e. Diiodomethane) liquids will be required for calculation as follow [42-44]:

$$\gamma sl = \left(\sqrt{\gamma_{sv}^{LW}} - \sqrt{\gamma_{lv}^{LW}}\right)^2 \left(\sqrt{\gamma_{sv}^{AB}} - \sqrt{\gamma_{lv}^{AB}}\right)^2 \qquad (5)$$

$$\Delta Gadh = \Delta G_{adh}^{LW} + \Delta G_{adh}^{AB} \qquad (6)$$

$$\Delta G_{adh}^{LW} = -2\left(\sqrt{\gamma_{mv}^{LW}} - \sqrt{\gamma_{lv}^{LW}}\right)\left(\sqrt{\gamma_{sv}^{LW}} - \sqrt{\gamma_{lv}^{LW}}\right)dA \tag{7}$$

$$\Delta G_{adh}^{AB} = -2\left(\sqrt{\gamma_{mv}^{AB}} - \sqrt{\gamma_{lv}^{AB}}\right)\left(\sqrt{\gamma_{sv}^{AB}} - \sqrt{\gamma_{lv}^{AB}}\right)dA \tag{8}$$

3. The third approach separates the acid –based component to an electron-donating γ^{-} and an electron – accepting γ^{+}. So two polar and one non-polar component will be required for the calculation as follow [42-44]:

$$\gamma sl = \left(\sqrt{\gamma_{sv}^{LW}} - \sqrt{\gamma_{lv}^{LW}}\right)^{2} + 2\left(\sqrt{\gamma_{sv}^{+}\gamma_{sv}^{-}} + \sqrt{\gamma_{lv}^{+}\gamma_{lv}^{-}} - \sqrt{\gamma_{sv}^{-}\gamma_{lv}^{+}} - \sqrt{\gamma_{sv}^{+}\gamma_{lv}^{-}}\right) \tag{9}$$

$$\Delta Gadh = \Delta G_{adh}^{LW} + \Delta G_{adh}^{AB} \tag{10}$$

Equation of state is considered relatively easy method to compute ΔG_{adh} (Eq. 4). It becomes more complicated when surface free energy components γ^{LW}, γ^{AB} and parameters γ^{-} and γ^{+} are included as shown in the second and third approaches [40-42].

Furthermore and based on the thermodynamic model, Power et al. [45] developed a novel model that calculates Gibbs free energy (ΔG_{adh}) of adhesion for the initial bacterial attachment process. The merit of this model is that it eliminates the need to calculate interfacial free energies and instead relies on measurable contact angles. In their work, they were able to calculate the ΔG_{adh} of adhesion for a *Pseudomonas putida* bacterium interacting with a mercaptoundecanol and dodecanethiol self-assembled monolayer. They developed the following Gibbs free energy:

$$\Delta G_{adh} = \frac{1}{2}\gamma_{l}(1 - \cos\theta_{bl})(1 - \cos\theta_{sl})dA \tag{11}$$

The term γ_{l} is free energy of liquid, θ_{bl} and θ_{sl} are contact angles measured from bacteria/air-liquid and substrate-air-liquid interfaces, respectively.

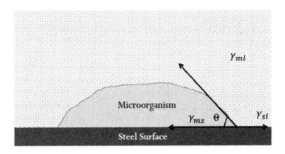

Figure 3. Illustration of the different interfacial energies involved during bacterial adhesion

DLVO Approach

The drawback of thermodynamics approach is that it ignores the electrical double layer interaction with the bacteria as illustrated by Figure 4. This assumption is invalid as the bacterial cells have surface negative or positive charge. In contrast, the DLVO approach displays a balance between attractive Lifshitz-van der Waals (ΔG^{LW}) and repulsive or attractive electrostatic forces (ΔG^{EL}) . These two forces are function of distance (d) between the bacteria and surface. In order to calculate the adhesion free energy (ΔG_{adh}), the electrostatic interactions between surfaces should be included. The inclusion of electrostatic interactions requires that the zeta potentials of the interacting surfaces be measured, in addition to measuring contact angles [41-45, 46]. So the total free energy expression is:

$$\Delta G_{adh} = \Delta G^{LW}(d) + \Delta G^{EL}(d) \tag{12}$$

The attractive Lifshitz- van der Waals ΔG^{LW} is calculated by:

$$\Delta G^{LW} = \frac{-AR}{6d} \quad \text{and} \quad A = 24\,\pi\, d_0^2\, \gamma_i^{LW} \tag{13}$$

The repulsive or attractive electrostatic forces ΔG^{EL}:

$$\Delta G^{EL}(d) = \pi \varepsilon_r \varepsilon_o \pi a \left\{ 2\Phi_b \Phi_s \ln\left[\frac{1+\exp(-kd)}{1-\exp(-kd)} \right] \left(\Phi_b^2 + \Phi_s^2 \right) \ln[\, 1 + \exp(-\kappa d)] \right\} \tag{14}$$

The term A is the Hamakar constant , $\Phi 1$ and $\Phi 2$ are the zeta potentials of the bacteria and the flat surface, R is the sphere radius assuming the bacteria is sphere shapes, ε_o and ε_r is the electrical permittivity of the vacuum and medium respectively, κ is Debye- Hu^ckel parameter and d is the distance in nm [41, 46].

It has been found that the medium ionic strength has no influence on the Lifshitz-van der Waals attraction, whereas both the range and the magnitude of the electrostatic interactions decrease with increasing ionic strength due to shielding of surface charges. In case of high ionic strengths, electrostatic interactions have lost their influence [41].

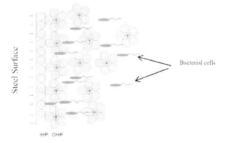

Figure 4. Illustrations of bacterial cell interactions with the electrical double layers: electrical double layer depicting the Inner Helmhotz Plane (IHP) formed by a layer of solvent and the Outer Helmhotz Plane (OHP) determined by the alignment of hydrated cations.

Extended DLVO Approach

The extended DLVO theory relates the origin of hydrophobic interactions in microbial adhesion and includes four fundamental interaction energies: Lifshitz-van der Waals, electrostatic, Lewis acid-base and Brownian motion forces:

$$\Delta G_{adh} = \Delta G^{LW}(d) + \Delta G^{EL}(d) + \Delta G^{AB}(d) + \Delta G^{BW} \tag{15}$$

The effect of acid-based interaction is higher than those for the electrical and the Lifshitz- van der Waals energies; however, it is short range and requires a close distance (< 5nm) between the bacteria and the surface. On the other hand, the Brownian motion comprise (1/2) kT per degree of freedom and the ΔG^{BW} of adhered bacteria to a surface equals 1kT =0.414 *10^{-20} J [41, 46].

SUBSEQUENT PHYSICAL-CHEMICAL INTERACTION BETWEEN THE BIOFILM IN PIPELINE

The subsequent influence of the biofilm on linepipe steel is the development of MIC or Biofouling. MIC is not new type of corrosion process, but it incorporates the role of bacteria and resulted biofilm in the corrosion processes. There are diverse types of corrosion resulting from MIC. Generally, MIC produces localized type of corrosion that exhibits pitting. Other types of corrosion include crevice corrosion, under deposit corrosion, cracking, enhanced erosion corrosion and dealloying [2-4].

Corrosion is classified as an interfacial process. The thermodynamics and kinetics of the corrosion process is strongly influenced by the physico-chemical environment at the interface including the pH, oxygen concentrations, salt, conductive, developed oxides and redox potentials. It is well established that the metabolic activities and the biofilm has the ability to alter these factors [2-4, 47]. The type and extent of damage depend on the bacterial type and associated environment. The main types of bacteria associated with metals in pipeline systems are sulfate reducing bacterial (SRB), iron reducing bacterial and iron and manganese oxidizing bacteria [5]. Among them, SRB has been recognized to be the major MIC causative microorganisms in pipeline systems. According to Iverson's estimation, 77% of the corrosion in the producing oil wells in the United Stated is introduced by SRB [47]. Therefore, the following discussion will be limited to the influence of the physical-chemical interactions between the SRB, biofilm and linepipe surface.

Sulfate reducing bacteria (SRB) is related to the Bacterial domain. SRB are anaerobic and do not need oxygen to survive; rather, they use sulfate ions as a terminal acceptor and produce hydrogen sulfide (H_2S). Furthermore, this type of bacteria has the ability to reduce nitrate and thiosulfate. SRB can manage to stay alive in an aerobic environment until the environment becomes suitably anaerobic for them to grow. In this case, the aerobic type (i.e IRB) of bacteria consumes the oxygen faster than the oxygen diffusion towards the biofilm, so the environment deeper in the biofilm will become anaerobic and, in turn, SRB will thrive. SRB obtains their energy from organic nutrients, such as lactate. They can grow in a pH range from 4 to 9.5 and tolerate pressure up to 500 atmospheres. Most SRB exist in temperature ranges of 25 – 60°C. SRB can be found everywhere in the oil and gas production facilities, deeper in the well, and extent to the treatment facilities. The environment inside the pipeline systems has anaerobic or low oxygen concentration, considering SRB as the main contributor to bio-corrosion [2-4, 8, 48, 49].

There are different ways that SRB and the resulting biofilm produce MIC damage in the pipeline. According to the classical theory, SRB consume the cathodic hydrogen by an enzyme known as hydrogenase to obtain the electron required for metabolic activities. Therefore, the removal of hydrogen from the metal surface will catalyze the revisable activation of hydrogen and in turn will force the iron to dissolve at the anode [2-4, 47]. In 1934, Khur and Vlugt [3] had proposed that the reactions that govern the classical theory as follow:

$$Fe \rightarrow 4Fe^{2+} + 8e \qquad \text{anodic Reaction} \tag{16}$$

$$8H^+ + 8e \rightarrow 8H \qquad \text{cathodic reaction} \tag{17}$$

$$SO_4^{2-} + 8H \xrightarrow{SRB} S^{2-} + 4 H_2O \qquad \text{SRB metabolism} \qquad (18)$$

$$\left.\begin{array}{l} Fe^{2+} + S^{2-} \to FeS \\ Fe^{2+} + \ 6OH^- \ \to 3Fe(OH)_2 \end{array}\right\} \qquad \text{corrosion products} \qquad (19)$$

Miller and King [3] related the corrosion effects of SRB to both the hydrogenase and iron/iron sulfide galvanic cell. As proposed, the iron sulfide will act as a cathode and absorb the molecular hydrogen, and the area beneath will be the anode sites. In anaerobic conditions, oxygen free environment that are a prerequisite for SRB growth, the concentration of hydrogen ions will be extremely low and will not be able to form a layer of atomic hydrogen in this type of environment. For this reason, additional cathodic reaction has been considered, such as H_2S reduction, as follows:

$$H_2S + e \to HS^- + \frac{1}{2} H_2 \qquad (20)$$

Furthermore, the biofilm forms on the metal surface is heterogeneous in nature and form community centers of bacteria. Those sites may be chosen based on chemical and metallurgical profiles, such as inclusions and roughness that induce attachment sites for the bacteria. These colonies produce EPS that attract more bacteria and organic materials to these sites. Subsequently, the conditions under these colonies such as oxygen level, ion concentrations and pH, will be different than those in the bulk stream and, in turn, leads to the formation of concentration cells, pitting and crevice corrosion. Other literature proposed that the area beneath the biofilm will act as anodic sites while the outside region will be cathodic. The fixing community center will form fixing anodic sites that are affected by the immobile bacteria growth, their activities and the biofilm developed under these colonies. This behavior will initiate pits under those colonies and will become fixed anodic sites under an immobile community; as a consequence those pits will grow with time [3, 47, 50].

Some strains of SRB, such as *Desulfovibrio,* use the organic carbon source in the nutrition system such as lactate to produce the hydrogen necessary as electron donor and yields pyruvate or acetate, which is excreted to the bulk as these bacteria are nonacetate oxidizers as follow:

$$4CH_3CHOHCOO^- + SO_4^{2-} \to 4CH_3COO^- + 4HCO_3^- + H_2S + HS^- + H^+ \qquad (21)$$

$$2CH_3CHOHCOO^- + SO_4^{2-} \to 2CH_3COO^- + 2HCO_3^- + H_2S + HS^- + CO_2 \qquad (22)$$

Therefore, the deposit of acetic acid as a result of the above reaction will form an aggressive environment to the linepipe steel when concentrated under colony or other corrosion product and leads to localized metal dissolution beneath it [3, 48]. Figure 5 shows extensive pitting resulting from MIC caused by SRB on low alloy carbon steel surfaces [19].

 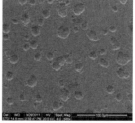

Figure 5. Extensive pitting induced by SRB, *Desulfovibrio africanus sp,* on API 5L X65 carbon linepipe steel coupons [19].

CONCLUSIONS

The MIC process starts with the attachment of planktonic bacteria to linepipe surfaces, which leads to the formation of the biofilm and subsequently results in metal deterioration. Bacterial, surface and medium characteristics play significant roles in the extent of bacterial transfer rate, adhesion and resulting biofilm size. The bacteria characteristics include surface charge, cell size and hydrophobicity. Surface properties include chemical compositions, roughness, inclusions, crevices, oxides and zeta potential, whereas the aqueous medium properties include pH, ionic strength and flow regime of the system. The tendency of a bacterium to adhere to the surface can be evaluated using different approaches via interaction energies. These approaches include: Thermodynamics, DLVO and Extended DLVO. These approaches are based on the fundamentals of interaction forces between the bacteria and the surface, and in order to have an adequate description of this interaction, both long range and short range forces should be considered.

The main types of bacteria associated with MIC in pipeline systems are sulfate reducing bacteria (SRB), iron reducing bacteria and iron and manganese oxidizing bacteria. Among them, SRB has been recognized to be the major MIC causative bacteria in oil and gas operations. The biofilm and the active metabolisms of SRB alter the electrochemical process and subsequently change the pH level, produce more H_2S and introduce multiple cathodic side reactions, which enhance the reduction quality of the system and accelerate the anodic dissolution. Moreover, the accumulation of iron sulfide on the steel surfaces forms a galvanic cell with iron, resulting in localized galvanic attack of the iron surface adjacent to deposits of iron sulfide. Mostly, the nature of SRB damage is localized extensive pitting attacks.

REFERENCES

[1]Florian Mansfeld, The interaction of bacteria and metal surfaces,*Electrochimica Acta*, 52: 7670-7680, (2007)

[2]B. Little and J. Lee, Microbiologically Influenced Corrosion, *Wiley- Interscience* (2007).

[3]R. Javaherdashti, Microbiologically Influenced Corrosion- An Engineering Insight, *Springer-Verlag London* (2008).

[4]R. Javaherdashti, MIC and Cracking of Mild and Stainless Steel, *VDM Verlage, Germany* (2010).

[5]E Miranda, M Bethencourt, F J Botana and M J Cano, Biocorrosion of carbon steel alloys by anhydrogenotrophic sulfate-reducing bacterium *Desulfovibrio capillatus* isolated from a Mexican oil field separator, *Corrosion Science*, 48:2417-2431, (2006).

[6]Flemming, H.C., Economical and technical overview. In: Heitz, E.,Flemming, H.C., Sand, W. (Eds.), Microbially Influenced Corrosion of Materials, *Springer Verlag, Berlin, New York*, pp. 5–141 (1996).

[7]Jacobson, G.A., Corrosion at Prudhoe Bay – a lesson on the line.*Materials Performance*, 46:27–34 (2007).

[8]M. Madigan, Brock Biology of Microorganisms, *12th Edition, Pearson Benjamin Cummings, San Francisco, USA* (2009).

[9]I.B. Beech and J. Sunner, Biocorrosion towards understanding interactions between biofilms and metals, *Curr. Opin.Biotechnol.* 15:181–186(2004).

[10]M. Walch, Corrosion, microbial, in: J. Lederberg (Ed.), Encyclopedia of Microbiology,*Academic Press, New York*, 1:585–59 (1992).

[11]I.B. Beech and C.C. Gaylarde, Recent advances in the study of Biocorrosion-An overview, *Rev. Microbiol.* 30:177–190(1999).

[12]D.A. Jones and P.S. Amy, A Thermodynamic Interpretation of Microbiologically Influenced Corrosion, *Corrosion* 58:938-945(2002).

[13]Tingyue. Gu, K. Zhao and S. Nesic, A New Mechanistic Model For MIC Based On A Biocatalytic Cathodic Sulfate Reduction Theory, Paper No. 09390, *NACE Conference*, , Houston, TX (2009).

[14]Xiaoxia Sheng a, Yen Peng Ting a and SimoOlaviPehkonen, Force measurements of bacterial adhesion on metals using a cell probe atomic force microscope, *Journal of Colloid and Interface Science* 310:661–669 (2007).

[15]Y. Ong, A. Razatos, G. Georgious and M.M. Sharma, Adhesion forces between E. coli bacteria and biomaterial surfaces, *Langmuir* 15:2719–2725(1999).

[16]C.J. van Oss, R.J. Good and M.K. Chaudhury, The role of van der Waals forces and hydrogen bonds in "hydrophobic interactions between biopolymers and low energy surfaces, *J. Colloid Interface Sci.* 111: 378–390(1986).

[17]Morgan, J. W., Evison and L. M. and Forster, C. F., Examination into the composition of extracellular polymers extracted from anaerobic sludges polymers seem to be responsible for microbial attachment either to a support medium or within the sludge granule, *Process Safety and Environmental Protection.* 69:231-236 (1991).

[18]Arundhati Pal and A. K. Paul, Microbial extracellular polymeric substances: central elements in heavy metal bioremediation" *Indian J. of Microbiology*, 48:49-64 (2008).

[19]F. Al-Abbas, A. Kakpovbia, B. Mishra , D. Olson and J. Spear, Utilization of Nondestructive Electrochemical Techniques in Characterizing Microbiologically Influenced Corrosion (MIC) of API-5l X65 Carbon Linepipe Steel: Laboratory Study, *Proceeding QNDE Conference*, , Burlington, VT (2011).

[20]DR Korber, A Choi, GM Wolfaardt, SC Ingham and DE Caldwell, Substratum topography influences susceptibility of Salmonella enteritidis biofilms to trisodium phosphate, *Appl. Environ. Microbiol.*63:3352-3358(1997).

[21]Kurissery R Sreekumaria, KanavillilN and akumaraand Yasushi, Bacterial attachment to stainless steel welds: Significance of substratum microstructure, *Biofouling*, 17:303-316 (2001).

[22]Little, Brenda ,Wagner Pat and Jacobus, John, The Impact of Sulfate-Reducing Bacteria on Welded Copper-Nickel Seawater Piping Systems, *Material Performance*, 27:57-61 (1988).

[23]Edi Medilanskia, Karin Kaufmanna, Lukas Y Wicka and Oskar Wanner, Influence of the Surface Topography of Stainless Steel on Bacterial Adhesion, *Biofouling*, 18:193-203(2002).

[24]R Armona, J Starosvetskyb, M Dancygierband D Starosvetsky, Adsorption of flavobacteriumbreve and pseudomonas fluorescens on different metals: Electrochemical polarization effect, *Biofouling*, 17:289-301 (2001).

[25]Matilde F. de Romero, Junire Parra and Rosemary Ruiz, Cathodic polarization effect on Sessile SRB growth and iron protection, Paper # 06526, *NACE Conference*, Huston , TX (2006).

[26]Baikun Li a and Bruce E. Loganb, Bacterial adhesion to glass and metal-oxide surfaces, *Colloids and Surfaces B: Biointerfaces*, 36:81–90(2004).

[27]W.P. Johnson, M.J. Martin, M.J. Gross and B.E. Logan, Facilitation of bacterial transport through porous media by changes in solution and surface properties, *Colloid Surface*, 17:263(1996).

[28]Whitehead KA and Verran J., The effect of substratum properties on the survival of attached microorganisms on inert surfaces, *Marine and Industrial Biofouling*, Springer-Verlag Berlin, pp13–34 (2008).

[29]Avery SV, Howlett NG and Radice S, Copper toxicity towards Saccharomyces cerevisiae: dependence on plasma membrane fatty acid composition, *Appl. Environ Microbiol*, 62: 3960–3966(1996).

[30]Starr TJ and, Jones ME, The effect of copper on the growth of bacteria isolated from marine environments, *Limnol Oceanog*, 2:33–36(1957).

[31]Jonas RB, Acute copper and cupric ion toxicity in an estuarine microbial community, *Appl. Environ Microbiol*, 55: 43–49(1989).

[32]Keevil CW, Antibacterial properties of copper and brass demonstrate potential to combat toxic E. coli outbreaks in the food processing industry, *Paper presented at the symposium on copper and health, CEPAL*, Santiago, Chile(2001).

[33]Gerchakov,Sol M.,Marszalek, Donald S., Roth,Frank J. ,Sallman, Bennett and Udey Lanny R ,Observations on Microfouling Applicable to OTEC Systems, *Proceeding OTEC biofouling and Corrosion Symposium*, WA, pp 63-75 (1977)

[34]Malte Hermansson, The DLVO theory in microbial adhesion, *Colloids and Surfaces B: Biointerfaces*, 14: 105-119 (1999).

[35]Xiaoxia Sheng a, Yen Peng Ting and Simo OlaviPehkonen, The influence of ionic strength, nutrients and pH on bacterial adhesion to metals,*Journal of Colloid and Interface Science*, 2:256–264(2008).

[36]M Fletcher, Attachment of *Pseudomonas fluorescens* to glass and influence of electrolytes on bacterium-substratum separation distance, *J. Bacteriol.*, 170: 2027-2030 (1988).

[37]Marjorie M Cowanab1, Tessie M Warrena and Madilyn Fletcher, Mixed-species colonization of solid surfaces in laboratory biofilms, *Biofouling*, 3:23-34 (1991)

[38]Shunichi Ishii, Jun Koki, Hajime Unno and andKatsutoshi Hori, Two Morphological Types of Cell Appendages on a Strongly Adhesive Bacterium, *Acinetobacter sp.* Strain Tol 5, *Applied and Environmental Microbiology*, 8: 5026-5029 (2004)

[39]D. Davies, Small-scale systems for in vivo drug delivery, *Nat. Rev. Drug Discov.* 2: 114–122(2003).

[40]Rolf Bos , Henny C. van der Mei and Henk J. Busscher, Physico-chemistry of initial microbial adhesive interactions its mechanisms and methods for study,'' *FEMS Microbiology Reviews*, 23:179-230(1999).

[41]Neumann, A.W., Good, R.J., Hope and C.J. and Sejpal, M., An equation-of-state approach to determine surface tensions of low-energy solids from contact angles, *J. Colloid Interface Sci.* 49: 291-304 (1974).

[42]Dann, J.R.,Forces involved in the adhesive process, I: critical surface tensions of polymeric solids as determined with polar liquids, *J. Colloid Interface Sci.*, 32: 302-320(1970).

[43]Dann, J.R., Forces involved in the adhesive process. II: Non-dispersion forces at solid-liquid interfaces, *J. Colloid Interface Sci.*,32: 321-331(1970).

[44]Owens, D.K. and Wendt, R.C., Estimation of the surface free energy of polymers, *J. Appl. Polymer Sci.*, 13: 1741-1747(1969).

[45]Laura Power, Sophie Itier, Margaret Hawton, and Heidi Schraft, Time Lapse Confocal Microscopy Studies of Bacterial Adhesion to Self-Assembled Monolayers and Confirmation of a Novel Approach to the Thermodynamic Model, *Langmuir*, 23:5622-5629 (2007).

[46][17] Sonia Bayoudh, Ali Othmane, Laurence Mora and Hafedh Ben Ouada, Assessing bacterial adhesion using DLVO and XDLVO theories and the jet impingement technique, *Colloids and Surfaces Biointerfaces*, 73 1–9(2009).

[47]Fei Kuang , Jia Wang , Li Yan and Dun Zhang, Effects of sulfate-reducing bacteria on the corrosion behavior of carbon steel, *Electrochimica Acta*, 52:6084–6088(2007).

[48]H. Videla, C. Swords, and R. Edyvean, Corrosion Products and Biofilm Interactions in the SRB Influenced Corrosion of Steel, Paper No 2557, *NACE conference*, Houston, TX (2002).

[49]H.A. Videla, Manual of Biocorrosion, CRC Lewis Publishers, Boca Raton, Florida, 1996

[50]H. Castaneda and X. Benetton, SRB-biofilm influence in active corrosion sites formed at the steel-electrolyte interface when exposed to artificial seawater conditions, *Corrosion Science*, 50:1169–1183(2008).

ELECTROPHORETIC DEPOSITION OF SOFT COATINGS FOR ORTHOPAEDIC APPLICATIONS

Sigrid Seuss, Alejandra Chavez
Institute of Biomaterials, University of Erlangen-Nuremberg
91058 Erlangen, Germany

Tomohiko Yoshioka
Department of Metallurgy & Ceramic Science, Tokyo Institute of Technology
Tokyo, Japan

Jannik Stein, Aldo R. Boccaccini[*]
Institute of Biomaterials, University of Erlangen-Nuremberg
91058 Erlangen, Germany

ABSTRACT
    Electrophoretic deposition is gaining increasing interest in the field of coatings for biomedical implants. Especially soft coatings, which functionalize the surface of metallic substrates, are in the focus of current research activities. A previous comprehensive review has shown the large variety of biopolymers and bioceramics that can be conveniently deposited and codeposited by electrophoretic deposition. This short review summarizes the latest research activities in this field and discusses, as an example, the application of EPD to obtain composite coatings in the system TiO_2 nanoparticles/ polylactic acid. This system combines the good biocompatibility of TiO_2 with the film forming properties of the biodegradable polymer. It was shown that the material combination enhances the flexibility and the adhesion of the coatings to the substrate, leading to soft coatings that should enable better connection between the rigid implant and the surrounding bone tissue.

INTRODUCTION
    Electrophoretic deposition (EPD) is an attractive processing method for the production of bioactive ceramic and ceramic/polymer coatings for a range of biomedical applications[1]. The electrophoretic coating process is divided into two steps. During electrophoresis, which is the first step, charged particles or polymer macromolecules in suspension move under the influence of an electric field. In the second step, particles or macromolecules are deposited on the oppositely charged electrode to form a dense coating[2]. Figure 1 shows a schematic diagram of a typical EPD cell with vertical planar electrodes. To obtain homogeneous and highly adhesive coatings, well dispersed suspensions are necessary. In comparison to other coating processes, EPD offers the possibility to coat complex three dimensional or porous structures and to produce free standing objects[1, 2]. Another advantage of this process is the possibility to deposit a wide range of materials and to co-deposit various material combinations to obtain composite coatings, multilayer structures or functionally graded materials. EPD is a cost-effective process since it requires only simple equipment consisting basically of a two-electrode cell and a power supply (Figure 1). In a recent review paper[1] the high versatility of EPD for the fabrication of coatings for applications in the biomaterial sector was discussed. In this context one of the most interesting fields is the combination of organic with inorganic materials to exploit the favorable properties of both materials in composite coatings. As EPD is a colloidal process, coatings made with ceramic materials usually require a further heat-treatment (sintering) step to densify the

---

[*] Corresponding author. Email: aldo.boccaccini@ww.uni-erlangen.de

coatings. These sintering processes are carried out at high temperatures, which can lead to degradation of the metallic substrate. It has been suggested that combination of ceramic materials with polymers, which act as binder or "glue" for the ceramic particles and for improving the adhesion to the substrate material, leads to robust coatings without the need for post-EPD sintering[3]. The combination of polymers with ceramics in the field of orthopedic coatings also offers the possibility to produce (nano-) structured coatings to mimic for example the structure of bone[4]. Two kinds of polymers for biomedical coatings are being investigated: synthetic polymers and natural polymers. Stable synthetic polymer coatings include materials like PEEK, which has been used for composite coatings in combination with bioactive glasses[5] or alumina[6]. On the other hand, it is also possible to use degradable synthetic polymers, like polycaprolactone (PCL), to fabricate soft coatings, however research on EPD of these polymers is limited. In one of the few studies available, Xiao et al. reported on the codeposition of hydroxyapatite and PCL[4]. Much more research has been carried out on the EPD of natural polysaccharides like chitosan, alginate or hyaluronic acid. For example, Zhitomirsky et al. successfully deposited the stable polysaccharide hyaluronic acid, which was also combined with hydroxyapatite to develop composite films[7]. EPD of chitosan and chitosan composites has been studied extensively[8-14]. Chitosan is a biocompatible polymer, a linear cationic polysaccharide, which exhibits antimicrobial activity, accelerated angiogenesis, little fibrous encapsulation and improved cellular attachment. Chitosan has been successfully deposited from acidic aqueous suspensions[9]. Moreover, Zhitomirsky et al.[10] have carried out significant research on the production of composite coatings based on chitosan, e.g. chitosan with HA, and they also used chitosan-ceramic composites to develop drug delivery devices[11]. Another degradable polysaccharide is alginate, which can be also deposited successfully by EPD from aqueous suspensions[15].

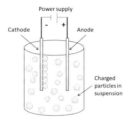

Figure 1. Schematic representation of the EPD setup showing positively charged particles depositing on the negative electrode.

In this paper, the latest published research in the field of biomedical soft coatings by EPD is briefly discussed and, additionally, results on the electrophoretic codeposition of biopolymer-TiO$_2$ nanocomposite coatings are presented as an example to demonstrate the effectiveness of EPD to develop this class of biopolymer-bioceramic composite coatings.

ELECTROPHORETIC POLYMER/BIOCERAMIC COATINGS: LATEST DEVELOPMENTS
It is well known that bioceramic-based hard coatings based on bioactive glasses, hydroxyapatite, and titanium oxide exhibit bioactivity, which is the property required to bind to bone directly. However, this kind of coatings often needs high temperatures to sinter and to increase adhesion to the substrates. In addition, rigid coatings may not be the optimal approach to enable effective bonding of the implant to host tissue[14]. Therefore the co-deposition of polymers with bioceramics is an effective way to develop soft coatings which can provide a better connection between the implant and bone tissue, serving as a soft interface between vascularised bone and the hard, inert metallic implant.

Table I. Summary of recent research developments about EPD of biopolymer-ceramic composite coatings

Coating composition	Deposition conditions				Ref.
	Bath composition	Solvent	Electric field	Substrate	
HA-Bioglass-hyaluronic acid	0.6 g/L HA (150 nm) 0.15 g/L Bioglass (5 μm) 0.5 g/L hyaluronic acid	70% ethanol	15-25 V/cm		16
HA-PLL	0.5-2 g/L HA (150 nm) 1-2 g/L PLL	70-80% ethanol	3-20 V/1.5 cm	platinized Si wafer	17
HA-PLO	0.5-2 g/L HA (150 nm) 1-2 g/L PLO	70-80% ethanol	3-20 V/1.5 cm	platinized Si wafer	17
HA-PPY-heparin	0-0.2 g/L HA 27 g/L PPY 1 g/L heparin	water	1 mA/cm^2 0-12 min	graphite, Pt, 304 stainless-steel	18
Bioglass-chitosan	0.4-2 g/L Bioglass (6 μm) 0.5 g/L chitosan	1vol% acetic acid	10-30 V/1.5 cm 200-600 s	AISI 316L stainless- steel	13
TiO$_2$-hyaluronic acid	0.1-2 g/L TiO$_2$ (0.5 μm) 0.5-1 g/L hyaluronic acid	70% ethanol	15-25 V/cm		16
TiO$_2$-hyaluronic acid	0-1.2 g/L TiO$_2$ (25 nm) 0-1 g/L hyaluronic acid	70% ethanol	4-20 V/1.5 cm 1-10 min	Pt, graphite, 304 stainless-steel	19
TiO$_2$-PAA	0-4 g/L TiO$_2$ 0.5-2 g/L PAA	water (pH 8)	3-5 V/1.5 cm 1-10 min	Pt, 304 stainless-steel	20

*HA; hydroxyapatite, PLL; poly-L-lysine, PLO; poly-L-ornithine, PPY; polypyrrole, PPA; polyacrylic acid

In the last 2 years, several electrophoretic composite coatings of interest for biomedical applications have been reported[13, 14, 16-20], which are summarized in Table I. Hydroxyapatite was co-deposited with hyaluronic acid[16], poly-L-lysine (PLL)[17], poly-L-ornithine (PLO)[17], or polypyrrole[18]. Hyaluronic acid is a naturally occurring anionic polysaccharide that has been associated with several cellular processes including angiogenesis and the regulation of inflammation. Ma et al.[16] reported the EPD of hyaluronic acid and bioceramic nano- or micro-particles. The adsorption of hyaluronic acid on the surface of the bioceramics resulted in the generation of negative charges for anodic EPD. The coatings were crack-free. PLL and PLO are chiral polymers commonly used as coating agents to promote cell adhesion. Wang et al.[17] deposited hydroxyapatite/PLL and hydroxyapatite/PLO composite coatings on platinized silicon wafers by EPD. PLL and PLO provided both stabilization and charging of hydroxyapatite nanoparticles in suspensions and cathodic deposition occurred at constant voltages in the range 3-20 V in an ethanol-water mixed solvent. Ma et al.[18] developed hydroxyapatite/polypyrrole composite coatings on a 304 stainless-steel substrate by EPD. The deposition was performed galvanostatically in an aqueous solution. However, because the composite deposits showed low adhesion and tendency to peel off from the substrate, the addition of salicylate was necessary as an anionic dopant.

Bioglass® has been also electrophoretically deposited in combination with hyaluronic acid[16], or chitosan[13]. Pishbin et al.[13] reported EPD of chitosan/45S5 Bioglass® composite coatings on AISI 316L stainless-steel substrates. The microstructure of the coatings showed a uniform distribution of the Bioglass® microparticles in the chitosan matrix. Titanium oxide that induces apatite formation in SBF has been co-deposited with hyaluronic acid[16, 19], or polyacrylic acid (PAA)[20]. Ma and Zhitomirsky[19] prepared electrochemically deposited titanium oxide/hyaluronic acid nanocomposites on 304 stainless-steel substrates. The thickness of the coatings could be varied in the range 0-10 μm. PAA is a well-known biocompatible polymer[21]. Wang et al.[20] showed the EPD of titanium oxide/PAA composites. The deposition occurred on the anode at pH 8. The PAA molecules were negatively charged at pH 8, and the suspensions of PAA and titanium oxide nanoparticles were well-dispersed and stable. It was suggested that PAA macromolecules adsorbed on the surface of the titanium oxide particles and could thus provide efficient stabilization of the suspension. The microstructure observation of the coating indicated that PAA can act as a binder in these composite coatings, preventing composite cracking and improving the adhesion to the substrate. These polymer-bioceramic composite coatings by EPD are suitable for applications in orthopedics.

A CASE IN POINT: BIODEGRADABLE POLYMER - TiO₂ NANOCOMPOSITE COATINGS

Materials and Methods

In this work, TiO_2-polylactic acid (PLLA) composite coatings were developed by a combination of electrophoretic deposition (EPD) and dip coating process. The aim was to infiltrate electrophoretically deposited TiO_2 porous coatings with PLLA. Titania is mainly used as coating for orthopedic implants due to its biocompatibility and antibacterial properties[22, 23] and synthetic polymers, in particular polyesters, are interesting for biomedical coatings because of their excellent processing characteristics as well as biocompatibility and biodegradability[24]. The addition of polymers to porous inorganic coatings allows low temperature processing of the materials while keeping the mechanical stability (rigidity) provided by the inorganic phase, which can be kept to a relatively high volume fraction. Similar composites, involving PDLLA as matrix, have been reported in the literature and successfully tested for biocompatibility. For example, it has been reported that the addition of TiO_2 nanoparticles to PDLLA can enhance cell attachment and can provide interfacial bonding to the surrounding tissue by formation of a hydroxyapatite layer[25,26]. PDLLA films with different concentrations of TiO_2 nanoparticles were prepared by solvent casting technique and tested with MG-63 osteoblasts-like cells[25]. The results showed no cytotoxicity and good cell adhesion and cell proliferation. Similarly, Gerhardt et al,[26] demonstrated the bioactivity of TiO_2-PDLLA films (30wt% TiO_2) in simulated body fluid (SBF) showing the formation of non-stoichiometric HA nanocrystals with an average diameter of 40 nm after two weeks of immersion in SBF. Also, the release of TiO_2 nanoparticles from the polymer matrix was studied assessing the response of nanoparticles to MG-63 osteoblast-like cells. No significant effect on MG-63 cell mitochondrial respiratory rates was found when using low concentrations of nanoparticles, confirming that TiO_2 nanoparticle concentrations up to a certain value do not affect cell osteoblast viability.

For the present EPD experiments, suspensions of TiO_2 nanoparticles were prepared based on our previous work[27]. Commercially available TiO_2 nanoparticles (P25, Evonik Industries, particle size of 21 nm) in concentration of 1 wt% were suspended in acetylacetone with addition of iodine. The suspensions were magnetically stirred for 5 minutes followed by 15 minutes ultrasonification and 5 minutes of stirring. The substrates used were planar 316 L stainless steel foils. For EPD tests, voltages of 35 V and deposition time of 2 minutes were used. The deposits were dried in air at room temperature for a day. After EPD, the coatings were dipped in PLLA dissolved in dichloro-methane (DCM). A concentration of 5 wt% PLLA in DCM was used and the dip coating process lasted for 3 minutes. After extraction of the coatings from the suspensions they were dried in open air at room temperature. The surface and cross sections of the TiO_2 coatings before and after infiltration with PLLA were analyzed by SEM. Also, the wetting behavior of these coatings was measured by contact angle using a Surftens-Universal device. For the measurements a drop of deionized water was placed with a syringe on the coating and the left and right contact angles were measured optically.

Results and Discussion

Figure 2 (a and b) shows the surface and cross section morphology of TiO_2 coatings before and after infiltration with PLLA. Fig. 2(a) is a low magnification SEM image of the surface of the TiO_2 coating obtained at 35 V and 2 minutes deposition time while Fig. 2(a') shows the cross section morphology of the same coating. The deposited layer is homogeneous, but there is evidence of microcracking on the surface due to the fast evaporation of the solvent used. It was observed that the thickness of the layer is approximately 25 μm and it is well adhered to the substrate (which was manually assessed by bending the substrate). Fig. 2(b and b') shows the surface and cross section of the TiO_2 coating after infiltration with PLLA (5 wt%). The microstructure is very similar to the original coating and the main differences are the presence of bubbles due to the dipping in the PLLA-DCM suspension and the presence of PLLA filling the cracks, as expected. The cross section fracture surface presents a well-detached layer from the substrate which could be related to the complete infiltration of PLLA into the porous TiO_2 coating. The presence of bubbles originates during PLLA dissolution in DCM and then after dipping, these bubbles remain on the coating surface. This simple experiment has shown that combination of electrophoretic deposition and dip coating is favorable to obtain composite coatings, which could be applied also to other inorganic nanoparticles and biopolymers. The addition of PLLA to TiO_2 coatings, which are otherwise brittle and prone for microcracking leads to robust soft coatings exhibiting enhanced flexibility, which can be proposed to act as the "soft component and hard device interconnect" as discussed in the literature[14, 28, 29].

Figure 3 (a and b) shows the images of contact angle measurements on the surface of TiO_2 coating and TiO_2-PLLA nanocomposite, respectively. Fig. 3(a) indicates that the TiO_2 surface is highly hydrophilic with an average contact angle of 2.15°. On the other hand, Fig. 3(b) shows a contact angle of 100° for the TiO_2-PLLA nanocomposite. The behavior of the nanocomposite is hydrophobic due to the presence of PLLA indicating that an effective surface modification of the TiO_2 nanoparticles is possible by the addition of PLLA. The main advantage of the TiO_2-PLLA composites is that no sintering is required and this type of nanocomposites can be of interest due to better mechanical performance, room temperature processing, biodegradability, bactericidal properties or combination of these properties. For example, it should be possible to add biomolecules or therapeutic drugs (antibiotics) to the starting polymer suspension to incorporate them in the TiO_2 coating, which could thus exhibit controlled drug delivery ability.

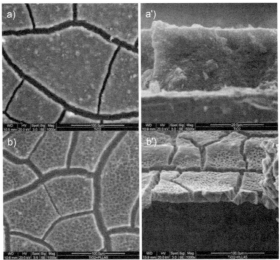

Figure 2. Surface and cross-section SEM images of TiO_2 coatings obtained by EPD (a, a') and TiO_2-PLLA nanocomposite coatings (b, b').

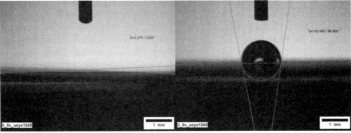

Figure 3. Images illustrating the wetting behavior of (a) TiO_2 coatings and (b) TiO_2-PLLA nanocomposite.

CONCLUSIONS

Electrophoretic deposition offers the possibility of combining soft coating materials, e.g. biodegradable polymers such as PCL or chitosan with ceramic materials like bioactive glasses or TiO_2, leading to interesting material combinations for biomedical applications. The favorable properties of both materials are combined to form a functional composite material in which the hard ceramic particles are embedded in soft polymer matrices, which act as a glue, so there is no need for high sintering temperatures which could damage the coatings and substrates. The present review of the recent literature indicates that in the past years considerable effort has been done to optimize the coating process for various polymer/ceramic combinations for various applications. Results about the EPD of TiO_2/ PLLA composite coatings show the suitability of combining EPD with dip coating to obtain composite coatings with high ceramic concentration. The flexibility of the TiO_2 coatings obtained by EPD was increased by infiltrating the coatings with PLLA. Further investigations are planned to incorporate drugs or bioactive biomolecules in the composite coatings at the EPD stage to increase the coating functionality.

ACKNOWLEDGEMENTS

We thank Ms Eva Springer, Institute of Glass and Ceramics, University of Erlangen-Nuremberg for SEM pictures. A. Chavez gratefully acknowledges Conacyt, Mexico, for financial support. T. Yoshioka thanks the "Institutional Program for Young Researcher Overseas Visits" of the Japan Society for the Promotion of Science (JSPS) for financial support to visit the Institute of Biomaterials at the University of Erlangen-Nuremberg.

REFERENCES

[1] A.R. Boccaccini, S. Keim, R. Ma, Y. Li, and I. Zhitomirsky, Electrophoretic Deposition of biomaterials, *Journal of the Royal Society*, 7, 581-613 (2010).

[2] O.O. Van der Biest, and L.J. Vandeperre, Electrophoretic deposition of materials, *Annual Review of Materials Science*, **29**, 327-52 (1999).

[3] K. Grandfield, F. Sun, M. FitzPatrick, M. Cheong, and I. Zhitomirsky, Electrophoretic deposition of polymer-carbon nanotube-hydroxyapatite composites, *Surface & Coatings Technology*, **203(10-11)**, 1481-7 (2009).

[4] X.F. Xiao, R.F. Liu, and X.L. Tang, Electrophoretic deposition of silicon-substituted hydroxyapatite/poly(epsilon-caprolactone) composite coatings, *Journal of Materials Science-Materials in Medicine*, **20(3)**,691-7 (2009).

[5] A.R. Boccaccini, C. Peters, J.A. Roether, D. Eifler, S.K. Misra, and E.J. Minay, Electrophoretic deposition of polyetheretherketone (PEEK) and PEEK/Bioglass (R) coatings on NiTi shape memory alloy wires, *Journal of Materials Science*, **41(24)**, 8152-9 (2006).

[6] I. Corni, M. Cannio, M. Romagnoli, and A.R. Boccaccini, Application of a neural network approach to the electrophoretic deposition of PEEK-alumina composite coatings, *Materials Research Bulletin*, **44(7)**, 1494-501 (2009).

[7] F. Sun, and I. Zhitomirsky, Electrodeposition of hyaluronic acid and composite films, *Surface Engineering*, **25(8)**, 621-7 (2009).

[8] F. Gebhardt, S. Seuss, M.C. Turhan, H. Hornberger, S. Virtanen, and A.R. Boccaccini, Characterization of electrophoretic chitosan coatings on stainless steel, *Materials Letters*, **66**, 302–4 (2012).

[9] L.-Q. Wu, A.P. Gadre, H. Yi, M.J. Kastantin, G.W. Rubloff, W.E. Bentley, G.F. Payne, and R. Ghodssi, Voltage-dependent assembly of the polysaccharide chitosan onto an electrode surface, *Langmuir*, **18(22)**, 8620-5 (2002).

[10] X. Pang, and I. Zhitomirsky, Electrodeposition of composite hydroxyapatite-chitosan films, *Materials Chemistry and Physics*, **94(2-3)**, 245-51 (2005).

[11] F. Sun, X. Pang, and I. Zhitomirsky, Electrophoretic deposition of composite hydroxyapatite-chitosan-heparin coatings, *Journal of Materials Processing Technology*, **209(3)**, 1597-606 (2009).

[12] A. Simchi, F. Pishbin, and A.R. Boccaccini, Electrophoretic deposition of chitosan, *Materials Letters*, **63**, 2253-6 (2009).

[13] F. Pishbin, A. Simchi, M.P. Ryan, and A.R. Boccaccini, Electrophoretic deposition of chitosan/45S5 Bioglass® composite coatings for orthopaedic applications, *Surf. Coat. Technol.*, **205**, 5260-8 (2011).

[14] Z. Zhang, T. Jiang, K. Ma, X. Cai, Y. Zhou, and Y. Wang, Low temperature electrophoretic deposition of porous chitosan/silk fibroin composite coating for titanium biofunctionalization, *Journal of Materials Chemistry*, **21(21)**, 7705-13 (2011).

[15] M. Cheong, and I. Zhitomirsky, Electrodeposition of alginic acid and composite films, *Colloids and Surfaces a-Physicochemical and Engineering Aspects*, **328(1-3)**, 73-8 (2008).

[16] R. Ma, Y. Li, and I. Zhitomirsky, Electrophoretic deposition of hyaluronic acid and composite films for biomedical applications, *JOM*, **62**, 72-5 (2010).

[17] Y. Wang, X. Pang, and I. Zhitomirsky, Electrophoretic deposition of chiral polymers and composites, *Colloids Surf. B*, **87**, 505-9 (2011).

[18] R. Ma, K. N. Sask, C. Shi, J. L. Brash, and I. Zhitomirsky, Electrodeposition of polypyrrole-heparin and polypyrrole-hydroxyapatite films, *Mater. Lett.*, **65**, 681-4 (2011).

[19] R. Ma, and I. Zhitomirsky, Electrophoretic deposition of silica-hyaluronic acid and titania-hyaluronic acid nanocomposites, *J Alloys Compounds*, **509**, 510-3 (2010).

[20] Y. Wang, I. Deen, and I. Zhitomirsky, Electrophoretic deposition of polyacrylic acid and composite films containing nanotubes and oxide particles, *J. Colloid Interface Sci.*, **362**, 367-74 (2011).

[21] G. S. El-Bahy, E. M. Abdelrazek, M. A. Allam, A. M. Hezma, Characterization of In Situ Prepared Nano-Hydroxyapatite/Polyacrylic Acid (HAp/PAAc) Biocomposites, *J. Appl. Polym. Sci.* **122**: 3270-3276 (2011).

[22] T. Kokubo, H.M. Kim, and M. Kawashita, Novel bioactive materials with different mechanical properties, *Biomaterials*, **24(13)**, 2161-75 (2003).

[23] C.X. Cui, H. Liu, Y. Li, J. Sun, R. Wang, S. Liu, and A.L. Greer, Fabrication and biocompatibility of nano-TiO2/titanium alloys biomaterials, *Materials Letters*, **59(24-25)**, 3144-8 (2005).

[24] J.C. Middleton, and A.J. Tipton, Synthetic biodegradable polymers as orthopedic devices, *Biomaterials*, **21(23)**, 2335-46 (2000).

[25] J. Wei, Q.Z. Chen, M.M. Stevens, J.A. Roether, A.R. Boccaccini. Biocompatibility and bioactivity of PDLLA/TiO2 and PDLLA/TiO2/Bioglass® nanocomposites. *Materials Science and Engineering C* **28**, 1–10 (2008).

[26] L.-C. Gerhardt, G. M. R. Jell, A. R. Boccaccini. Titanium dioxide (TiO2) nanoparticles filled poly(D,L lactid acid) (PDLLA) matrix composites for bone tissue engineering. *J Mater Sci: Mater Med* **18**,1287–1298 (2007).

[27] A.R. Boccaccini, P. Karapappas, J.M. Marijuan, and C. Kaya, TiO2 coatings on silicon carbide and carbon fibre substrates by electrophoretic deposition, *Journal of Materials Science*, **39(3)**, 851-9 (2004).

[28] G.F. Payne, and S.R. Raghavan, Chitosan: A soft interconnect for hierarchical assembly of nano-scale components, *Soft Matter*, **3**, 521-7 (2007).

[29] L.T. de Longe, S.C.G. Leeuwenburgh, J.G.C. Wolke, and J.A. Jansen, Organic-inorganic surface modifications for titanium implant surfaces, *Pharm Res.*, **25(10)**, 2357-69 (2008).

GLUTAMIC ACID-BIPHASIC CALCIUM PHOSPHATES: IN VITRO BONE CELL-MATERIAL INTERACTIONS

Solaiman Tarafder[a], Ian McLean[b], and Susmita Bose[a*]

[a]W. M. Keck Biomedical Materials Research Lab, School of Mechanical and Materials Engineering
[b]The Gene and Linda Voiland School of Chemical Engineering and Bioengineering
Washington State University
Pullman, WA 99164-2920
E-mail: sbose@wsu.edu

ABSTRACT
Glutamic acid (Glu) can act in bone cell regulation through glutamate signaling. We have investigated the adsorption and release behavior of Glu on biphasic calcium phosphates (BCPs) consisted of 1:1 ratio hydroxyapatite (HA) and β-tricalcium phosphate (β-TCP). These BCPs showed high affinity for Glu adsorption. Very low Glu release from these BCPs indicated strong interactions between Glu and Ca^{2+} probably through complex formation. Increased osteoblasts proliferation by MTT assay was observed with increased Glu on these BCPs.

INTRODUCTION
Glutamic acid (Glu), an amino acid, in addition to being a neuro-transmitter can also regulate bone cell activities by glutamate signaling [1]. Recent studies show that the amino acid, L-glutamic acid (L-Glu), which is mostly recognized for its principal role as an excitatory neurotransmitter in the vertebrate central nervous system [2], also plays a significant function in bone cells signaling through glutamate signaling pathway [3] [4]. Both osteoblasts (bone forming cells) and osteoclasts (bone resorbing cells) express glutamate receptors (GluRs). GluRs take part in the regulation of the bone formation and resorption by osteoblasts and osteoclasts, respectively. Glutamate transporters (GluTs), presented in the cell membrane, are also expressed by bone cells. Mechanical loading has been attributed as an important factor in maintaining adequate bone mass density of the skeleton system [5]. It has been reported that osteocytes show a mechanoresponsive expression of GluTs [4].

Both hydroxyapatite (HA) and β-tricalcium phosphate (β-TCP) are very widely used for hard tissue repair and augmentation in orthopedics and dentistry because of their osteoconductivity, excellent bioactivity, and compositional similarities to bone mineral [6] [7] [8]. HA has lower solubility than β-TCP in the physiological environment. High solubility of β-TCP makes it to be known as bioresorbable ceramic [6]. Thus, the development of a biphasic calcium phosphate (BCP) based biomaterial consisting of HAp and β-TCP has some benefits over only HA or β-TCP in terms of degradation behavior.

In this study, glutamic acid (GA) was adsorbed on BCPs. Purpose of this study was to understand the chemistry of adsorption and release kinetics of GA in HA and β-TCP based BCP ceramics to investigate the potential use of GA as an osteogenic agent. The influence GA on bone cell-material interactions were studied using human osteoblast cells. Effect of GA on osteoblast proliferation was also investigated by MTT assay.

MATERIALS AND METHODS

Sample preparation
Biphasic calcium phosphate samples containing HA and β-TCP, 1:1 ratio, were prepared. Hydroxyapatite and β-tricalcium phosphate with an average particle size of 550 nm were obtained from Berkeley Advanced Biomaterials Inc., Berkeley, CA. The samples were prepared by mixing 50 g

powder with appropriate amounts of HA and β-TCP in 250 ml polypropylene Nalgene bottles containing 75 ml of anhydrous ethanol and 350 g of 5 mm diameter zirconia milling media. The mixtures were then milled for 24 h at 70 rpm to minimize the formation of agglomerates, and increase the homogeneity of the powders. After milling, powders were dried in an oven at 60° C for 72 h. The resulting dried powders were uniaxially pressed to obtain disk samples [12 mm (φ) X 1 mm (h)] and then sintered at 1150 °C for 2 h.

GLUTAMIC ACID ADSORPTION AND RELEASE STUDY
To investigate the adsorption behavior of glutamic acid (Glu) (Sigma-Aldrich, St. Louis, MO) on BCPs, samples were kept separately in a 10 ml GA solution containing 5 mg/mL, 7.5 mg/mL, and 10 mg/mL GA aqueous solution. Adsorption was carried out at room temperature for two different time intervals, 6 h and 12 h. BCP samples obtained after 6 h GA adsorption were used to study GA release behavior. Release study was carried out at pH 7.4 phosphate buffer. Samples from each concentration were placed in 10 mL of pH 7.4 phosphate buffer in individual vials. These vials were then kept at 37 °C under 150 rpm constant shaking. Buffer solutions were changed at 2, 4, 8 and 12 hours. At each time point, 2mL of buffer solution was replaced with a fresh 2 mL buffer solution. GA measurement was carried out using a previously described method [9]. Briefly, 1.0 mL of 0.2 M sodium bicarbonate (NaHCO$_3$) solution and 1.0 mL of 1% dinitrofluorobenzene (DNFB) (prepared by mixing 1.0 DNFB in 100 mL 1,4-dioxane) were added with 2 mL GA released solution. The resulting mixture was heated in dark at 60 °C for 40 minutes in sealed condition. After that, reactions were stopped by adding 0.5 mL of 1 mol/L HCl. The resulting dinitrophenyl (DNP) derivatives were extracted with 5 mL ethyl acetate, kept at room temperature for 10 minutes, and then the absorbance was measured at 420 nm against ethyl acetate as blank.

IN VITRO BONE CELL INTERACTIONS
In vitro bone cell interactions on the surfaces of the three BCP samples were investigated by culturing human fetal osteoblast cell (hFOB) for 1, 3, and 5 days of incubation period. The cells used in this study were immortalized osteoblastic cell line, which were derived from human bone tissue. All samples were sterilized by autoclaving at 121^{0}C for 30 min prior to cell culture. Aliquot of 150 μL cell suspension containing 2.4×10^5 cells were seeded directly on each samples placed in the wells of 24-well plates. Samples were then incubated at 37 °C under 5 % CO$_2$ / 95 % humidified air atmosphere for 1 h. After 1 h of cell seeding, a 1 mL aliquot of DMEM media enriched with 10% fetal bovine serum was added to the surrounding of each sample in the well. Cultures were maintained at 37 °C under 5 % CO$_2$ / 95 % humidified air atmosphere. The culture media was changed every alternate day for the duration of the experiment.

CELL MORPHOLOGY
Cell morphology was assessed by SEM observation. All samples for SEM observation were fixed with 2% paraformaldehyde/2% glutaraldehyde in 0.1 M cacodylate buffer overnight at 4^{0}C. Post fixation for each sample was performed with 2% osmium tetroxide (OsO$_4$) for 2 h at room temperature. Fixed samples were then dehydrated in an ethanol series (three times for each of 30%, 50%, 70%, 95%, and 100%), followed by a hexamethyldisilane (HMDS) drying procedure. Dried samples were then gold coated (Technics Hummer V, San Jose, CA, USA), and observed under an FESEM (FEI Inc., Hillsboro, OR, USA).

CELL PROLIFERATION USING MTT ASSAY
The MTT assay (Sigma, ST. Louis, MO) was performed for 1, 3, and 5 days of incubation to assess hFOB cell proliferation on the bare BCP and Glu adsorbed BCP samples. The MTT solution of

5 mg mL^{-1} was prepared by dissolving MTT in PBS, and filter sterilized. The MTT was diluted (50 µL into 450 µL) in serum free, phenol-red free Dulbeco's Minimum Essential (DME) medium. 100 µL diluted MTT solution was then added to each sample in 24-well plates. After 2 h incubation samples were then taken into new well plate and 500 µL solubilizer (made up of 10% Triton X-100, 0.1 N HCl, and isopropanol) was added to each well to dissolve the formazen crystal. 100 µL of the resulting supernatant was transferred into a 96-well plate, and read by a plate reader at 570 nm. Triplicate samples per group were evaluated and three data points were measured from each sample.

STATISTICAL ANALYSIS

Statistical analysis was performed on MTT assay results using student's t-test, and P value < 0.05 was considered significant.

RESULTS AND DISCUSSION

Glu adsorption and release

Figure 1 shows the Glu adsorption after 6 h and 12 h. There was no significant difference observed in Glu adsorption between these two time points. A maximum of 15 mg/cm2 Glu is adsorbed on BCP samples after 6 h from 7.5 and 10 mg/mL Glu solution. High Glu adsorption

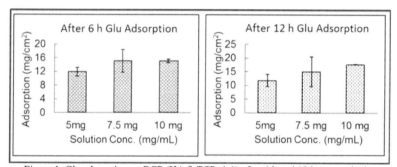

Figure 1. Glu adsorption on BCP (HA:β-TCP: 1:1) after 6 h and 12 h, respectively.

indicates BCPs has very high affinity for Glu. Figure 2 shows the FTIR spectra after Glu adsorption on a BCP sample. The presence of carbonyl (C=O) peak at 1631 cm^{-1} on the Glu adsorbed BCP sample clearly indicates the Glu adsorption. Figure 3 shows the release behavior

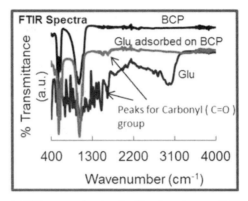

Figure 2. FTIR spectra showing the Glu adsorption on a BCP sample.

of Glu from these BCP samples up to 12 hours. There was no significant detectable Glu release observed from these BCP samples. In all cases, a decreasing trend in Glu release was observed. This was probably due to re-adsorption of the released Glu. Glu is a metal chelating

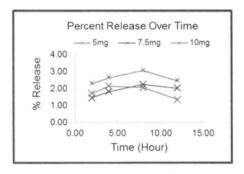

Figure 3. Release profile of Glu from BCPs.

agent, which makes it capable for strong chemical interactions with metallic ions. A very low percent release of GA from BCPs might be due to strong chemical interaction between Glu and Ca^{2+} cation of these BCPs [10].

EFFECT OF Glu ADSORPTION ON OSTEOBLASTS CELL PROLIFERATION
hFOB cells proliferation on the bare BCP and Glu adsorbed BCPs were studied using MTT assay. Figure 4 shows a comparison of cell densities on all samples after 1, 3, and 5 days of cell culture. With the increasing culture time, cell density was increased on all surfaces. There were not

very large differences in cell density between control and Glu adsorbed BCP samples; although a significant difference (*p < 0.05) in cell density was always observed between control and 100 μg Glu adsorbed BCP samples. A significant differences was always observed between 100 μg and 500 μg Glu adsorbed BCP samples with higher cell density on BCP samples with 500 μg Glu adsorbed samples.

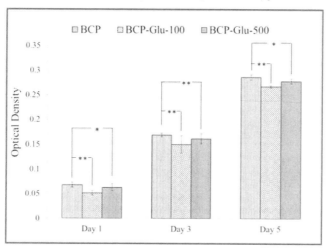

Figure 4. MTT assay of hFOB on bare BCP sample (control), 100 μg Glu adsorbed BCP samples (BCP-Glu-100), and 500 μg Glu adsorbed BCP samples (BCP-Glu-500) after 3, 5, and 7 days. Statistical analysis shows the significant (**$P < 0.05$, n = 3) and nonsignificant (*$P > 0.05$, n = 3) differences.

SEM morphology of hFOB cells on the Glu adsorbed BCP samples, as shown in Figure 5, did not show any deleterious effect of Glu on osteoblast cells. Though in this study, we always observed higher cell density in bare BCP samples compared to Glu adsorbed samples; it cannot be concluded that there are some adverse effects of Glu on osteoblast cell proliferation. It is not determined at this point that what Glu concentration is optimum for osteoblast cell proliferation or differentiation. If complex formation between Glu and Ca^{2+} ion is the probable reason for very low release of Glu from these BCP samples, then, it is also unknown whether Glu-Ca^{2+} complex can influence on osteoblast proliferation and differentiation. Glu release from these BCP or any calcium phosphate (CaP) surfaces might be the determinant factor if Glu can play its role in cell signaling only in its free form. Higher osteoblast cell density in 500 μg Glu adsorbed BCP samples compared to 100 μg Glu adsorbed BCP sample could probably due to relatively higher Glu release from 500 μg Glu adsorbed BCP sample.

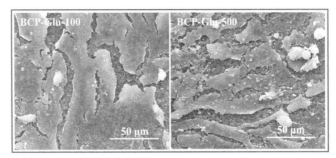

Figure 5. SEM micrographs showing the morphology of hFOB cells on 100 μg (BCP-Glu-100) and 500 μg (BCP-Glu-500) Glu adsorbed BCP sample surfaces after 5 days of culture.

Osteogenic drug or growth factors delivery from calcium phosphates (CaP) are widely used in bone tissue engineering [11]. One of the major challenges is timely and effective delivery of the drug at the site of interest. A potential complex formation or strong interaction between the drug molecules and CaPs is a hindrance for effective sustained delivery. Delivery of the drug from the polymeric matrix coated on CaPs could be one alternative to avoid the complex formation [11-13]. Delivery from the surface of the polymeric coating can also be applied instead. In that case, functionalization of the polymeric surface needs to be done for inducing interactions between polymeric surface and drug molecules [13]. A future study focusing to controlled and sustained Glu delivery from CaP based scaffolds would guide us design biomaterials of specific needs for orthopedic and dental applications.

CONCLUSIONS

Bihasic calcium phosphates (BCPs) consisted of hydroxyapatite (HA) and β-tricalcium phosphate showed very high affinity for glutamic acid (Glu). A very low release of Glu could probably due to strong interactions through complex formation between Glu and Ca^{2+}. Control BCPs always showed higher osteoblast cell density than the Glu adsorbed BCPs. Nevertheless, a higher osteoblast cell density was observed with increased Glu on BCPs. In summary, a further study is needed, to understand the effect of Glu on cell signaling pathway, if there is any in presence of TCP, for BCP-GLU based materials to be used in bone tissue engineering.

ACKNOWLEDGMENTS

The authors like to acknowledge the financial support from the National Institute of Health, NIBIB for financial support (Grant # NIH-R01-EB-007351).

REFERENCES
[1] E. P. Seidlitz, M. K. Sharma, and G. Singh, "Extracellular glutamate alters mature osteoclast and osteoblast functions," *Canadian Journal of Physiology and Pharmacology*, vol. 88, no. 9, pp. 929-936, 2010.
[2] M. BS, "Glutamate as a neurotransmitter in the brain: review of physiology and pathology.," *The Journal of nutrition*, vol. 130, no. 4, p. 1007S.
[3] E. Hinoi, T. Takarada, and Y. Yoneda, "Glutamate signaling system in bone," *Journal of Pharmacological Sciences*, vol. 94, no. 3, pp. 215-220, 2004.
[4] D. J. Mason, "The role of glutamate transporters in bone cell signalling," *Journal of Musculoskeletal & Neuronal Interactions*, vol. 4, no. 2, pp. 128-131, Jun. 2004.

[5] D. J. Mason, "Glutamate signalling and its potential application to tissue engineering of bone," *European Cells Materials*, vol. 7, pp. 12–25, 2004.

[6] A. Bandyopadhyay, S. Bernard, W. Xue, and S. Bose, "Calcium Phosphate-Based Resorbable Ceramics: Influence of MgO, ZnO, and SiO2 Dopants," *Journal of the American Ceramic Society*, vol. 89, no. 9, pp. 2675-2688, Aug. 2006.

[7] S. Tarafder, S. Banerjee, A. Bandyopadhyay, and S. Bose, "Electrically Polarized Biphasic Calcium Phosphates: Adsorption and Release of Bovine Serum Albumin," *Langmuir*, vol. 26, no. 22, pp. 16625-16629, Nov. 2010.

[8] S. Tarafder, S. Bodhak, A. Bandyopadhyay, and S. Bose, "Effect of electrical polarization and composition of biphasic calcium phosphates on early stage osteoblast interactions," *Journal of Biomedical Materials Research Part B: Applied Biomaterials*, vol. 97, no. 2, pp. 306-314, May 2011.

[9] L. Chen, Q. Chen, Z. Zhang, and X. Wan, "A novel colorimetric determination of free amino acids content in tea infusions with 2,4-dinitrofluorobenzene," *Journal of Food Composition and Analysis*, vol. 22, no. 2, pp. 137-141, Mar. 2009.

[10] M. H. Sousa, J. C. Rubim, P. G. Sobrinho, and F. A. Tourinho, "Biocompatible magnetic fluid precursors based on aspartic and glutamic acid modified maghemite nanostructures," *Journal of Magnetism and Magnetic Materials*, vol. 225, no. 1-2, pp. 67-72, 2001.

[11] S. Bose and S. Tarafder, "Calcium phosphate ceramic systems in growth factor and drug delivery for bone tissue engineering- A review," *Acta Biomaterialia*, vol. 8, no. 4, pp. 1401-1421, April 2012.

[12] W. Xue, A. Bandyopadhyay and S. Bose, " Polycaprolactone coated porous tricalcium phosphate scaffolds for controlled release of protein for tissue engineering," *Journal of Biomedical Materials Research Part B: Applied Biomaterials*, vol. 91B, no. 2, pp. 831-838, November 2009.

[13] W. Mattanavee, O. Suwantong, S. Puthong, T. Bunaprasert, V. P. Hoven and P. Supaphol," Immobilization of Biomolecules on the Surface of Electrospun Polycaprolactone Fibrous Scaffolds for Tissue Engineering, *ACS Applied Materials and Interfaces*, vol. 1, no. 5, pp. 1076-1085, April 2009.

DETONATION SPRAYING OF TiO$_2$-Ag: CONTROLLING THE PHASE COMPOSITION AND MICROSTRUCTURE OF THE COATINGS

Dina V. Dudina^a, Sergey B. Zlobin^b, Vladimir Yu. Ulianitsky^b, Oleg I. Lomovsky^a, Natalia V. Bulina^a, Ivan A. Bataev^c, Vladimir A. Bataev^c

^{a} Institute of Solid State Chemistry and Mechanochemistry SB RAS, Novosibirsk, Russia
^{b} Lavrentiev Institute of Hydrodynamics SB RAS, Novosibirsk, Russia
^{c} Novosibirsk State Technical University, Novosibirsk, Russia

ABSTRACT
TiO$_2$ has become the subject of numerous investigations owing to its photocatalytic activity, biocompatibility and ability to form oxygen-deficient phases of interesting physical properties. This work is aimed at studying the influence of the detonation spraying parameters on the phase composition and microstructure of the TiO$_2$-Ag coatings. Through changing the composition of the explosive gaseous mixture, it is possible to deposit coatings under reducing conditions or under excess of oxygen thereby controlling reduction/oxidation reactions in the sprayed material. Ti$_3$O$_5$ and oxygen-deficient rutile form in the coatings deposited under a reducing atmosphere. With larger volumes of the explosive mixture filling the detonation gun, higher temperatures of the particles leaving the gun can be achieved. In the coatings formed by molten and semi-molten particles, some flattened re-solidified areas on the surface are observed; however, these can be avoided at certain combinations of particle velocity/degree of overheating. Silver inclusions experience melting and coalescence or dispersion during spraying; dispersion of silver particles is favored when titanium dioxide in the sprayed particles is a molten or semi-molten state.

INTRODUCTION
Titanium dioxide, TiO$_2$ has become the subject of numerous investigations owing to a host of exciting properties, such as photocatalytic activity under UV illumination [1-3], biocompatibility and bioactivity [4], ability to form oxygen-deficient phases showing catalytic activity in visible light [5-6] and interesting physical properties [7]. The addition of metallic silver particles to TiO$_2$ improves its photocatalytic behavior [8] and imparts antibacterial properties to the composite material [9]. The subject of this work is TiO$_2$-Ag coatings deposited on copper substrates from composite powders prepared by mixing titanium dioxide (rutile) and elemental silver. Such coatings may protect copper coils of air-conditioners preventing accumulation and growth of bacteria on the coil surfaces.
The choice of the deposition technique for the TiO$_2$-Ag coatings is dictated by a lack of inherent plasticity in titanium dioxide making it necessary to use thermal spraying to efficiently deposit TiO$_2$ layers well-bonded to substrates. Detonation spraying is based on heating and acceleration of the sprayed particles by gaseous detonation products [10-11]. Powder particles are fed into the barrel of a detonation gun filled with an explosive gaseous mixture. As the particles move inside the barrel, they are influenced by the chemistry of the gaseous environment. Detonation spraying is suitable to form dense coatings of good adhesion. To our best knowledge, physical and chemical transformations of titanium dioxide during detonation spraying have not been previously studied. Here, we present the phase evolution of titanium dioxide sprayed under different atmospheres and address microstructural features of the detonation sprayed coatings. In order to rationalize the observed phase and microstructure changes, we involve calculations of temperatures and velocities of the sprayed particles.

MATERIALS AND METHODS
Titanium dioxide (rutile, 99.999%, 30 µm) and metallic silver (99.99%, 0.5 µm) powders were used to prepare powder mixtures for the detonation spraying. The mixtures corresponding to the TiO$_2$-

2.5vol.%Ag composition were ball milled in a high energy ball mill using a ball acceleration of 200 m·s^{-2}. The purpose of ball milling was to form a composite structure in the powders with a uniform distribution of Ag particles. The highest efficiency of deposition in the detonation spraying is achieved when the sprayed particles are several tens of microns in size [10]. Since the milled TiO$_2$-2.5vol.%Ag powders contained a large fraction of very fine particles (less than 5 μm in size) and were unsuitable for the detonation spraying, a procedure was developed to agglomerate the milled powders using a polyvinyl alcohol water solution to produce composite particles ranging between 10 and 60 μm.

A Computer Controlled Detonation Spraying (CCDS) facility [11] was used to deposit the coatings. The barrel of the detonation gun was 850 mm long and 20 mm in diameter. Acetylene C$_2$H$_2$ was used as a fuel. Two O$_2$/C$_2$H$_2$ ratios were used to create reducing (1.05) and oxidizing (2.5) conditions of deposition. The fraction of the barrel volume filled with an explosive mixture was varied between 30 and 60%. The powder injection point was located 350 mm from the open end of the barrel. The injection was performed by a powder feeder. While an average distance traveled by the powder particles in the detonation gun was 350 mm, some particles could start their travel 300 mm or 400 mm from the end of the barrel. The spraying distance was set at 150 mm. The coatings were deposited on copper substrates 1 mm thick. The substrates were sand-blasted before the deposition of coatings. The weight of the substrates before and after the deposition was accurately measured.

In order to estimate the velocities and temperatures of the particles leaving the detonation gun, a computer numerical model was used [12-14]. Reduction processes in TiO$_2$ were not taken into account in the calculations. The calculations were performed for 3 particles sizes: 20, 40 and 60 μm.

The microstructure of the coatings was studied by Scanning Electron Microscopy (SEM) and Energy Dispersive Spectroscopy (EDS) using a Carl Zeiss EVO50 Scanning Electron Microscope and EDS X-Act (Oxford Instruments). The XRD phase analysis of the coatings was performed using an X-ray diffractometer (D8 ADVANCE, Bruker) with Cu Kα radiation. The surface roughness of the coatings was measured using a Laser Scanning Confocal Microscopy (LEXT OLS4000, Olympus). The roughness was measured in two modes: across a line of 625 μm (R$_a$) and across an area of 646·646 μm^2 (S$_a$).

RESULTS AND DISCUSSION

When the barrel of a detonation gun is filled with a mixture of C$_2$H$_2$+O$_2$ with an O$_2$/C$_2$H$_2$ molar ratio equal to 1.05, the atmosphere of spraying is reducing due to the presence of hydrogen and carbon monoxide in the detonation products. We have sprayed the TiO$_2$-2.5vol.%Ag powders under a reducing atmosphere varying the volume the C$_2$H$_2$+O$_2$ mixture (modes 1-4) and under conditions of full conversion of acetylene (mode 5) as is summarized in Table I.

Table I. Detonation spraying parameters of the TiO$_2$ (rutile) - 2.5vol.%Ag powders.

Detonation spraying mode	Fraction of the barrel volume filled with O$_2$/C$_2$H$_2$ mixture, %	O$_2$/C$_2$H$_2$ molar ratio
1	30	1.05
2	40	1.05
3	50	1.05
4	60	1.05
5	50	2.50

The calculated temperatures and velocities of the TiO$_2$ - 2.5vol.%Ag particles leaving the barrel of the detonation gun (Table II) show that the particle size, temperature and velocity are interrelated.

The temperature of particles of all sizes increased as larger volumes of the C$_2$H$_2$+O$_2$ mixture were used. The mass of the material deposited per 1 shot of the gun was 2.0, 5.0, 8.0 and 8.8 mg in modes 1, 2, 3 and 4, respectively, which shows that more efficient spraying occurs in high-temperature modes 3 and 4. Comparing the calculation results for modes 3 and 5, we conclude that increasing oxygen content in the explosive mixture while keeping its volume constant leads to higher particle temperatures.

In high-temperature conditions of modes 3, 4 and 5, smaller particles are heated up to a higher extent than larger particles. In these modes, rutile experiences partial melting. In modes 1 and 2, the hottest particles are those of 40 μm in size. The calculations show that melting of rutile was not reached in mode 1. As was calculated using the average distance traveled by the particles of 350 mm, rutile does not melt in mode 2. However, if some 20 μm-sized particles start their travel in the gun barrel 400 mm from its open end, they partially melt (Table III). Indeed, as will be shown below, the microstructural features of the surface of the coatings sprayed in mode 2 are direct evidence of melting and re-solidification of titanium dioxide. The difference in the location of the starting point is a source of significant deviations in the calculated particle temperature. The temperature difference can be several hundred degrees and is greater for smaller particles and higher-temperature modes. Despite this uncertainty in the particle temperature, the calculations help explain the observed microstructures and determine the role of the molten phases in the microstructure formation. Particle velocities are less affected by the length of the distance traveled by the particles inside the gun.

Table II. Calculated temperatures (T) and velocities (V) of the TiO$_2$ - 2.5vol.%Ag particles leaving the barrel of the detonation gun in modes 1-5. The calculations were performed for 3 particle sizes (20, 40 and 60 μm) and the average distance traveled by the particles of 350 mm. T$_m$ is the melting temperature of TiO$_2$, 2123 K.

Detonation spraying mode	20 μm			40 μm			60 μm		
	V, m s^{-1}	T, K	T/T$_m$	V, m s^{-1}	T, K	T/T$_m$	V, m s^{-1}	T, K	T/T$_m$
1	550	780	0,37	430	1320	0,62	360	1120	0,53
2	610	1600	0,76	510	·1700	0,80	430	1400	0,66
3	690	2123	1,00	560	1980	0,93	460	1640	0,77
4	680	2650	1,25	530	2123	1,00	420	2040	0,96
5	580	2440	1,15	480	2125	1,00	390	2123	1,00

Smaller particles have higher velocities at the exit of the gun barrel than larger particles. The calculated particle velocities increase as the explosive charge increases from 30 to 50%; at 60% there is a slight drop in the particle velocity. This non-monotonous dependence of the particle velocity on the explosive charge has been previously elucidated and is related to the relative position of the powder injection point and the end of the barrel volume filled with an explosive mixture [11].

The phase composition of the coating sprayed in mode 1 does not show any significant differences from that of the sprayed powder (Fig.1). Due to a limited thickness of the coatings, some XRD patterns contain diffraction lines of copper from the substrate. In the XRD pattern of the coating deposited in mode 2, the intensity of the main reflection of rutile (110) decreases compared to that in the pattern of the coating sprayed in mode 1. The (101) and (211) rutile reflections shift to lower angles in the patterns of the coatings sprayed in modes 2, 3 and 4. The reason of such changes is the formation of oxygen-deficient rutile in the reducing conditions with increased lattice parameter *a*.

A peculiar feature of the coatings sprayed in high-temperature reducing conditions is the presence of significant amounts of the Ti_3O_5 phase, as can be concluded from the XRD patterns. The analysis of the peak positions of the Ti_3O_5 phase shows that the observed phase is the monoclinic λ-phase. At room temperature, λ-Ti_3O_5 is metastable and can be stabilized by doping [15]. In the detonation sprayed coating, this phase can form as a result of rapid cooling of the detonation splats.

Table III. Calculated temperatures (T) and velocities (V) of the TiO₂ - 2.5vol.%Ag particles leaving the barrel of the detonation gun in modes 2 and 3. The calculations were performed for 3 distances traveled by the particles (300, 350 and 400 mm) and 3 particle sizes (20, 40 and 60 μm).

Detonation spraying mode	Distance traveled by the particles, mm	20 μm		40 μm		60 μm	
		V, m s⁻¹	T, K	V, m s⁻¹	T, K	V, m s⁻¹	T, K
2	400	630	2123	520	1930	430	1550
	350	610	1600	510	1700	430	1400
	300	630	810	510	1460	420	1250
3	400	670	2440	520	2123	410	1890
	350	690	2123	560	1980	460	1640
	300	680	1780	560	1710	470	1440

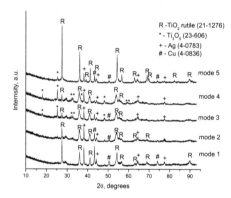

Fig.1. XRD patterns of the detonation sprayed coatings (JCPDS card numbers of the identified phases are given in parentheses).

The amount of the Ti_3O_5 suboxide phase and oxygen-deficient rutile in the deposited coatings is determined by the kinetics of reduction reactions. The amount of the reducing agents in the gaseous mixture is excessive relative to the amount of titanium dioxide, such that, given a sufficient time, TiO_2 can be fully reduced to Ti_3O_5 or to other low-oxygen content oxides. However, the sprayed particles are in a reducing environment of the gaseous detonation products for a limited time and only partial reduction occurs. In mode 5, well-crystallized rutile forms in the coating and only small amounts of the

Ti$_3$O$_5$ phase are present. The positions of the reflections from the rutile phase are restored to correspond to stoichiometric rutile. So, the reduction product in the conditions of mode 5 is the Ti$_3$O$_5$ phase and not the oxygen-deficient rutile.

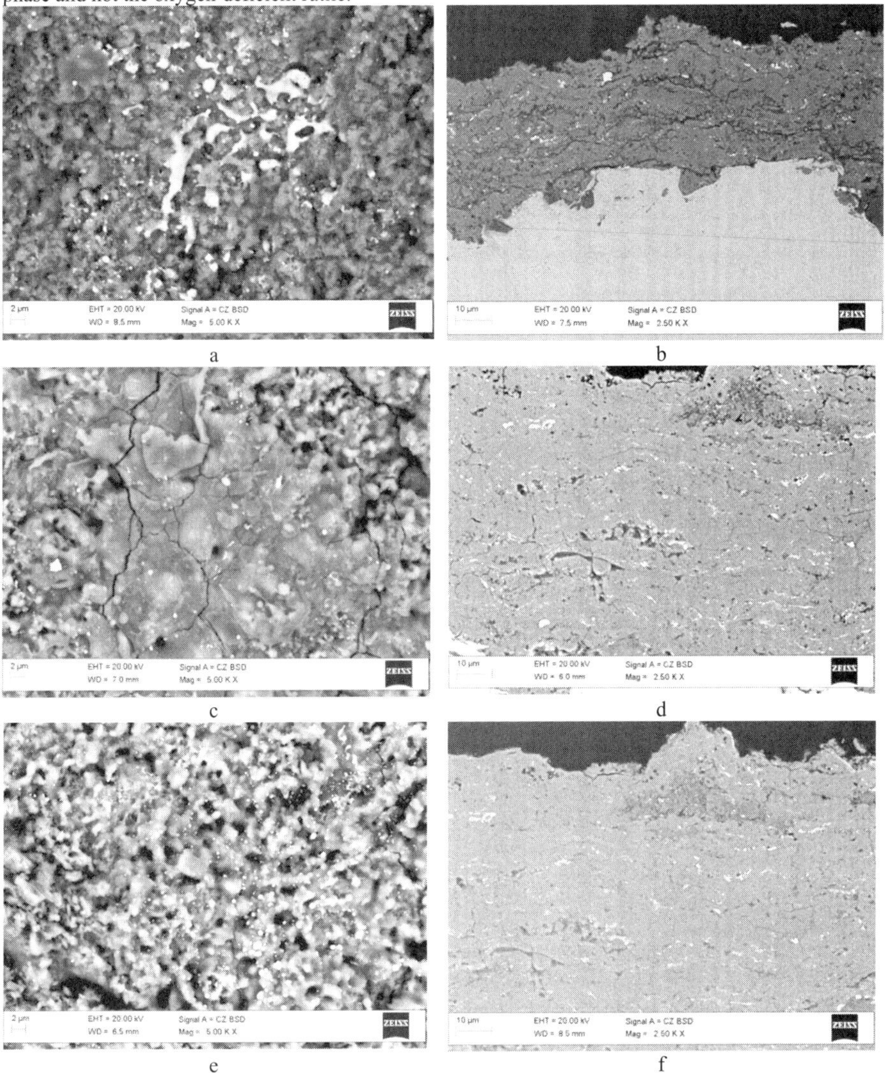

Fig.2. Back-scattered electron images of the surface and cross-section of the detonation sprayed coatings: mode 1 – a, b; mode 3 – c, d; mode 4 – e, f.

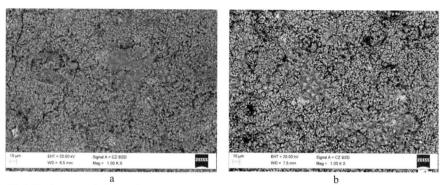

a   b

Fig.3. Back-scattered electron images of the surface the detonation sprayed coatings showing flattened areas (splats): mode 2 – a, mode 3 – b.

SEM images of the surface and cross-section of the detonation sprayed coatings are shown in Fig.2. Solid particles in mode 1 form thin porous coatings (Fig.2, a-b). The evidence of melting can be found on the surface of coatings sprayed in modes 2, 3, 4 and 5. Under reducing conditions, the coatings become denser and thicker from mode 1 to mode 4. The growing thickness of the coatings agrees well with an increased mass gain in the samples after the deposition. The coatings sprayed in modes 3 and 4 possess a lamellar structure typical for detonation sprayed coatings formed in the presence of liquid phases.

Table IV. Surface roughness of the detonation sprayed TiO$_2$-2.5vol.%Ag coatings.

Detonation spraying mode	Roughness R_a, µm	Roughness S_a, µm
1	3.147	4.554
3	2.703	3.713
4	1.529	3.147
5	2.224	4.147

In the coatings formed by molten and semi-molten particles, some flattened areas on the surface are observed (Fig.2, c and Fig.3), which are direct indications of melting and solidification processes occurring in the sprayed material. They were not found in the coating sprayed in mode 1, but occasionally observed in mode 2. In higher-temperature modes 3 and 5, flattened areas covered up to 40-50% of the coating surface. Interestingly, such flattened areas were not characteristic of coatings formed in mode 4. Li ct al [16] observed a disappearance of frozen splats on the surface of the HVOF-sprayed TiO$_2$ with increasing propane fuel flow and explained it by higher velocities of the sprayed particles and higher kinetic energies forcing particles to disperse upon impact with the substrate or previously deposited layers. In our case, the sprayed particles in mode 4 move faster than in mode 5 but more slowly than in mode 3. At the same time, the particles in mode 4 are overheated to a higher extent than the particles sprayed in mode 3. The degree of overheating of the melt influences the viscosity of the molten droplets. A droplet of a lower-viscosity melt is more likely to disperse upon impact with the substrate. Therefore, at certain temperature/velocity combinations, a sprayed particle will break into smaller droplets and the formation of flattened areas on the surface will be avoided. Flattened areas on the surface of the coatings are not desirable, as they cause surface cracking. Interestingly, the surface roughness of the coatings sprayed in mode 4 is lower than that of the coatings sprayed in modes 1, 3 and 5 (Table IV), which shows that melt dispersion upon impact favors formation of a layer with a more uniform structure.

Bright particles distinguished in the back-scattered electron images of the surface and cross-section of the coatings correspond to the silver inclusions as was confirmed by the EDS analysis. On the surface of the coatings, silver particles are in a shape of frozen droplets. In mode 1, titanium dioxide does not melt while silver particles experience melting. The molten silver droplets coalesce, which results in the formation of Ag agglomerates in the coatings several microns in size. When the sprayed particles are in a molten or semi-molten state and they impact on the substrate with high velocities, molten silver inclusions disperse into smaller droplets leading to the formation of submicron and nanoparticles uniformly distributed in the coatings, which is best achieved in the coatings sprayed in mode 4.

The development of TiO$_2$-based coatings by detonation spraying as an alternative method to widely used plasma spraying is promising for biomedical applications among others. TiO$_2$ coatings on implants can facilitate the formation of bonelike apatite on their surfaces [4], which is vital for successful bonding between the implants and bone tissues. Furthermore, due to its corrosion resistance, TiO$_2$ acts as a chemical barrier preventing the release of implant components into the body [17]. The plasma sprayed TiO$_2$ coatings in the as-sprayed state are usually bio-inert and are to be rendered bioactive by laser treatment [18], UV illumination [19], hydrogen plasma immersion ion implantation [4] or acid treatment [20]. In detonation spraying, the temperatures of the particles are lower than those in plasma spraying; the solidification conditions of titanium oxides are, therefore, different in these two spraying methods. Those can be crucial in predetermining the bioactivity of solidified deposits. The tailorable phase composition and porosity of the detonation sprayed coatings are of interest as potentially influencing their bioactivity. Further research is needed in order to evaluate the bioactivity of the detonation sprayed titanium oxide coatings of varied phase composition and microstructure. The possibility of obtaining nanoparticles of Ag during detonation spraying of the TiO$_2$-Ag composite coatings opens up a perspective of using them as antimicrobial coatings. The remaining porosity of the detonation sprayed TiO$_2$-Ag coatings may be a positive factor allowing access to silver particles located in the inner layers of the coatings.

CONCLUSIONS

We have shown that TiO$_2$-2.5vol.%Ag composite powders can be efficiently deposited on copper substrates by the detonation spraying. Through careful adjustment of the spraying parameters, it is possible to deposit coatings under reducing conditions or under excess of oxygen thereby controlling the amount of titanium suboxides in the coatings. The calculations of the temperatures and velocities of the particles help shed light on how the microstructure forms during spraying. The amount of the C$_2$H$_2$+O$_2$ mixture in the gun barrel and its stoichiometry influence both temperatures and velocities at which the particles leave the detonation gun. Under reducing conditions and sufficient heating of the sprayed particles - up to a molten or semi-molten state – titanium suboxide Ti$_3$O$_5$ forms. The Ti$_3$O$_5$ phase observed in the coatings corresponds to λ-phase, which is metastable at room temperature. Silver particles experience melting and coalescence or dispersion during spraying and remain well preserved in the coatings. Dispersion of molten silver droplets dominates when the particles are heated to reach the melting temperature of the oxide matrix. The tailorable phase composition, porosity and dispersion of silver particles are of interest as factors potentially influencing bioactivity and antimicrobial properties of the TiO$_2$-Ag detonation sprayed coatings.

ACKNOWLEDGEMENTS

The work was supported according to the Program of the Siberian Branch of Russian Academy of Sciences V.36.4. "Controlling chemical processes using high-pressures, radiation, and electrical and magnetic fields in stationary and impulse modes", Integration Projects № 57 and 82 of the Siberian Branch of the Russian Academy of Sciences and Scientific School № 5770.2010.1.

The Authors are grateful to LLC "Melytec" for providing a Laser Scanning Confocal Microscopy (LEXT OLS4000, Olympus) for surface roughness measurements of the coatings.

REFERENCES

1. O.Carp, C.L.Huisman and A.Reller, Photoinduced reactivity of titanium dioxide, Progress Solid State Chem., **32**, 33-177 (2004).
2. F.L.Toma, L.M.Berge, D.Jacquet, D.Wicky, I.Villaluenga, Y.R.de Miguel and J.S.Lindelov, Comparative study on the photocatalytic behaviour of titanium oxide thermal sprayed coatings from powders and suspensions, Surf. Coat. Tech., **203**, 2150-56 (2009).
3. Y.Paz, Application of TiO$_2$ photocatalysis for air treatment: Patents' overview, Appl. Catalysis B: Environmental, **99**, 448-60 (2010).
4. X.Liu, X. Zhao, R. K.Y. Fu, J. P.Y. Ho, C. Ding and P. K. Chu, Plasma-treated nanostructured TiO$_2$ surface supporting biomimetic growth of apatite, Biomater. **26**, 6143–50 (2005).
5. I.N.Martyanov, S.Uma, S.Rodrigues and K.J.Klabunde, Structural defects cause TiO$_2$-based photocatalysts to be active in visible light, Chem. Comm., **21**, 2476-77 (2004).
6. I.N.Martyanov, T.Berger, O.Diwald, S.Rodrigues and K.J.Klabunde, Enhancement of TiO$_2$ visible light photoactivity through accumulation of defects during reduction-oxidation treatment, J .Photochem. Photobiology A: Chemistry, **212**, 123-41 (2010).
7. D.B.Strukov, G.S.Snider, D.R.Stewart and R.S.Willams, The missing memristor found, Nature, **453**, 80-83 (2008)
8. P.D.Cozzoli, E.Fanizza, R.Comparelli, M.L.Curri and A.Agostiano, Role of metal nanoparticles in TiO$_2$/Ag nanocomposite-based microheterogeneous photocatalyst, J.Phys.Chem. B, **108**, 9623-9630 (2004).
9. B.S.Necula, L.E.Fratila-Apachitei, S.A.J.Zaat, I.Apachitei and J.Duszczyk, In vitro antibacterial activity of porous TiO$_2$-Ag composite layers against methicillin-resistant *Staphylococcus aureus*, Acta Biomater., **5**, 3573-3580 (2009).
10. Y.A.Nikolaev, A.A.Vasiliev and V.Yu.Ulianitsky, Comb. Expl. Shock Waves, **39**, 382-410, (2003).
11. V.Ulianitsky, V.Shtertser, S.Zlobin and I.Smurov, Computer-controlled detonation spraying: from process fundamentals toward advanced applications, J. Thermal Spray Technol., **20**, 791-801 (2011).
12. T.P.Gavrilenko, Y.A.Nikolaev, V.Y.Ulianitsky, M.C.Kim and J.W.Hong, Computational Code for Detonation Spraying Process, Proc. Intl. Thermal Spray Conf., Nice, 1475-83 (1998).
13. I.Smurov, D.Pervushin, Y.Chivel, B.Laget, V.Ulianitsky and S.Zlobin, Measurements of particle parameters at detonation spraying, Proc. Intl. Thermal Spray Conf. Singapore, 481-95 (2010).
14. S.B.Zlobin, V.Y.Ulianitsky, A.A.Shtertser and I.Smurov, High-velocity collision of hot particles with solid substrate under detonation spraying: detonation splats, Proc. Intl. Thermal Spray Conf., USA, 728-34 (2009).

15. M.Onoda, Y.Ogawa and K.Taki, Phase transitions and the doping effect in Ti$_3$O$_5$, J Phys: Cond. Matter, **10**, 7003-13 (1998).
16. C.J.Li, G.J.Yang, Y.Y.Wang, C.X.Li, F.X.Ye and A.Ohmori, Phase formation during deposition of TiO$_2$ coatings through high velocity oxy-fuel spraying, Mater. Trans., **47**, 1690-96 (2006).
17. X.Nie, A.Leyland and A.Matthews. Deposition of layered bioceramic hydroxyapatite/TiO$_2$ coatings on titanium alloys using a hybrid technique of micro-arc oxidation and electrophoresis. Surf. Coat. Technol., **125**, 407-414 (2000).
18. N.Moritz, M.Jokinen, T.Peltola, S.Areva and A.Yli-Urpo. Local induction of calcium phosphate formation on TiO$_2$ coatings on titanium via surface treatment with a CO$_2$ laser. J. Biomed. Mater. Res. A, **65**, 9-16 (2003).
19. X.Liu, X.Zhao, C.Ding and P.K.Chu. Light-induced bioactive TiO$_2$ surface. Appl. Phys. Lett., **88**, 013905-1 – 013905-3 (2006).
20. X.Zhao, X.Liu and C.Ding. Acid-induced bioactive titania surface. J. Biomed. Mater. Res. A, **75**, 888-894 (2005).

SiO$_2$ AND SrO DOPED β-TCP: INFLUENCE OF DOPANTS ON MECHANICAL AND BIOLOGICAL PROPERTIES

Gary Fielding, Johanna Feuerstein, Amit Bandyopadhyay, Susmita Bose*
W.M Keck Biomedical Materials Research Laboratory , School of Mechanical and Materials Engineering, Washington State University
Pullman, Wa United States

*Contact author e-mail: sbose@wsu.edu

Keywords: β-tricalcium phosphate, Bioresorbable, Dopants, strength degradation, simulated body fluid

ABSTRACT

Metal ion substitution in β-tricalcium phosphate (β-TCP) ceramics, a commonly used bone substitute material, has been shown to significantly alter the material's mechanical and biological properties without adversely affecting its biocompatibility. Previous studies have shown that SiO$_2$ addition causes increased densification and that strontium (Sr) plays a positive role in bone formation. To further understand the effects of SiO$_2$ and SrO doping on mechanical properties and mineralization characteristics of β-TCP ceramics, dense β-TCP compacts of two different compositions (pure TCP, 1.0 wt. % SrO + 0.5 wt. % SiO$_2$ TCP) were prepared by uniaxial and cold isostatic pressing followed by sintering at 1250°C. X-ray diffraction (XRD) analysis of sintered compacts revealed that dopants impede the β- to α-TCP transformation process. Doping with SiO$_2$/SrO decreased the compressive strength of the samples to 53% of the value measured for pure TCP compacts. TCP compacts were immersed in simulated body fluid (SBF) for 12 weeks to imitate the impact on mechanical properties and mineralization characteristics in vivo. Significant apatite formation was observed on both pure and doped samples. The doped samples, however, exhibited a greater affinity for mineralization than their pure counterparts. Pure TCP samples demonstrated a 40% strength loss over the course of the 12 week study, while the doped samples retained their initial strength. These results indicate that tailorable strength and strength loss behavior can be achieved in β-TCP via compositional modifications.

INTRODUCTION

Calcium phosphate synthetic biomaterials have been extensively studied over the past 20 years because of their compositional similarity to the inorganic constituent of natural bone. β-tricalcium phosphate (β-TCP) is of particular interest due to its excellent biocompatibility, bioresorbability and time dependent strength degradation [1-3]. It is currently used in many clinical applications ranging from dental reconstruction to bone cement. Contemporary research has been focused on the development of β-TCP for use as a scaffolding material for critical size defects, where the critical size defect is the smallest intraosseous wound that will not heal naturally during the lifespan of the patient. Presently, autografts and allografts are the gold standard in practice, but they leave much to be desired. Autografts require a second site surgery, where bone mass is usually taken from the iliac crest, and can result in donor site morbidity and high surgery costs. There is also a relatively small amount of bone available for harvesting. Allografts may be less effective than autografts and often have longer healing times. They also run the risk of host rejection.

The quintessential scaffold for bone tissue engineering needs to provide enough mechanical support to maintain stability while new tissue ingrowth occurs. Ideally, the scaffold will degrade at the same rate that ingrowth ensues. While β-TCP does exhibit time dependent strength degradation and dissolution behavior, it may not be suitable for all applications. There are many factors that affect the bone growth rate that occurs during the healing process, some of which include the size, type and location of defect, as well as the age of patient and disease [4-6].

Metal ion substitution has been studied by our lab and others as a way to influence the mechanical and biological properties of β-TCP. Results have shown that different combinations of dopants as well as varying concentrations of dopants can have significant effects on the overall mechanical strength, strength degradation profiles and cell-material interactions of β-TCP compacts [7-10]. Silicon (Si), as a trace element, has been shown to play an essential role in bone developmental biology, where deficiencies result in decreased bone mass density and decreased bone turnover rates [11-13]. It is also a well-studied sintering additive to calcium phosphate materials. Pure β-TCP undergoes phase transformation at about 1125° C to α-TCP. The α-phase signifies a rapid grain growth stage and is considered deleterious to the mechanical properties of the sintered body due to spontaneous microcracking in the sintered body due to the expansion-contraction cycle generated by the differences in density between β-TCP (3.07 g/cm3) and α-TCP (2.86 g/cm3) [14]. When SiO$_2$ is added, the β to α phase transformation is stunted at high temperatures, helping to eliminate microcracking and increasing overall density [15]. Strontium is a non-essential element that has bone seeking behavior. In small amounts, less than 2g per day, strontium has been shown to have beneficial effects on osteoporotic patients by increasing bone mass density and providing sustained anti-fracture efficacy [16]. Strontium acts by displacing Ca^{2+} ions in osteoblastic calcium-mediated processes. Namely, researchers have noted that strontium may stimulate bone formation by activating the calcium sensing receptor [17-18], while impeding bone resorption by increasing osteoprotegerin (OPG) and inhibiting receptor activator of nuclear factor kappa B ligand (RANKL) expression [19]. OPG is a protein produced by osteoblasts that inhibits RANKL induced osteoclastogenesis by operating as a decoy receptor for RANKL[20]. The OPG/RANKL ratio, then, is an important balance for regulating bone resorption and osteoclastogenesis.

The aims of the present study are to produce SiO$_2$/SrO binary doped β-TCP compacts and investigate the effects of the dopants on the mechanical strength, strength degradation profiles, and biomineralization profiles over a 12 week period in simulated body fluid.

MATERIALS AND METHODS

Synthetic β-tricalcium phosphate nanopowder (β-TCP) with an average particle size of 550 nm was obtained from Berkley Advanced Biomaterials Inc. Berkeley, CA. High purity strontium oxide (SrO, 99.9% purity) was purchased from Aldrich, MO, USA and silicon oxide (SiO$_2$, 99.998% purity) was procured from Alfa Aesar, MA, USA. All other chemicals were of analytical grade and used without further purification.

Concentrations for dopants, 0.5% SiO$_2$ and 1% SrO, were chosen based on previous work [8, 21-22]. The powders were prepared by adding 50 g of β-TCP nanopowder to 250 mL polypropylene Nalgene bottles containing 100 g of 5 mm diameter zirconia milling media. The appropriate amount of SiO_2 and SrO dopants were added to the Nalgene bottle to achieve the desired concentration and the mixture was then ball milled with ethanol for 6 h at 65 rpm. This time period and rate were chosen to minimize the formation of agglomerates and increase the homogeneity of the powders [8]. After milling, the powders were dried in an oven at 60°C for 72 h followed by further drying at 210°C for 3 h. Dried powders were pressed into disc compacts and compression cylinders for further analysis. Disc compacts were used for density, dissolution and mineralization studies, while compression cylinders were used for mechanical property evaluation. The disc compacts were formed by pressing 500 mg of powder at 145 MPa for one minute to produce samples 12 mm in diameter and 2.5 mm thick. Compression cylinders were formed by pressing 750 mg of powder at 58 MPa for one minute to produce samples 6.5 mm in diameter and 12 mm in height.

Following uniaxial pressing, all green compacts were isostatically pressed at 413 MPa and sintered at 1250°C for two hours in a Thermolyne muffle furnace using a three step heating cycle. In the first step, the samples were heated at 3°C/min to 300°C and held at this temperature for 1 h to remove any unwanted moisture from the samples. Next the samples were heated at the same rate to 600°C and held at this temperature for 1 h to reduce residual stresses in the green sample. Finally, the samples were heated to 1250°C at 1°C/min and sintered at this temperature for 2 h.

Phase analysis of the sintered samples was performed by X-ray diffraction using a Siemens D500 Kristalloflex system (Siemens D500 Kristalloflex, Madison WI) at room temperature using Cu-K$_\alpha$ radiation with a Ni-filter. Each run was performed with 2θ values of 10° to 60° at a step size of 0.02° and a count time of 0.5 s per step. The peak positions for doped samples were compared to those of undoped β- and α- TCP (JCPDS No's 09-169 and 09-0348). The surface morphology of disc compacts of all compositions was observed under a scanning electron microscope (SEM) (FEI Inc., OR, USA) following gold sputter-coating (Technics Hummer V, CA).

To evaluate *in vitro* bioactivity of undoped and doped β-TCP samples, disc compacts of all compositions were immersed in a simulated body fluid (SBF) buffered at pH 7.35 ±0.05 with tris(hydroxymethyl)aminomethane/HCl. SBF is an acellular solution with an ionic composition (in units of mM, 142 Na^+, 5.0 K^+, 1.5 Mg^{2+}, 2.5 Ca^{2+}, 103 Cl^-, 27.0 HCO_3^-, 1.0 HPO_4^{2-} and 0.5 SO_4^{2-}) almost equal to that of human plasma and buffered at a similar pH[23]. The samples were exposed to the SBF solution under static conditions at 37°C for 2, 4, 8 and 12 weeks, respectively. Over the course of the study, the SBF was changed every 3-4 days and a portion of the used SBF from a set of samples was taken for Ca^{2+} concentration analysis using an atomic absorption spectrophotometer (AAS) (Shimadzu, Kyoto, Japan). At each time point, five samples of each composition were taken out, washed with distilled water and then dried at 65°C in an oven for 7 days. Once dry, the weight of each sample was carefully recorded and compared with its dry weight before immersion. Surface microstructures of the samples were observed using a scanning electron microscope (SEM) (FEI Inc., OR, USA) following gold sputter coating (Technics Hummer V, CA) to determine formation of an apatite layer. Attenuated total reflection infrared (ATR-IR) analysis was done on disk surfaces using a Fourier transform infrared spectrometer (FTIR, Nicolet 6700, ThermoFisher, Madison, WI) to confirm the presence of the apatite layer.

Uniaxial compressive strength measurements were done for each composition using a screw-driven universal testing machine with a constant crosshead speed of 0.33 mm/min. Compressive strength was calculated using the maximum stress before failure. At least five replicate samples were tested at each data point.

RESULTS AND DISCUSSION

XRD analysis was performed to determine possible phase transformation due to the addition of dopants and relatively high sintering temperature. Figure 1 shows the XRD patterns of undoped and SiO₂/SrO doped TCP samples. Spectra of both pure and doped samples showed β-TCP as the primary phase (JCPDS No. 09-0169). Both samples also exhibited the presence of α phase TCP (JCPDS No. 09-0348). This is expected as the phase transformation from β to α occurs at 1125° C. The doped samples, however, indicated a significant reduction in α phase formation when compared to pure TCP samples, suggesting that SiO₂ and SrO dopants stabilize the β-TCP phase at 1250° C. In calcium phosphate (CaP) ceramics, sintering additives have been used to mitigate the phase transformation of β-TCP to α-TCP at temperatures above 1125° C. The presence of α phase formation in TCP ceramics has been shown to be closely related with the expansion of sample volume and declining shrinking rate and is generally considered to prevent TCP from further densification [15, 24]. The XRD patterns also confirm, with a shift in 2θ and d-spacing values, the substitution of Si^{4+} and Sr^{2+} for Ca^{2+} into the TCP structure. The change in lattice parameters can be attributed to the difference in ionic radius of Si^{4+} and Sr^{2+} compared to Ca^{2+} [25-26]. The ionic radius of Si^{4+} (0.4 Å) is much smaller than that of Ca^{2+} (1 Å), while the ionic radius of Sr^{2+} is slightly larger than Ca^{2+}. The result is an increase in the unit cell parameters where Sr^{2+} is substituted and a decrease where Si^{4+} is substituted. The average combined substitution of Sr^{2+} and Si^{4+} in a 1:1.5614 mole ratio (ratio derived from weight percent used for mixing samples) can be found by a simple calculation: (1.5614(0.4 Å) + 1.18 Å)/2 = 0.9 Å. This yields an overall decrease in lattice parameters and, therefore, a shift in peak to higher 2θ values is observed [27].

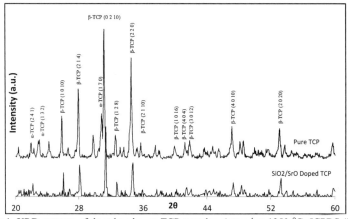

Figure 1. XRD patterns of doped and pure TCP samples sintered at 1250 °C. JCPDS # 09-0169 (β-TCP) and 09-0348 (α-TCP)

Bulk density of disc compacts sintered at 1250°C was determined for the pure and doped TCP compositions. The results were then normalized to the theoretical density of pure β-TCP (3.07 g/cm^3). Pure TCP had a relative bulk density of 91.72 ± 2.32% and that of SiO$_2$/SrO doped TCP was 91.4 ±1.49%. Pure and doped samples were also analyzed using SEM to understand the influence of dopants on their microstructure. The SEM micrographs of sintered sample surfaces, demonstrated in Figure 2, showed that both TCP sample compositions exhibited a highly dense structure. The pure samples exhibited distinctly faceted grains, while the SiO$_2$/SrO grains were more fluid in shape, evidencing liquid phase sintering in the doped samples. The presence of a liquid phase during sintering has the benefit of resulting in decreased porosity when compared to solid state sintering [28]. Increased wettability during liquid phase sintering leads to an increase in capillary action. Increased capillary forces during sintering results in more efficient particle rearrangement and increased density. The densities of the samples were not significantly different, but the doped TCP exhibited larger and non-uniform grain size when compared to pure TCP samples. When everything is taken into consideration, this makes sense. The absence of α-TCP and the liquid phase sintering increases densification of doped samples, while the expansion and contraction of lattice parameters due to Sr^{2+} and Si^{4+} substitution leads to non-uniform grain growth and, in turn, overall decreased densification and larger grain size. The pure TPC had α phase present, which leads to decreased, but more uniform densification and grain size. The end results are samples that have similar densities, but different grain sizes.

Figure 2. SEM micrographs of pure TCP (a) and SiO$_2$/SrO doped TCP (b) sintered at 1250 °C for 2h. Insets show an expanded view of sample microstructure.

In order to better understand the influence of these dopants on the mineralization and strength degradation of TCP, this study was carried out in a 3 pronged approach: 1) Apatite formation was measured and confirmed by ATR-IR and SEM. 2) The dissolution and mineralization kinetics were examined by use of AAS for Ca^{2+} ion release and by monitoring the change in weight of the samples over the 12 week study. 3) Samples were tested for compressive strength at 2, 4, 8 and 12 weeks in SBF.

The weight change of samples after submersion in SBF for 2, 4, 8 and 12 weeks was recorded and is presented in Figure 3. All samples gained weight over the 12 week period, indicating an affinity for apatite formation rather than dissolution. The doped samples presented the greatest weight gain at 1.41 ±0.09%, while the pure samples exhibited a gain of 0.34 ±0.15%. The weight change in TCP ceramics is controlled by two primary mechanisms: (i) weight loss due to dissolution and (ii) weight gain due to apatite formation [8]. SEM micrographs of the samples after 12 weeks in SBF are shown in Figure. 4. Significant apatite formation was observed on the surface of both compositions, however a greater degree of mineralization was observed in the doped samples. It can be seen that the surface of the doped sample is completely covered in near homogenous apatite layer, while the pure samples still show evidence of rod like apatite formation. Figure 5 shows the result of the ATR-IR analysis after 12 weeks in SBF. All samples show typical P-O bending bands at 469, 567, 600, 961 and 1041. Both samples also showed weak C-O bands at 873 and 1417. These bands confirm the presence of an apatite layer on the surface of the compacts after 12 weeks in SBF. The ATR-IR spectra showed higher intensities for $[PO_4]^{4-}$ at wavenumbers 1041, 961, 600, 567 and 469 for doped samples when compared to their pure counterparts, indicating a higher degree of mineralization.

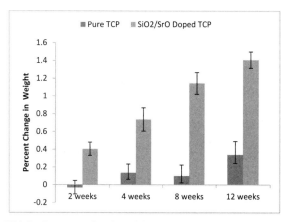

Figure 3. Weight change as a function of time for pure TCP and SiO$_2$/SrO doped TCP in simulated body fluid.

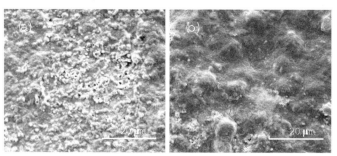

Figure 4. SEM micrographs showing in vitro mineralization of pure TCP (a) and SiO$_2$/SrO doped TCP (b) after 12 weeks in SBF

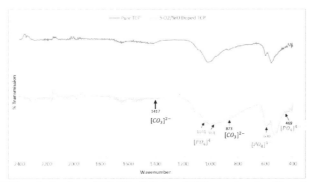

Figure 5. FTIR analysis of pure TCP SiO$_2$/SrO doped TCP samples after 12 weeks in simulated body fluid.

Figure 6 shows the collective change in Ca^{2+} concentration in the SBF as a function of incubation time. Over the first two weeks of the study, pure TCP exhibited rapid dissolution followed by a three week period of near steady state behavior and then a gradual increase in dissolution for the duration of the study. The SiO$_2$/SrO samples demonstrated significantly slower resorption kinetics than its pure TCP counterparts and had two short periods of equilibrium at 4 weeks and 5 weeks. Overall, the dissolution over the 12 week period was steady and nearly linear. AAS for total Ca^{2+} ions released further confirms all outcomes that apatite formation favors the doped samples. Results indicated that more Ca^{2+} ions were released from pure TCP samples in the SBF solution over the course of 12 weeks when compared to the doped samples. The difference in release kinetics can be attributed to the distortion in the crystal structure caused by the addition of the dopants [28]. The effect of adding dopants was an overall lattice contraction effect, which results in the hindering of Ca^{2+} release from the samples [27,29,30].

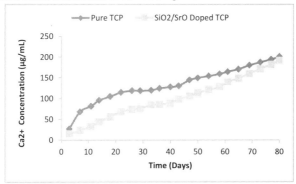

Figure 6 Total Ca^{2+} release as a function of time from pure TCP and SiO$_2$/SrO doped TCP in simulated body fluid.

Addition of SiO_2 and SrO decreased the compressive strength of the samples. Initially, as demonstrated in Figure 7, at week 0, the compressive strength of pure TCP (318 ± 30 MPa) was nearly double that of the SiO_2/SrO doped TCP (170 ± 17 MPa). The drastic difference in initial strength can be credited to the combination of increased grain size, grain size non-uniformity and amorphous glassy phase formation due to liquid phase sintering of the doped samples. At week 2, the pure TCP samples decreased in compressive strength (197 ± 41 MPa), while the doped samples remained constant. The weight of the pure samples only decreased at week 2, suggesting that the samples were favoring dissolution rather than mineralization kinetics. This is also confirmed by the AAS readings for Ca^{2+} ion concentration, where the pure TCP samples exhibited rapid release of calcium over the course of the first two weeks before reaching a state of quasi-equilibrium at weeks 3 and 4. The weight of the doped samples increased at week 2, offsetting any effects of dissolution and allowing a stable strength degradation profile. At week 4, the pure TCP samples had a significant increase in mechanical strength from week 2 (316 ± 41 MPa). There was an increase in weight of the samples while AAS reveals that apatite formation kinetics were favored over weeks 3 and 4. Combined, this can explain the strength increase. The SiO_2/SrO doped samples demonstrated no significant strength loss over the course of the experiment. Strength degradation profiles are controlled in large part by the balance of dissolution and mineralization kinetics. Increased apatite formation favors slow strength degradation as it helps to fill pores and voids in the sample, provide support to the structure and increase failure stresses [9]. AAS showed a fairly constant release of Ca^{2+} and the weight of the pure samples only gradually increased indicating dominating dissolution kinetics for these weeks. The doped samples continued to show significant weight increase and demonstrated a constant Ca^{2+} ion release over weeks 4, 8 and 12. The Ca^{2+} ion concentration was consistently much less than measured in the pure TCP samples. The compressive strength between weeks 2 and 12 did not very, signifying that there was a good balance between the rate at which the compact was dissolving and the rate at which new apatite was forming.

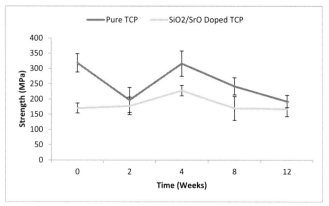

Figure 7 Compressive strength of pure TCP and SiO_2/SrO doped TCP over the course of 12 weeks in SBF

Conclusions

In this study, the influence of the addition of SiO_2 and SrO on the biodegradability of TCP ceramics was investigated. The presence of dopants in β-TCP suppressed the phase transition from β- to α-TCP. Doped TCP showed highest weight gain among the compositions, suggesting good apatite growth. SEM micrographs and ATR-IR confirmed enhanced apatite formation on doped samples when compared to pure TCP samples. Pure TCP underwent initial rapid dissolution behavior and had strength loss by week 2. AAS and weight change measurements confirmed that during this period the dissolution kinetics were favored over mineralization kinetics. At week 4 the strength of pure samples increased as a result from high mineralization rates. The strength degradation profile of pure TCP for weeks 8 and 12 showed a gradual decrease in overall strength, confirming that dissolution kinetics was favored during the time frame. Doped TCP samples had less strength at all time points, but remained constant throughout the 12 week study, suggesting that dissolution and mineralization kinetics were closely matched.

Acknowledgments
Authors would like to acknowledge financial support from the National Institutes of Health, NIBIB (Grant # NIH-R01-EB-007351).

References

[1] S. Raynaud , E Champion, J.P. Lafon, D. Bernache-Assollant, Calcium phosphate apatites with variable Ca/P atomic ratio III. Mechanical properties and degradation in solution of hot pressed ceramics, *Biomaterials*, **23**, 1081-89, (2002).

[2] F. Wua, J. Weia, H. Guoa, F. Chena, H. Honga, C. Liu, Self-setting bioactive calcium-magnesium phosphate cement with high strength and degradability for bone regeneration, *Acta Biomateriala*, **4**, 1873-84, (2008).

[3] M. Nagano, T. Nakamura, T. Kokubo, M. Tanahashi, M. Ogawa. Differences of bone bonding ability and degradation behaviour in vivo between amorphous calcium phosphate and highly crystalline hydroxyapatite coating, *Biomaterials*, **17**, 1771-77, (1996).

[4] M. Mehta, P. Strube, A. Peters, C. Perka, D. Hutmacher, P. Fratzl, G.N. Duda. Influences of age and mechanical stability on volume, microstructure, and mineralization of the fracture callus during bone healing: Is osteoclast activity the key to age-related impaired healing?, *Bone*, **47**, 219-28, (2010).

[5] N. Little, B. Rogers, M. Flannery. Bone formation, remodeling and healing, Surgery (Oxford) **29**, 141-5, (2011).

[6] Y.X. He, G. Zhang, X.H. Pan, Z. Liu, L.Z. Zheng, C.W. Chan, K.M. Lee, Y.P. Cao, G. Li, L. Wei, L.K. Hung, K.S. Leung, L. Qin, Impaired bone healing pattern in mice with ovariectomy-induced osteoporosis: A drill-hole defect model, *Bone*, **48**, 1388-1400, (2011).

[7] A.M. Pietak, J.W. Reid, M.J. Stott, M. Sayer, Silicon substitution in the calcium phosphate bioceramics, *Biomaterials*, **28**, 4023-4032, (2007).

[8] A. Bandyopadhyay, S. Bernard, W. Xue, S. Bose, Calcium phosphate-based resorbable ceramics: Influence of MgO, ZnO, and SiO2 dopants, *Journal of the American Ceramic Society*, **89**, 2675-2688, (2006)

[9] W. Xue, K. Dahlquist, A. Banerjee, A. Bandyopadhyay, S. Bose, Synthesis and characterization of tricalcium phosphate with Zn and Mg based dopants, *J Mater Sci: Mater Med*, **19**, 2669-2677, (2008).

[10] K. Qiu, X.J. Zhao, C.X. Wan, C.S. Zhao, Y.W. Chen, Effect of strontium ions on the growth of ROS17/2.8 cells on porous calcium polyphosphate scaffolds, *Biomaterials*, **27**, 1277-1286, (2006).

[11] R. Jugdaohsingh, K. Tucker, N. Qiau, L. Cupples, D. Kiel, J. Powell, Dietary silicon intake is positively associated with bone mineral density in men and premenopausal women of the Framingham Offspring cohort, *J Bone Miner Res*, **19**, 297–307, (2004).

[12] F. Nielson, R. Poellot, Dietary Si affects bone turnover and differentiation in overiectomized and sham operated growing rats, *J Trace Elements Exp Med*, **17**, 137–49. (2004).

[13] M. Hott, Short term effects of organic silicon on trabecular bone in mature ovariectomized rats, *Cal Tiss Inter*, **53**, 174–9, (1993).

[14] F.H. Perera, F.J. Martı́nez-Va´zquez, P. Miranda, A.L. Ortiz, A. Pajares, Clarifying the effect of sintering conditions on the microstructure and mechanical properties of β-tricalcium phosphate, *Ceramics International*, **36**, 1929-35, (2010).

[15] T. Kanazawa, T. Umegaki, K. Yamashita, H. Monma, T. Hiramatsu. The effecs of additives on sintering and some properties of calcium phosphates with various Ca/P ratios, *J. Mater. Sci.*, **26**, 417-22, (2010).

[17] J.Y. Reginster, O. Bruyère, A. Sawicki, A. Roces-Varela, P. Fardellone, A. Roberts, J.P. Devogelaer, Long-term treatment of postmenopausal osteoporosis with strontium ranelate: Results at 8 years, *Bone*, **45**, 1059-64, (2009).

[18] E.M. Brown, Is the calcium receptor a molecular target for the actions of strontium on bone? *Osteoporos Int*, **14**, S25–34, (2003).

[18] J. Coulombe, H. Faure, B. Robin, M. Ruat, In vitro effects of strontium ranelate on the extracellular calcium-sensing receptor, *Biochem Biophys Res Commun*, **323**, 1184–90, (2004).

[19] T. Brennan, M.S. Rybchyn, A.D. Conigrave, R.S. Mason, Strontium ranelate effect on proliferation and OPG expression in osteoblasts, *Calcif Tissue Int*, **78**, 129, (2006).

[20] P.J. Kostenuik, V. Shaloub, Osteoprotegerin: a physiological and pharmacological inhibitor of bone resorption, *Curr Pharm Des*, **7**, 613–35, (2001).

[21] W. Xue, J.L. Moore, H.L. Hosick, S. Bose, A. Bandyopadhyay, W.W. Lu. Osteoprecursor cell response to strontium-containing hydroxyapatite ceramics. *J Biomed Mater Res*, **79A**, 804–14, (2006).

[22] S.S. Banerjee, M.S. Tarafder, N.M. Davies, A. Bandyopadhyay, S. Bose, Understanding the Influence of MgO and SrO Binary Doping on Mechanical and Biological Properties of β-TCP Ceramics, *Acta Biomaterialia*, **6**, 4167–74, (2010).

[23] W. Suchanek, M. Yashima, M. Kakihana, M. Yohimura. Hydroxyapatite ceramics with selected sinting additives, *Biomaterials*, **18**, 923-33, (1997).

[24] H.S. Ryu, H.J. Youn, K.S. Hon, B.S. Chang, C.K. Lee, S.S. Chung, An improvement of sintering property of B-tricalcium phosphate by addition of calcium pyrophosphate, *Biomaterials*, **23**, 909-14, (2002).

[25] G. Renaudin, P. Laquerriere, Y. Filinchuk, E. Jallotd, J.M. Nedelec, Structural characterization of sol–gel derived Sr-substituted calcium phosphates with anti-osteoporotic and anti-nflammatory properties, *J Mater Chem*, **18**, 3593–600, (2008).

[26] S.R. Kim, J.H. Leeb, Y.T. Kimb, D.H. Riua, S.J. Junga, Y.J. Leea. Synthesis of Si, Mg substituted hydroxyapatites and their sintering behaviors, *Biomaterials*, **24**, 1389–98, (2003).

[27] X. Yin, L. Calderin, M.J. Stott, M. Sayer. Density functional study of structural, electronic and ibrational properties of Mg- and Zn-doped tricalcium phosphate biomaterials, *Biomaterials*, **23**, 4155–63, (2002)

[28] S. Nath, K. Biswas, K. Wang, R.K. Bordia, B. Basu. Sintering, phase stability, and properites of calcium phosphate-mullite composites, *J. Am. Ceram. Soc.*, **93**, 1639-49, (2010).

[29] J. Christoffersen, M. Christoffersen, N. Kolthoff, O. Barenholdt. Effects of strontium ions on growth and dissolution of hydroxyapatite and on bone mineral detection, *Bone*, **20**, 47-54, (1997).

[30] G. Renaudin, P. Laquerriere, Y. Filinchuk, E. Jallot, J. Nedelec, Structural characterization of sol-gel derived Sr-substituted calcium phosphates with anti-osteoporotic and anti-inflammatory properties, *Journal of Materials Chemistry*, **18**, 3593-3600, (2008).

INHIBITION OF LOW-TEMPERATURE DEGRADATION AND BIOCOMPATIBILITY ON SURFACE OF YTTRIA-STABILIZED ZIRCONIA BY ELECTRIC POLARIZATION

Naohiro Horiuchi, Norio Wada, Miho Nakamura, Akiko Nagai, Kimihiro Yamashita

Institute of Biomaterial & Bioengineering, Tokyo Medical and Dental University,
2-3-10 Kanda-Surugadai, Chiyoda-ku, Tokyo, Japan

ABSTRACT

We found that electric polarization decelerate the tetragonal to monoclinic phase transformation on yttria-stabilized zirconia (YSZ) under wet environment. The electric polarization was confirmed via measurements of thermally stimulated depolarization currents, and stored charges were increased with increasing applied voltages of polarization procedure. After exposure to hot water, the monoclinic phase was formed on the surfaces of both polarized and non-polarized YSZ. The monoclinic ratios on the polarized surfaces were less than half of the ratios of non-polarized surfaces. Meanwhile, contact angles of water on the surface are decreased with increasing applied voltages. The decrease of contact angles indicates the surface becomes hydrophilic. This deceleration of degradation is due to some electric barrier layer consist with the hydroxyl groups formed at the surface of YSZ via the electric polarization. In addition, the polarized surfaces have been proved to have same biocompatibility as the non-polarized surfaces.

INTRODUCTION

Zirconia ceramics are one of the most important materials in its broad usage from structural materials to electronic materials. Zirconia is also used as biomedical materials: artificial hip joint cup and heads due to its toughness and biocompatibility.[1] Zirconia ceramics are used for dental restoration, because of the high mechanical toughness and biocompatibility.[2,3] In the 1990s, partially stabilized zirconia was developed for usage in a material of all-ceramic restorations, accompanying with development of CAD/CAM technique. However, so-called low temperature degradation (LTD) is well known; Tetragonal zirconia become deteriorate in water present conditions.[4,5] Stabilized tetragonal phase gradually transforms to monoclinic phase in wet condition at relatively moderate temperature e.g. room temperature to 200 °C. The transformation begins from the surface and penetrates into inside of the tetragonal zirconia, and the formed monoclinic phase impairs the strength of the tetragonal zirconia.

Improvement of surfaces is a crucial and inevitable issue in biomedical material science. Few elements are allowed to use for biomaterial in respect to biocompatibility, because many elements are toxic for human body. Obviously, properties of surfaces are generally dominated by properties of bulk materials roughly determined by the chemical composition. Thus, studies trying to improve the surface properties by changing the composition are difficult for the restriction of selecting the compositions. Our group have developed an innovative surface modification method via formation of electric polarization in hydroxyapatites.[6-9] In the procedure of electric polarization, any chemicals or doping of elements had not been used. In present study, we have found that the electric polarization improved the durability against the LTD. The electric polarization decelerated the tetragonal to monoclinic phase transformation. The huge surface charges were induced by the polarization, which was confirmed by thermally stimulated depolarization current. Measurement of contact angles of water indicated that the surface charges generate the electrostatic force.

EXPERIMENTAL

Sintered disks of yttria stabilized zirconia (YSZ) were fabricated from yttria doped (3 mol%) zirconia powders (Tosoh, TZ-3YB). The powders were uniaxially pressed into disks and sintered at 1450 °C for 2 hours. Thermally stimulated depolarized currents (TSDC) were measured with Keithley 6514. The rate of temperature elevation was 5 °C min^{-1}. The charges induced by polarization were calculated from TSDC spectra using the equation:

$$Q = (1/\beta) \int J(T) dT \qquad (2)$$

where $J(T)$ the measured current density at temperature T, and β is the heating rate. Sintered disks of 10 mm in diameter and 1 mm of thickness are polarized as follows: disks are sandwiched between a pair of platinum films. DC voltages were applied between opposite electrodes at elevated temperature of 200 °C for 30 minutes. The voltage was applied until the specimen was cooled to room temperature. The measurements of the contact angles were performed using commercial contact angle meter (CA-X, Kyowa Interface Science, Japan) with drops of de-ionized water placed on the disk surface. The contact angle was measured at least 10 times with the same sample in an ambient air, and the average value was used. Phase transformation, tetragonal to monoclinic, was evaluated with X-ray diffraction measurement (D8 advance, BrukerAXS). The fractions of monoclinic phases were calculated using Garvie-Nicholson's formula.[10]

Adhesion assay was performed using the osteoblastic cells (MC3T3-E1 cell line) which obtained from the RIKEN Cell Bank (Tsukuba, Japan). The cells were cultured until reaching 70 % confluency. The cells were detached and seeded at a density of 1×10^4 cells onto the polarized and non-polarized surfaces, which were sterilized with 70% ethanol. At 45 min after seeding, the cells on the surfaces were fixed with 4% paraformaldehyde and permeabilized with 0.1% Triton X-100 in phosphate buffered saline (PBS). Following washing with PBS, the YSZ were incubated in Alexa-conjugated goat immunoglobulin in a blocking solution containing rhodamine phalloidin solution for 30 min at RT. After 4,6-diamino- 2-phenylindole (DAPI) staining, the fluorescence were observed using a fluorescence microscope (OLYMPUS IX71).

RESUTLS AND DISCUSSIONS

TSDC spectra of the polarized YSZ are shown in Figure 1(a). These spectra were measured after fabrication of sputtered platinum electrodes on the surfaces. The results indicate the dependence on the applied electric field in polarization procedure. The increase of the electric field gives higher current peaks, suggesting charges induced on the surfaces are increased with increasing polarization voltage. Each spectrum has two peaks located at high and low temperature: 300 °C and 400-500°C. Separated peaks in TSDC spectra indicate the polarization of YSZ consists of at least two kinds of depolarization elements. Figure 1(b) shows the heights of peaks located at high and low temperature as function of the applied electric field in polarization procedure. The peak height located at low temperature is linearly increasing with increasing polarization voltages, while the peak height of peaks located at high temperature are exponentially increasing. The linear dependence on the applied voltages suggests that the depolarization currents are attributed to electrical dipole produced by displacement of ions.[11] On the other hand, exponential increase indicated that the current is due to electrons trapped electrons.[12] The trapped electrons are also confirmed visually. Portions of the positively charged surface (P-surfaces: which were contacted to the negative electrodes in the polarization procedure) become black after polarization. The blackened area increased with the applied voltages in polarization. The blackening is caused by the electrons induced from electrode to YSZ.[13,14] The trapped electrons absorb the visible light. Consequently, the second peaks located at high temperature are considered to be attributed to the trapped charges. Although the polarization consists of electrical dipole and trapped electrons, only the dipole element contributes to the formation of surface charges. This is because the trapped charges do not exist on the negatively charged surfaces (N-surfaces). In addition, in the P-surfaces, the trapped electrons decrease the positive charges, suggesting that trapped electrons do not contribute. The monoclinic fraction on P surfaces was larger than that of N surfaces. This is explained that the trapped electron injected from the negative electrode in polarization procedure decreases the surface charges. The surface charges should be calculated from only the peaks located at lower temperature in the TSDC spectra. The surface charges for applied voltage of 500 V/cm are estimated to be 2.0 ± 0.1 mC cm^{-1}. The value is extremely large, comparing with several tens of μC cm^{-1} of spontaneous polarization of ferroelectric material with perovskite structure.

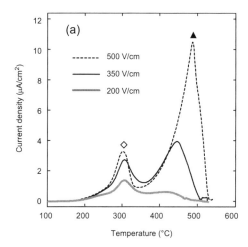

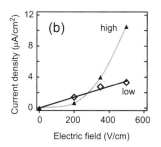

Figure 1. (a) Thermally stimulated depolarization current (TSDC) spectra for different voltages. (b) Peak heights as a function of applied electric fields. Gray and black lines indicate peaks located at high and low temperature.

The electrical dipoles are formed with the migration of oxygen vacancies. YSZ is well known as an oxygen ion conductor. The oxygen ions migrate via the oxygen vacancies formed by substitution of trivalent ions Y^{3+} for Zr^{4+}:

$$Y_2O_3 \xleftrightarrow{\quad ZrO_2 \quad} 2Y'_{Zr} + 3O_O^X + V_O^{\bullet\bullet}. \tag{2}$$

These notations express an ion or a vacancy at a lattice site with effective charge according to Kroger-Vink notation. Under a dc field of polarization procedure, O^{2-} ions migrate via oxygen vacancies. The migration gives rise to a local distribution of concentration of O^{2-} ions. As a result, electrical dipoles of space charge polarization are formed.

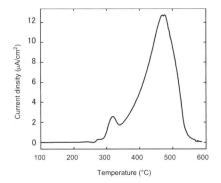

Figure 2. TSDC spectrum; Sample was polarized under 10 kV/cm. TSDC was measured after fabrication of Au electrode with vapor deposition.

In case of applying for a surface modification of biomaterials, we cannot fabricate the sputtered electrodes. The polarization procedure was performed without electrodes.

TSDC spectra in Figure 1(a) are measured using samples of polarized and depolarized after fabrication of sputtered Pt electrodes. On the contrary, Figure 2 shows TSDC spectrum using sample of polarized at the electric field of 10 kV/cm without electrodes. After polarization procedure was performed, Au electrodes were fabricated with vapor deposition before the measurement of TSDC. The figure has similar shape and comparable current to the TSDC spectra in Figure 1, indicating that there are no differences between polarization with and without electrodes.

The LTD is occurred in presence of water. Therefore, an interaction between water and surfaces of YSZ is expected to be play big role in progress of transformation. Wettability of the surfaces was characterized by contact angle measurement. Figure 3 shows contact angles as a function of applied electric fields in polarization treatment. The caption of P-surface denotes positively charged surfaces which contacted to negative electrodes in the polarization procedure, and N-surface denotes negatively charged surfaces which contacted to positive electrodes. The contact angles decreased as the applied electric fields were increased. The angles on N-surface started to decrease at lower electric fields than that on P-surface. The decreases of contact angles with increasing the applied polarization is explained by electrowetting:[15,16] The electric charges on the surface decreases the contact angles. The more charges on the surfaces give more small contact angles. Consequently, it is concluded that the surfaces polarized at higher voltage have more surface charges and electrostatic forces.

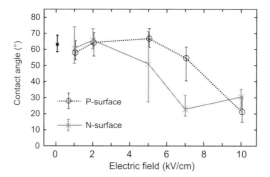

Figure 3. Contact angel as a function of applied voltage. The caption of P-surface denotes positively charged surfaces which contacted to negative electrodes in the polarization procedure, and N-surface denotes negatively charged surfaces which contacted to positive electrodes.

Accelerated degradation tests were performed in order to evaluate the inhibition of LTD by electric polarization. Sintered YSZ disks of 0.1 mm in thickness were prepared for the tests. The prepared disks are electrically polarized without electrodes and annealed in water at 120 °C as accelerated degradation. The fractions of produced monoclinic phase on the surfaces were calculated using Garvie-Nicholson's formula[10] from measured X-ray diffraction spectra. Figure 3 shows monoclinic fractions on the surfaces after accelerated test. The horizontal axis in Figure 1 indicates applied voltage in polarization treatment. The caption of P-surface denotes positively charged surfaces which contacted to negative electrodes in the polarization procedure, and N-surface denotes negatively charged surfaces which contacted to positive electrodes. The monoclinic fractions of no-polarized surfaces are indicated at 0 V in applied voltages. The monoclinic fractions are decreased with increasing the applied voltages. While monoclinic fractions of no-polarized surfaces are about 0.6, that of N-surface applied voltage of over 7 kV/cm^{-1} are less than 0.3. The phase transformation from tetragonal to monoclinic phase was decelerated in surfaces polarized at high voltage. Especially in N-surface, the transformation was significantly inhibited.

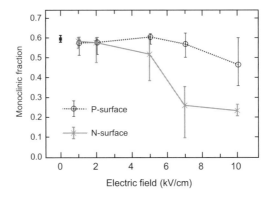

Figure 4. Monoclinic phase fractions as a function of applied voltage.

Figure 3 and 4 have quite similar appearance, indicating the LTD and contact angles have some correlation. The relationship between the measured contact angles and monoclinic phase fractions are shown in Figure 5. The horizontal and vertical axes indicate contact angels and monoclinic fractions, respectively. They have strong correlation in both of surfaces. Especially in the N-surfaces, the correlation is clearly observed. The lower contact angle gives lower monoclinic fraction. As mentioned above, lower contact angle indicates the higher electrostatic forces. This correlation suggests that the higher electrostatic force on the surface inhibits the more effectively phase transformation.

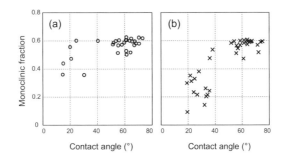

Figure 5. Correlation between contact angles and monoclinic fractions. (a) P-surface, (b) N-surface.

Although the mechanism of LTD is not all revealed, prevailing explanation in the literature is the species originated from waters invading into the YSZ causes the transformation from metastable tetragonal phase to monoclinic phase.[4,5] The process of the LTD is proposed as follows: First, H_2O reacts with O^{2-} on the YSZ surface and forms hydroxyl ions. Second, OH^- ions penetrate into the inner part of YSZ by grain boundary diffusion, and diffused hydroxyl ions fill the oxygen vacancies. The decrease of oxygen vacancies result in the unstability of tetragonal phase, because the stability of tetragonal phase is originated from formation of oxygen vacancies due to substitution of trivalent ions Y^{3+} for Zr^{4+}.

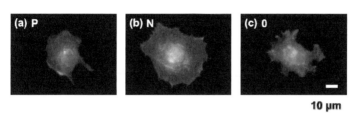

10 µm

Figure 6. Morphology of the adhered osteoblastic cells on YSZ pellets. Merged fluorescence images of the cells cultured on the YSZ pellets for 45 min. (a) P-surfaces, polarized at 700 V, (b) N-surfaces, polarized at 700 V, (c) non-polarized.

Figure 6 shows the morphology of the adhered osteoblastic cells on YSZ pellets. The images in Figure 6 is merged fluorescence images of the cells cultured for 45 min: (a) P-surfaces, polarized at 700 V, (b) N-surfaces, polarized at 700 V, and (c) non-polarized. Typical adhered cells had a spindle or fanlike spreading shape on all kinds of the surfaces of YSZ. The results indicate that the cells well adhere on the polarized surfaces and the polarized surfaces have comparable biocompatibility to the non-polarized surfaces.

We propose two kinds of mechanism of inhibiting the progress of LTD. First, electrostatic repulsive forces induced by the polarization keep away from the surface of YSZ. The electrostatic force interferes with the diffusion of OH^- ions. As a result, the progress of the LTD is inhibited. Next hypothesis is that modified surface by the induced surface charges have high chemical stability and the reaction of H_2O into OH^- ions are decelerated. For examples of the hydrophilic in TiO_2, the hydroxyl ion concentration on surfaces is increased and rearranged by UV irradiation, which are metastable state in thermodynamics.[17,18] On the surface of polarized YSZ, it is assumed that hydroxyl ion rearrangement is induced by the polarization. The metastable state prevents invasion of OH^- ions into the inner of YSZ, resulting in inhibition of the LTD.

CONCLUSION

We demonstrated electric polarization inhibited the low-temperature degradation in YSZ. Huge surface charges were induced by the polarization, which were confirmed by the TSDC measurements. The induced surface charges resulted in the hydrophilic surfaces. In addition, the more hydrophilic surfaces have the more durability for degradation. We proposed the mechanism of the inhibition of the LTD: metastable surface charge layer, which is consisted of hydroxyl ions, prevented invasions of water into the inner of YSZ. Furthermore, the polarized surfaces showed similar biocompatibility to the non-polarized surfaces.

REFERENCES

[1]B. Cales, "Zirconia as a Sliding Material: Histologic, Laboratory, and Clinical Data", *Clin. Orthop. Relat. Res.*, **379**, 94-112 (2000).

[2]A. J. Raigrodski, "Contemporary materials and technologies for all-ceramic fizzed partial dentures", J. Prosthet. Dent., **92**, 557-562 (2004).

[3]I. Denry and J. R. Kelly, "State of the art of zirconia for dental applications", *Dent. Mater.*, **24**, 299-307 (2008).

[4]X. Guo, "Property degradation of tetragonal zirconia induced by low-temperature defect reaction with water molecules", *Chem. Mater.*, **16**, 3988-3994 (2004).

[5]J. Chevalier, L. Gremillardw, A. V. Virkar and D. R. Clarke, "The tetragonal-monoclinic transformation in Zirconia: Lessons learned and future trends", *J. Am. Ceram. Soc.*, **92**, 1901-1920 (2009).

[6]K. Yamashita, Noriyuki Oikawa and T. Umegaki, "Acceleration and deceleration of bone-like crystal growth on ceramic hydroxyapatite by electric poling", *Chem. Mater.*, **8**, 2697-2700 (1996).

[7]S. Nakamura, H. Takeda and K. Yamashita, "Proton transport polarization and depolarization of hydroxyapatite ceramics", *J. Appl. Phys.*, **89**, 5386 (2001).

[8]M. Ueshima, S. Nakamura and K. Yamashita, "Huge, millicoulomb charge Storage in ceramic hydroxyapatite by bimodal electric polarization", *Adv. Mater.*, **14**, 591-595 (2002).

[9]T. Kobayashi, S. Nakamura and K. Yamashita, "Enhanced osteobonding by negative surface charges of electrically polarized hydroxyapatite", *J. Biomed. Mater. Res.*, **57**, 477-484 (2001).

[10]R. C. Garvie and P. S. Nicholson,"Phase analysis in zirconia systems", *J. Am. Ceram. Soc.*, 55, 303-305 (1972).

[11]R. M. Neagu, E. R. Neagu, I. M. Kalogeras and A. Vassilikou-Dova, "Evaluation of the dielectric parameters from TSDC spectra: application to polymeric systems", *Mater. Res. Innov.*, **4**, 115-125 (2001).

[12]G.M. Sessler, "Electrets", 3rd Edition, Vol. **1**, Laplacian Press, (1999).

[13]J. Janek and C. Korte, "Electrochemical blackening of yttria-stabilized zirconia – morphological instability of the moving reaction front", *Solid State Ionics*, **116**, 181-195 (1999).

[14]X. Guo, "Effect of DC voltage on the microstructure and electrical properties of stabilized-zirconia", *Solid State Ionics*, **99**, 143-151 (1997).

[15]R. Digilov, "Charge-induced modification of contact angle: the secondary electrocapillary effect", *Langmuir*, 16, 6719-6723 (2000).

[16]M. W. J.Prins, W. J. J. Welters and J. W. Weekamp, "Fluid control in multichannel structures by electrocapillary pressure", *Science*, **291**, 277-280 (2001).

[17]A. Nakajima, S. Koizumi, T. Watanabe and K. Hashimoto, "Photoinduced amphiphilic surface on polycrystalline anatase TiO_2 thin films", *Langmuir*, **16**, 7048 (2000).

[18]N. Sakai, A. Fujishima, T. Watanabe and K. Hashimoto, "Quantitative evaluation of the photoinduced hydrophilic conversion properties of TiO_2 thin film surfaces by the reciprocal of contact angle", *J. Phys. Chem. B*, **107**, 1028 (2003).

BIOMATERIALS FOR THERAPEUTIC GENE DELIVERY

Eric N. James[1], Bret D. Ulery[1, 2], and Lakshmi S. Nair[1, 2, 3*]

[1]Department of Orthopedic Surgery, University of Connecticut Health Center, Farmington, CT 06030
[2]Institute of Regenerative Engineering, University of Connecticut Health Center, Farmington, CT 06030
[3]Department of Chemical, Materials and Biomolecular Engineering, University of Connecticut, Storrs, CT 06268
*To whom correspondence should be addressed:

ABSTRACT
        The most widely studied gene therapy strategies are based on traditional viral delivery methods. However, viral methods suffer several drawbacks, namely toxicity and insufficient delivery. Thus, non-viral gene transfer with low toxicity and more efficiency has been the focus of recent investigation. Recent studies showed the exceptional promise of biocompatible and biodegradable material delivery systems as non-viral vectors for gene therapy. Biomaterial-DNA complexes are currently been investigated for the treatment of a number of genetic diseases. The objective of this paper is to review the latest advances on biomaterials based gene therapeutic approaches.

INTRODUCTION
        Gene therapy is (defined as) a method to provide an individual with the genetic information required for producing specific therapeutic proteins to correct or modulate a disease or disorder through altered protein expression. Specifically, gene therapy offers the potential to overcome problems associated with the direct administration of protein-based therapies, namely low bioavailability, systemic toxicity, in vivo instability, high hepatic and renal clearance rates, and high cost of manufacturing products [1]. The success of gene therapy largely depends, however on the availability of suitable delivery vehicles. In past years research has focused on trying to introduce genes directly into cells, and has concentrated on diseases caused by single-gene defects, such as cystic fibrosis, hemophilia, muscular dystrophy and sickle cell anemia. This has proven difficult due to limited capacity to transport large sections of DNA to the correct site on the gene. Today, most gene therapy studies are aimed at cancer and hereditary diseases linked to a genetic defect. More specifically, research areas are focusing on angiogenesis, skeletal tissue regeneration, wound healing, several cancers and diabetes. Currently, gene transfer technologies or DNA delivery methods can be classified into three general types; mechanical transfections, electrical techniques, and vector assisted delivery systems.

CONVENTIONAL GENE DELIVERY TECHNIQUES

Mechanical and Electrical Based Delivery
        Mechanical and electrical strategies for introducing naked DNA into cells include microinjection, particle bombardment, pressure, and electroporation. Microinjection is a highly efficient technique in which one cell at a time is targeted for DNA transfer. This refers to the process of using a glass micropipette to insert foreign DNA into a living cell (a cell, egg, oocyte, embryos of animals) through a glass micropipette. It is a simple mechanical process in which a needle roughly 0.5 to 5 micrometers in diameter penetrates the cell membrane and/or the nuclear envelope. The desired contents are then injected into the sub-cellular compartment and the needle is removed. However, the precision needed requires significant time which is both laborious and expensive. Ironically, the major drawback of this approach is the very definition of the method; in other words because only one cell at

a time is transfected this limits its use as a therapeutic treatment strategy. Transfer of gene using gold microparticles is another method wherein particle bombardment equipment such as the gene gun is used. In this method, DNA-coated microparticles composed of metals such as gold or tungsten are accelerated to high velocity to penetrate cell membranes or cell walls. Bombardment is extensively used in DNA vaccination, in which limited local expression of delivered DNA (in cells of the epidermis or muscle) is adequate to achieve immune response. However, because of the difficulty in controlling the DNA entry pathway, this procedure is applied mainly to adherent cell cultures and has not yet been widely used systemically. Electroporation is a method that utilizes high-voltage electrical current to facilitate DNA transfer. This involves the application of high voltage pulses to induce skin perturbation, leading to transient pores which may account for the increase in skin permeability [2] . High voltages ($\geq$100 V) and short treatment durations (milliseconds) are most frequently employed. This technology has been mostly used to enhance the skin and muscle permeability of molecules with differing lipophilicity and size (i.e. small molecules, proteins, peptides and oligonucleotides) including biopharmaceuticals with a molecular weight greater than 7kDA [3] . Genetronics Inc. (San Diego, CA, USA) has developed a prototype electroporation transdermal device that has been tested with various compounds with a view to achieving gene delivery, improving drug delivery, and aiding the application of cosmetics. Other transdermal devices based on electroporation have also been proposed [2] however; more clinical information on the safety and efficacy of the technique is required to assess future commercial prospects of these devices. While electroporation has been widely successful in cell culture studies, high cell mortality associated with the process is a significant drawback that limits its clinical use [4] . Furthermore, though gene transfer has proven to be efficient for both mechanical and electrical transfer methods, such methods are limited to local targets, namely the skin, muscle, or mucosal tissue. Other tissues require invasive surgeries, are extremely difficult to standardize in a clinical setting and therefore may raise significant challenges.

Viral Delivery Systems

Vector-assisted DNA/gene delivery systems can be classified into two types based on their origin; biologically (viral DNA) based delivery systems and chemically (non-viral) based delivery systems. In viral delivery systems, non-pathogenic attenuated viruses are used as delivery systems for genes or DNA molecules, especially plasmid vectors. Plasmid vectors are circular DNA constructs that are used as vehicles to transfer foreign genetic material into a cell. Plasmid vectors usually consist of an insert transgene or delivery gene and a larger sequence that serves as the backbone of the vector. Viral DNA-delivery vectors can be either RNA or DNA viruses. A typical plasmid vector contains a polylinker that can recognize several different restriction enzymes, an ampicillin-resistance gene (amp^r) for selective amplification, and a replication origin (ORI) for proliferation in the host cell (Figure 1). The vector is made from natural plasmids by removing unnecessary segments and adding essential sequences. To clone a DNA sample, the same restriction enzyme must be used to cut both the vector and the DNA sample. Therefore, a vector usually contains a sequence (polylinker) which can recognize several restriction enzymes so that it can be used for cloning a variety of DNA samples. A plasmid must also contain a drug-resistance gene for selective amplification. After the vector enters into a host cell, it may proliferate with the host cell. However, since the transformation efficiency of plasmids in E. coli is very low, most E. coli cells that proliferate in the medium would not contain the plasmids. Typically, antibiotics are used to kill E. coli cells which do not contain the vectors. The transformed E. coli cells are protected by the ampicillin-resistance gene (amp^r) which can express the enzyme, b-lactamase, to inactivate the antibiotic ampicillin. The four common categories of gene therapy viral vectors are retroviruses (RV), adenoviruses (AD), adeno-associated viruses (ADA), and Herpes simplex viruses (HSV). Gene expression using viral vectors has been achieved with high transfection efficiencies in tissues such as kidney, heart, muscle, eye, and ovary, as well as in a number of cell lines [5] . Viruses are currently used in more than 70% [6] of human clinical gene therapy trials and they have shown tremendous potential in the treatment of diseases such as muscular dystrophy, AIDS, and cancer. The only currently approved gene therapy treatment is Gendicine® which delivers its transgene using a recombinant adenoviral vector to treat head and neck squamous cell carcinoma and other tumors caused by p53 gene mutations [5].

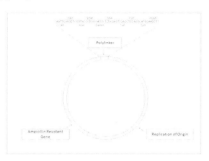

Figure 1. Diagram of a typical vector. Origin of Replication, Ampicillin Resitant gene and polylinker site are all present in most vectors

Retroviral vectors used in gene therapy are replication deficient, preventing their replication in the host cell and limiting their capacity to infect a single cell [7]. This characteristic, although essential for the safety of viral vectors in gene therapy, imposes restrictions on the amounts of virus that can be administered [7]. Retroviral-mediated delivery of therapeutic DNA has been widely used in clinical gene therapy protocols, including in the treatment of cancers (such as melanoma and ovarian cancer), adenosine deaminase deficiency-severe combined immune deficiency (ADA-SCID), and Goucher's disease [8]. Retroviral vectors are capable of transfecting high populations (45 - 95%) of primary human endothelial and smooth muscle cells which are generally extremely difficult to transfect [9].

Adenoviruses have been used in gene therapy strategies to treat metastatic breast, ovarian and melanoma cancers [10]. A strong host systemic immune response is a major issue with this therapy and contributes to the limited survival of adenoviral DNA in target cells resulting in a transient expression of the therapeutic gene. Interestingly, modified adenoviral vectors, which lack all viral genes, have been shown to facilitate delivery of up to 30 kilobase (Kb) pairs of a therapeutic DNA sequence with decreased toxicity [11]. This is an important advance since previous studies have been limited in the size of DNA segments to roughly ~4.7 kilobase that can be delivered to target cells. The use of adeno-associated viral (AAV) vectors provides an alternative to adenoviral vectors for gene therapy and a means for long-term gene expression with a reduced risk of adverse reactions upon administration. In the engineering of AAV vectors, most of the AAV genome can be replaced by the therapeutic gene, which significantly reduces potential adverse responses of the host to viral infection. However, the size of the therapeutic gene is limited to approximately 5 Kb [12]. Unfortunately, it has been demonstrated that adenoviruses in formulations may lose their potency after storage in commonly used pharmaceutical vials [1]. The HSV vector is a large and complex enveloped, double-stranded DNA virus that has the ability to carry large genes and, like AAV, can remain for prolonged amounts of time in transfected cells providing genetic material for long term expression of the therapeutic gene. Unfortunately, although HSVs are able to transfect many cell types. HSV offers a number of advantages over other viral vector systems. These advantages include: the broad host range that allows HSV to infect and replicate almost all cell lines; infectivity both in replicating and non-replicating cells such as neuronal cells; the potential for incorporating a large amount of foreign DNA; their control by anti-herpetic agents such as aciclovir (ACV) or ganciclovir (GCV); their ability to exist stably as episomes in neuronal cells, allowing long-lasting foreign gene expression; the use of mice, guinea pigs and monkeys as models due to the similarity in viral pathogenicity in these animals to that in humans; the determination of complete open reading frames and identification of disease-related viral genes. HSV vectors currently are limited in their use by vector toxicity. While strong immunogenicity and cell toxicity induced by HSV infection are major disadvantages for developing gene delivery vectors using HSV, they are beneficial when developing vaccine vectors or anti-cancer agents by HSV recombination. Although much more work is required to better understand the efficacy and safety issues of these recombinants before clinical use, the results from extensive pre-clinical and clinical trials have clearly demonstrated the potential of HSV recombinants for medical use, mainly as vaccine vectors and oncolytic viruses.

Current gene therapy is experimental and has not proved very successful clinically. Some have been life threatening. In 1999, gene therapy suffered a major setback when a patient died due to systemic organ failure caused by the adenovirus vector that had been used as a means of delivery [13]. In 2003 the FDA placed a temporary freeze on all gene therapy trials using retroviral vectors in blood stem cells when a child treated in a second experimental gene therapy trial developed leukemia-like symptoms. The FDA's Biological Response Modifiers Advisory Committee (BRMAC) removed the halt on retroviral gene therapy trials for the treatment of life-threatening diseases, adding more regulations to gene therapy approaches in 2003. Satisfactory translational delivery systems for patients have yet to be established. Future research is aimed at the development of low cost delivery technology with high transfection efficiency to do next generation gene therapies. In this review, we will examine the current progress and the clinical potential of biodegradable polymeric hydrogels as facilitators for gene delivery.

Non-viral Delivery Systems

Although much had been done to decrease the toxicity associated with viral gene delivery systems, non-viral delivery systems possess significant advantages over viral delivery systems. In particular, non viral delivery systems do not initiate a strong host immune response and are easy to formulate and assemble. These systems can be categorized as naked DNA delivery, lipid-based delivery, bioceramic-based delivery, and polymer-based delivery.

Naked DNA Delivery Systems

Naked DNA delivery consists of delivering DNA plasmids directly into the site of interest. These plasmids are designed to insert into the host cell genome allowing for the production of therapeutic products from the cells themselves. This approach is simple, cheap and has limited toxicity due to the lack of a secondary delivery vehicle. In addition, this strategy can be implemented through repeated injections similar to other pharmacological therapies. Naked DNA delivery has been used in gene therapy strategies for the treatment of bone cancer [14], diabetes [15], and heart disease [16]. While naked DNA delivery has shown promise, the limited uptake of plasmids, lack of trafficking to the nucleus and short retention of induced products severely limits the potential of this methodology for clinical practice. Moreover, naked DNA is highly susceptible to environmental degradation which greatly decreases its effectiveness. In order to protect DNA payload and enhance plasmid DNA transfection, a number of delivery vehicles have been studied including lipid-based, bioceramics-based and polymer-based vehicles.

Lipid-Based Delivery Systems

Lipids are naturally occurring fatty acids that are important structural elements of cells. Due to their natural occurrence, ability to be easily modified, compatibility with DNA and capacity to merge with cell membranes, they have been extensively used as gene therapy DNA delivery vehicles for nearly 25 years [17]. In order to effectively deliver DNA, lipids are often either complexed with DNA (lipoplexes) or formed into liposomal delivery vehicles. Lipoplexes are created by incubating cationic lipids with DNA causing condensation and complexation due to electrostatic interactions between them. Lipoplexes have been shown to be very effective in delivering DNA [18], but unfortunately a majority of cationic lipids have been found to be very toxic [19]. In order to decrease toxicity a number of methods have been used, including the creation of cyclic charged complexes [20], the inclusion of degradable bonds in the lipid backbone [21], and the use of chemical modifications such as galactosylation [22]. Another approach has been to trap plasmid DNA vectors in liposomes and this had been moderately successful in decreasing toxicity. Liposomes are lipid-based bilayer vesicles possessing a hydrophilic surface and a hydrophilic core separated by a hydrophobic membrane. DNA plasmid loaded liposomes are formed by incubating amphiphilic liposomes in solutions containing plasmids of interest. Liposomes have become a more popular gene delivery system lately since they can be formed with lower quantities of cationic lipids and in some studies have shown greater transfection efficiency than lipoplexes [23]. In addition, they can be easily surface modified to deliver payload to specific cell types. Lipid-based delivery systems have been used in the therapeutic treatment of cancer,[24] ischemia, [25] and bone disease [26]. While this method is promising, using lipids for DNA delivery does have its disadvantages. Lipids used for gene delivery can often be toxic via inhibition of normal cellular enzymatic function [27], induce inflammation through the complement cascade [28] and even be highly immunostimulatory [29].

Biocermaic-Based Delivery Systems

More recently biomaterials have shown particular promise as gene delivery vehicles in therapeutic solutions for bone disease, cancer and diabetes. Bioceramics due to their large surface area and affinity towards charged molecules exhibit much higher capacity of DNA loading than bulk materials and therefore have the potential to raise transfection efficiency [30, 31]. Calcium phosphate (CaP) was first used as a non-viral gene delivery vehicle by Graham and van der Eb [32], due to its biocompatibility and biodegradability. Briefly, CaP is generally precipitated with DNA to form the transfection complex [33–35]. The co-precipitate is obtained by supersaturating the medium containing the DNA with calcium and phosphate ions. The pH and temperature of the solution play significant roles in the properties of the resulting precipitate. Upon exposure to cells, the precipitate is endocytosed by the cells. The precipitate-containing vesicles will then fuse with lysosomes to release the DNA/calcium phosphate into the cytoplasm where the majority of the DNA is degraded [36, 37]. However, part of the DNA enters the nucleus and may be incorporated into the genetic material. The way the DNA crosses the nuclear membrane is still unclear. However, the preparation of DNA/CaP complex by the above co-precipitation method is difficult to control and spontaneously forms CaP bulk precipitate. The method of conventional in situ co-precipitation hence suffers drawbacks from low transfection efficiency [38, 39]. CaP nanoparticles are known to efficiently bind to bind DNA as complex for gene transfection [40, 41]. However, the amorphous nature of the particles and the difficulties in controlling the particle shape and preventing particle agglomeration are some of the reported limitations associated with them. Therefore, it is important synthesize shape-controllable CaP nanoparticles as the vector to elevate efficiency of gene transfection. Hydroxyapatite (HAp), the fundamental ceramic component of human tissue including bone and tooth has been extensively investigated as a biomaterial for bone tissue engineering and drug delivery vehicle due to its high affinity towards variety of biomolecules [40,41]. Recently, HAp nanoparticles are investigated as a gene delivery vehicle. Several other cationic ceramic nanoparticles have also been tested as gene delivery vehicles. The most common among them are cationic silica nanoparticles and carbon nanotubes [42,44, 45].

Although these nanoparticles have shown considerable gene transfer efficiency in vitro, not many reports exist concerning their use *in vivo*. One study tested calcium phosphate nanoparticles condensed with the reporter plasmid for β-galactosidase by directly injecting them into the substantia nigra region of the rat brain [43]. One month after injection moderate gene expression was observed in the midbrain area, thus showing the potential in CNS gene delivery. Another group has developed ceramic-coated liposome, called "cerasome" [44]. It has been reported that cerasome retains DNA integrity upon complexation with plasmid DNA and the size of the DNA complex of infusible or monomeric cerasome is ~70 nm (size of a virus) [45]. The complex also exhibits a remarkable transfection performance with high activity, minimized toxicity, and serum-compatibility. In summary, the past 10 years have seen significant interest in bioceramic-based gene delivery systems however, there is still much to investigate about its capacity to deliver efficiently with low toxicity levels *in vivo*.

Polymer-Based Delivery Systems

A promising alternative to using lipids for gene delivery has been the development of polymer-based delivery systems. Polymer-DNA complexes can be created that are smaller than lipid-based complexes which are desirable for intracellular delivery [46, 47]. They have been shown to be less toxic than lipid-based complexes, can mediate endosomal disruption allowing for efficient cytosolic delivery [48]. They have also been found to be less immunostimulatory [49]. Similarly to lipids, polymers have been used to form DNA/polymer complexes (polyplexes) and polymer bi-layer vesicles (polymersomes) for gene therapy applications. The most commonly used polymers are cationic and non-degradable and include polyethylenimine (PEI) and vinyl-based polymers [50-52] (Table 1). These gene delivery systems have shown promise in the treatment of lung disease [53], cancer, [54, 55] and liver disorders [56]. Due to moderate toxicity issues associated with non-degradable cationic polymers, other polymers have been investigated for their gene delivery potential. These include biodegradable synthetic cationic polymers such as poly(β-amino esters) [57] and poly(glycoamidoamines)[58] as well as naturally occurring cationic polymers like poly-L-lysine (PLL) [59] and chitosan [60]. Therapies including these polymers as delivery vehicles have potential in the treatment of cancer [61, 62] and bone disease [63 64]. While they are promising alternatives, polymer-based systems unfortunately still possess some of the same issues associated with the other delivery systems introduced so far. Using direct complexes, bi-layer vesicles or even particulate-based vehicles allows for rapid, whole-body dissemination of the constructs and exposes all associated DNA to the host at the time of injection.

Table 1. Summary of polymers discussed in this review.

Polymer	Characteristic	Interaction	Toxic	Chemical Structure
Poly(l-lysine)- (PLL)	Has a sufficient number of primary amines with positive charges to interact with the negatively charged phosphate groups of DNA	Has undesirable effect of aggregation of PLL/DNA complexes, but can be modified	Yes, glycosylated PLL is less toxic or with the addition of another polymer may reduce toxicity	
Polyethyleneimine (PEI)	Has a cationic polymer composed of 25% primary amines, 50% secondary amines and 25% tertiaryamines	Effectively condenses plasmids into colloidal particles that effectively transfect pDNA into a variety of cells both in vitro and in vivo	No	
Poly ethylene glycol (PEG)	Hydrophilic polymer is the covalent attachment of activated monomethoxy poly(ethylene) glycols (MPEGs) to free lysine groups of therapeutic	Enhances the physical stability of proteins and ablated humoral immune responses generated against them.	Low	

	proteins.	No	
Chitosan	Is a biodegradable poly-saccharide composed of two subunits, D glucosamine and N-acetyl-D-glucosamine, linked together by a b(1,4)glycosidic bond	Cationic-charged chitosan interacts with the negatively charged phosphate groups of DNA. The colloidal and surface properties depend on the molecular weight of chitosan, the ratio of plasmid to chitosan and the preparation medium.	
Gelatin	The protein fractions include various amino acids linked together by an amide linkage, with the major amino acids being glycine, proline, hydroxyproline, and glutamic	Cationic-charge interacts with the negatively charged phosphate groups of DNA.	

Poly(lactic-co-glycolic acid) (PLGA)		Low	Degrades by hydrolysis of its ester linkages in the presence of water. Degradation is related to the monomers' ratio used in production:	Hydrophilic properties synthesized by means of random ring-opening co-polymerization of) of glycolic acid and lactic acid.	acid.
Polyethylene glycol diacrylate (PEGDA)		No	Allows transfected cells with therapeutic gene to produce and release protein based on hydrogel pore size and diffusion	Bioinert and mimics many physical properties porous scaffold such as superporous hydrogels may increase surface area available for cellular attachment and encourage cell-cell and cell-matrix interactions in the absence
Poly(N-isopropylmethacrylamide) (pNIPMAm)		No	Has thermorespon sivity; it undergoes an entropically	Is an amphiphilic polymer that is strongly hydrated at physiological temperature and

	is likely therefore to resist protein adsorption relative to more hydrophobic carriers	driven coil-to-globule (swollen-to-collapsed) transition at 43 °C, which may have utility for thermally triggered delivery		
Silk-elastin-like protein polymers (SELPs)	Recombinant polymers composed of tandem repeats of a six amino acid sequence commonly found in silkworm silk fibroin (GAGAGS)12 and a five amino acidsequence commonly found in mammalian elastin(GVGVP)	These hydrogels display swelling and release properties for bioactive agents in a structure-dependent	No	

Biodegradable Polymeric Hydrogel-Based Delivery Systems

One possibility for improving the delivery issues associated with other systems is the use of localized delivery vehicles such as hydrogels. Hydrogels are three-dimensional hydrophilic polymeric networks that can absorb and retain a substantial amount of water while holding their shape due to their chemical or physical crosslinked structure [65, 66]. Chemically crosslinked hydrogels are prepared through photopolymerization, disulfide bond formation, or reaction between thiols and acrylates or sulfones. Physically crosslinked hydrogels are formed by the self-assembly of polymers in response to environmental stimuli, such as temperature or pH. Biodegradable hydrogels have been widely used in many other biomedical applications, such as drug or cell delivery and tissue engineering, and as we have detailed below, they have recently shown significant promise in gene therapy.

Bone Disease and Tissue Regeneration

Among the most promising developments in hydrogel-based gene delivery in recent years have been therapies focused on various bone diseases, disorders and fractures. Recently, a biodegradable cationic hydrogel composed of gelatin and chitosan was used to deliver an antisense oligonucletide targeting murine TNF-α for the treatment of endotoxin-induced osteolysis. An earlier investigation, silencing TNF-α by antisense nucleotide effectively alleviated osteolysis induced by particle wear from Co-Cr-Mo alloy materials used regularly in joint replacement [67]. As the body attempts to remove wear particles a systemic autoimmune reaction is triggered causing unwanted resorption of living bone tissue, known as aseptic loosening. Aseptic loosening is a local chronic inflammation, and thus this study created a localized controlled release system to enhance the antisense oligonucleotide (ASO) efficacy and extend its effect. The nucleotide was released after the cells adhered to the hydrogels. The released nucleotide was delivered efficiently into contacted cells and tissues during in vitro and in vivo studies. When tested in animal models of endotoxin-induced bone resorption, ASO delivered by such means showed effective suppression of TNF-α expression and subsequently the osteoclastogenesis in vivo [67]. This study is one of many that uses newly developed hydrogel systems in bone regeneration and is a successful attempt to apply localized gene delivery method to treat inflammatory diseases in vivo.

Another study used a slightly different transfection approach by evaluating the in vitro and in vivo viability, gene expression, and bone formation from pre-transfected fibroblasts encapsulated in polyethylene glycol diacrylate (PEGDA) microspheres [68]. Cytotoxicity assays showed greater than 95% viability in microencapsulated cells. Moreover, alkaline phosphatase and BMP-2 in serum demonstrated that BMP-2 secretion and specific activity from microencapsulated AdBMP2-transduced fibroblasts were not markedly different from monolayer. Furthermore, microencapsulated cells expressed BMP-2 longer than unencapsulated cells. Additional studies demonstrated that microencapsulated AdBMP2-transduced cells, upon intramuscular injection into mice, formed new heterotrophic bone formation, and had increased two fold the volume of unencapsulated cells . These data suggest that microencapsulation protects cells and prolongs and spatially distributes transgene expression. This study effectively demonstrated that incorporation of PEGDA hydrogels significantly advances current gene therapy for bone repair [68] . In a third study, DNA/PEI polyplexes were encapsulated inside poly (ethylene glycol) (PEG) hydrogels cross linked with MMP degradable peptides via Michael Addition chemistry. It was found that gene transfer to MSCs was possible in cells seeded both in tissue culture plates and three dimensional scaffolds. The use of hydrogel scaffolds that allow cellular infiltration to deliver DNA mimic a more physiologically conditioned environment and may result in long-lasting signals in vivo, which are essential for the regeneration of functional tissues [69] . Taken together these studies exemplify promising developments towards the treatment of various bone diseases and tissue regeneration via hydrogel gene delivery and transgenic cellular delivery.

Tumor and Cancer Treatment

Although traditional chemotherapeutic treatments have shown both preclinical and clinical effectiveness at inhibiting or eliminating cancerous cell growth, significant limitations exist due to toxic side effects of chemotherapeutic drugs needed to effectively eliminate malignant cells. This can also lead to chemoresistance and subsequent tumor recurrence, both major dilemmas in cancer treatment. Targeted therapies that can enhance cancer cell sensitivity to chemotherapeutic agents are of significant clinical benefit and could possibly augment drug efficacy while reducing toxic effects on untargeted cells. Targeted cancer therapy by RNA interference (RNAi) is a recently developed approach that can reversibly silence genes *in vivo* by selectively targeting genes. However, delivery represents the main hurdle for the broad development of RNAi therapeutics. A recent study targeted epidermal growth factor receptor (EGFR) [70], which has been shown to increase the sensitivity of cancer cells to taxane chemotherapy. This work demonstrated that core/shell hydrogel nanoparticles (nanogels) could be functionalized with peptides such as YSA that specifically target the EphA2 receptor to deliver small interfering RNAs (siRNAs) modulating EGFR. Treatment of EphA2 positive Hey cells with siRNA-loaded, peptide-targeted nanogels decreased EGFR expression levels and significantly increased the sensitivity of treated cells to docetaxel (Figure 2).

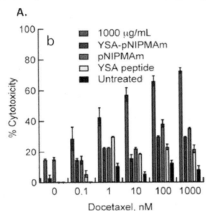

Figure 2. Chemosensitization of Hey cells treated with YSA-targeted, siRNA-loaded nanogels greatly increased the effectiveness of docetaxel therapy. Control: Unloaded YSA-conjugated nanogels (YSApNIPMAm), unloaded pNIPMAm nanogels (pNIPMAm), YSA peptide alone, and untreated cells. Reproduced from ref. 70, with permission from Biomedical Central.

Other investigators have developed a process (termed caged nanoparticle encapsulation or CnE) to load concentrated and unaggregated non-viral gene delivery nanoparticles into different hydrogels. It has been previously shown that PEG hydrogels loaded with DNA/PEI polyplexes through this process were able to deliver genes both in vitro and in vivo. Both hyaluronic acid and fibrin hydrogels loaded by CnE were able to deliver genes efficiently in vivo [71].

Recombinant silk-elastin-like protein polymers (SELPs), known for their highly tunable properties on both the molecular and macroscopic hydrogel levels, have also shown promise as gene delivery vehicles. Recent studies demonstrated that SELP-815K can be used as an injectable controlled delivery system in the treatment of head and neck cancer via a gene-directed enzyme prodrug therapy

(GDEPT) approach, a prodrug and virus combination delivery [71]. Due to its pore size and gelation properties in vivo, SELP restricts the distribution and controls the release of therapeutic viruses. It has been shown that SELP-mediated delivery significantly improves the therapeutic outcome of the herpes simplex virus thymidine kinase (HSVtk)/ganciclovir (GCV) system in xenograft models of human head and neck cancer. Unfortunately, there is still much to be learned about the potential benefits of this approach with regard to toxicity and the immune system. The studies presented here were designed to assess the change in toxicity of the SELP-mediated viral delivery compared to free viral injection in a non-tumor-bearing immune competent mouse model. Toxicity was assessed over a period of twelve weeks via body weight monitoring, complete blood count (CBC), and blood chemistry. In the acute and subacute phases there was a significant toxic response in groups combining the virus and the prodrug alone, whereas matrix-mediated gene delivery with SELP demonstrated a reduction in toxicity from the 2 week time point through the 4 week time point.

At the end of the twelve weeks, signs of toxicity had reduced in both groups. This treatment is ideal to prevent solid tumor growths, as these present a well-defined target and can easily be modeled in vivo. Based on these results, recombinant SELPs may offer a significant reduction in toxicity of virus-mediated GDEPT treatment compared to free virus injection in the acute and subacute phases [72].

Diabetes and Secondary Condition Treatments

Glucagon-like peptide-1 (GLP-1) is an incretin hormone that regulates blood glucose levels after food consumption. GLP-1 has the potential to be used in type 2 diabetes (T2D) mellitus treatment because of its insulinotropic activity. Despite its remarkable advantages, GLP-1 possesses an extremely short half-life owing to its degradation by dipeptidyl peptidase IV protease. New research is now focusing on GLP-1 gene delivery to overcome the issues associated with having to deliver multiple daily doses [73]. Novel approaches have shown promise in the effective and safe gene-based delivery of GLP-1 using chitosan/plasmid-DNA therapeutic nanocomplexes (TNCs) with the Zucker diabetic fatty (ZDF) serving as the animal model for T2D [74]. Animals injected with the TNC chitosan showed GLP-1 plasma levels roughly five fold higher than those in non-treated animals, and the insulinotropic effect of the treatment was demonstrated through a threefold increase in plasma insulin concentration when compared with untreated animals. Intraperitoneal glucose tolerance tests revealed an efficacious decrease in blood glucose compared with controls for up to 24 days after treatment, resulting in near-normalization of blood glucose levels. TNCs composed of specific chitosans and GLP-1-expressing plasmid constructs showed a remarkable ability to facilitate therapeutic potential of GLP-1 for the treatment of T2D mellitus.

Uncontrolled blood glucose levels in diabetics are the cause of many secondary conditions, leading to blindness, deafness, and kidney failure among other debilitating diseases. Diabetes is the leading cause of kidney failure, accounting for 44% of new cases in 2008. Hyperglycemia over time leads to many secondary conditions that affect various organs including the kidney. Diabetic Nephropathy (DN) is a progressive kidney disease caused by angiopathy (a weakening and thickening of the walls of the smaller blood vessels that causes them to bleed, leak protein, and slow the flow of blood through the body) of capillaries in the kidney glomeruli impairing their ability to filter toxins from blood for excretion in the urine. Thus there is a necessity to circumvent this issue with novel therapeutics. A study investigated matrix metalloproteinase (MMP) functions to prevent the occurrence of destructive fibrosis in progressive renal disease in streptozotocin (STZ)-induced diabetic mouse model [75]. As a sustained release carrier of plasmid DNA, biodegradable hydrogels and microspheres were formulated from cationized gelatin. Plasmid DNA was released from the cationized gelatin hydrogels as a result of hydrogel degradation. A plasmid DNA composed of a cytomegalovirus promoter and human recombinant (pCMV) MMP-1 gene pCMV-MMP was constructed. Gelatin microspheres with or without pCMV-MMP were injected into the renal subcapsule of C57BL/6 mice, which were intraperitoneally injected with STZ to induce diabetes. Immunofluorescence confirmed that MMP protein was expressed around the renal tissue injected with gelatin microspheres incorporating pCMV-MMP. When applied with cationized gelatin microspheres incorporating pCMV-MMP, the mice showed a level of blood urea nitrogen significantly lower than that of all other experimental groups. A reduced amount of collagen in the kidneys of mice administered gelatin microspheres incorporating pCMV-MMP was also shown. The hydroxyproline assay revealed a significantly decreased content of hydroxyproline in the kidneys. The sustained release of the MMP-1 gene is a potential prophylactic trial for kidney fibrolysis and dysfunction in the STZ-induced diabetic mouse model and possibly other diabetic mouse models, such as type 1 Non-Obese Diabetic mice.

Another notable study focused on diabetes and wound healing. Since there is a delayed healing process in diabetics [76], a diabetic mouse model was used in this investigation to examine the ability for TGF-1 gene accelerated diabetic wound healing using a thermosensitive hydrogel made of a triblock copolymer polyethylene glycol, and poly(lactic-co-glycolic acid (PEG-PLGA-PEG). Excision wounds on the back in parallel were made for each genetically diabetic mouse. The hydrogel containing plasmid TGF-1 was administered to the wound and formed an adhesive film in situ. The investigation found that thermosensitive hydrogel alone is slightly beneficial for improved wound healing during early stages of healing (day 1–5), but significantly accelerated re-epithelializaion, increased cell proliferation, and organized collagen was observed in the wound bed treated with thermosensitive hydrogel containing plasmid TGF-1. The healing was accompanied by enhanced collagen synthesis and more organized extracellular matrix deposition. This study demonstrated that thermosensitive hydrogel made of PEG-PLGA-PEG triblock copolymer provides excellent wound dressing activity and can serve as a possible model for gene therapy for other tissue regenerative treatments to promote wound healing in a diabetic mouse model.

Novel Biodegradable Hydrogel Nanoparticles as Gene Delivery Complexes

Efficient delivery of a therapeutic gene to target cells is the most challenging obstacle in gene therapy, and has been the focus of many research investigations. The use of viral vectors in vivo has been limited due to their latent activation of immune responses, their reduced capacity for packaging large genes, and the difficulty associated with their large-scale preparation acceptable for clinical testing. Although they provide superior efficiency of gene transfer, the outcomes of this method may not outweigh the risks. The use of nanoparticles as nucleic acid delivery vehicles has been a recent advancement in research over the past decade. More recently, investigators have developed new acid-degradable cationic nanoparticles created by using a monomer crosslinked to polymer, which enabled highly flexible nanoparticle fabrication to obtain controlled properties such as size, along with additional functionalities. The nanoparticles are designed to cause swelling and osmotic destabilization of the endosome, while cationic branches holding anionic DNA are cleaved from the polymeric backbone of the nanoparticles making plasmid DNA accessible for efficient gene expression. Efficient release of plasmid DNA upon hydrolysis of the nanoparticles at endosomal pH of 5.0 and transportation of the released DNA to the cell nucleus was shown. In vitro studies showed significantly higher transfection efficiency by degradable nanoparticles than polyethylenimine (PEI) polyplexes at very low concentrations. Size-dependent selective transfection of a phagocytic cell line (RAW 309 macrophages) and non-phagocytic cells (NIH 3T3 fibroblasts) was also shown via nanoparticles of two different sizes (240 nm and 680 nm in diameter), which is vital for adjustable gene therapy and DNA vaccination using the nanoparticle system. Preliminary pulmonary transfection of mice using the degradable nanoparticles demonstrated a remarkably higher expression of luciferase at a 70% lower concentration than using naked DNA alone. Properties of the nanoparticles such as acid degradability, surface charges, and types of cargo (e.g., siRNA, miRNA, and oligodeoxynucleotides) can be manipulated for specific applications and targeting and imaging modalities can also be coupled with the nanoparticles to achieve targeted delivery, as well as combined imaging for detection [77].

FUTURE PROSPECTS FOR ENHANCING BIOMATERIAL GENE DELIVERY

Gene therapy holds enormous promise for the treatment of acquired and inherited genetic diseases that are currently poorly treated, as well as for wound and fracture healing. Although viral transfection therapies have shown great improvement, their clinical potential is severely limited due to significant drawbacks associated with the use of delivery vehicles. Non-viral gene delivery systems are gaining recognition as a promising alternative to viral gene vectors due to their more limited immunogenicity and toxicity. In the past 10 years, the focus of gene delivery systems has shifted to biodegradable carrier systems. Compared to their non-degradable counterparts biodegradable carriers show reduced toxicity and avoid accumulation of the polymer. Moreover they offer the attractive option of being able to repeatedly deliver the therapy. In addition, their degradation can be used as a directed delivery tool to release nucleic acids into the target cell in a time-dependent manner. In the first wave of polycationic gene delivery studies, cytotoxicity was the most studied drawback associated with high molecular weight polycations used in gene delivery, followed by the capacity to release the DNA upon degradation of the carrier which permits the controlled intracellular delivery [5] . While much progress has been made in the field of biodegeradable hydrogels, bioceramics gene delivery system are only in its beginning stages of investigations and is gaining much attention while showing great promise for future applications. This review discussed several vital studies that demonstrate the enhancement of gene delivery efficiency and the sustained low toxicity of utilizing biocompatible ceramics and hydrogels as a means of transport. Currently, researchers are still in the early stages of *in vivo* studies and targeting strategies, but this will undoubtedly change in the near future. The potential advantage of these polymers is the pH triggered release of DNA. Biodegradable DNA nanogel microparticles have shown significant promise for DNA-based vaccination against viruses and tumors. These particles are selective for antigen presenting cells, which play a pivotal role in triggering immune responses. The challenges in future advancements in carrier design are rooted in the improvement of the release rate of intact DNA. The major intracellular barriers yet to be overcomed are sustained gene expression, endosomal escape and nuclear localization. A number of novel strategies have been approached including the addition of buffering elements or membrane destabilizing peptides, but still require further research. Although cationic lipids and non-degradable polymers investigated for DNA delivery have shown advancements in nucleic acid based therapies, bioceramics, biodegradable polymers and particles discussed in this review are ushering in a new treatment paradigm as promising candidates for suitable intracellular delivery therapeutics.

ACKNOWLEDGEMENTS

The authors greatly appreciate the funding from USAMRMI –W81WXH-10-1-0653

REFERENCES

[1] F.D. Ledley, Pharmaceutical approach to somatic gene therapy, Pharm. Res. 13 (1996) 1595-1614.

[2] J.C. Weaver, T.E. Vaughan, Y. Chizmadzhev, Theory of electrical creation of aqueous pathways across skin transport barriers, Adv. Drug Deliv. Rev. 35 (1999) 21-39. [3] A.R. Denet, R. Vanbever, V. Preat, Skin electroporation for transdermal and topical delivery, Adv. Drug Deliv. Rev. 56 (2004) 659-674.

[4] D.V. McAllister, M.G. Allen, M.R. Prausnitz, Microfabricated microneedles for gene and drug delivery, Annu. Rev. Biomed. Eng. 2 (2000) 289-313.

[5] X. Gao, K.S. Kim, D. Liu, Nonviral gene delivery: what we know and what is next, AAPS J. 9 (2007) E92-104.

[6] W. Walther, U. Stein, Viral vectors for gene transfer: a review of their use in the treatment of human diseases, Drugs. 60 (2000) 249-271.

[7] R. Mann, R.C. Mulligan, D. Baltimore, Construction of a retrovirus packaging mutant and its use to produce helper-free defective retrovirus, Cell. 33 (1983) 153-159.

[8] C. Bordignon, L.D. Notarangelo, N. Nobili, G. Ferrari, G. Casorati, P. Panina, E. Mazzolari, D. Maggioni, C. Rossi, P. Servida, A.G. Ugazio, F. Mavilio, Gene therapy in peripheral blood lymphocytes and bone marrow for ADA- immunodeficient patients, Science. 270 (1995) 470-475.

[9] K.J. Garton, N. Ferri, E.W. Raines, Efficient expression of exogenous genes in primary vascular cells using IRES-based retroviral vectors, BioTechniques. 32 (2002) 830, 832, 834 passim.

[10] R.D. Alvarez, J. Gomez-Navarro, M. Wang, M.N. Barnes, T.V. Strong, R.B. Arani, W. Arafat, J.V. Hughes, G.P. Siegal, D.T. Curiel, Adenoviral-mediated suicide gene therapy for ovarian cancer, Mol. Ther. 2 (2000) 524-530.

[11] G. Schiedner, N. Morral, R.J. Parks, Y. Wu, S.C. Koopmans, C. Langston, F.L. Graham, A.L. Beaudet, S. Kochanek, Genomic DNA transfer with a high-capacity adenovirus vector results in improved in vivo gene expression and decreased toxicity, Nat. Genet. 18 (1998) 180-183.

[12] J.Y. Dong, P.D. Fan, R.A. Frizzell, Quantitative analysis of the packaging capacity of recombinant adeno-associated virus, Hum. Gene Ther. 7 (1996) 2101-2112.

[13] S. Lehrman, Virus treatment questioned after gene therapy death, Nature. 401 (1999) 517-518.

[14] S. Rajendran, D. O'Hanlon, D. Morrissey, T. O'Donovan, G.C. O'Sullivan, M. Tangney, Preclinical evaluation of gene delivery methods for the treatment of loco-regional disease in breast cancer, Exp. Biol. Med. (Maywood). 236 (2011) 423-434.

[15] W.R. Hou, S.N. Xie, H.J. Wang, Y.Y. Su, J.L. Lu, L.L. Li, S.S. Zhang, M. Xiang, Intramuscular delivery of a naked DNA plasmid encoding proinsulin and pancreatic regenerating III protein ameliorates type 1 diabetes mellitus, Pharmacol. Res. 63 (2011) 320-327.

[16] G.M. Huang, G.S. Li, G.Y. Zhu, Y.M. Lai, H.X. Zhang, J. Wang, H.R. Wang, Safety and bioactivity of intracoronary delivery of naked plasmid DNA encoding human atrial natriuretic factor, Acta Pharmacol. Sin. 23 (2002) 609-611.

[17] P.L. Felgner, T.R. Gadek, M. Holm, R. Roman, H.W. Chan, M. Wenz, J.P. Northrop, G.M. Ringold, M. Danielsen, Lipofection: a highly efficient, lipid-mediated DNA-transfection procedure, Proc. Natl. Acad. Sci. U. S. A. 84 (1987) 7413-7417.

[18] C. Guillaume-Gable, V. Floch, B. Mercier, M.P. Audrezet, E. Gobin, G. Le Bolch, J.J. Yaouanc, J.C. Clement, H. des Abbayes, J.P. Leroy, V. Morin, C. Ferec, Cationic phosphonolipids as nonviral gene transfer agents in the lungs of mice, Hum. Gene Ther. 9 (1998) 2309-2319.

[19] L.T. Nguyen, K. Atobe, J.M. Barichello, T. Ishida, H. Kiwada, Complex formation with plasmid DNA increases the cytotoxicity of cationic liposomes, Biol. Pharm. Bull. 30 (2007) 751-757.

[20] M.A. Ilies, W.A. Seitz, B.H. Johnson, E.L. Ezell, A.L. Miller, E.B. Thompson, A.T. Balaban, Lipophilic pyrylium salts in the synthesis of efficient pyridinium-based cationic lipids, gemini surfactants, and lipophilic oligomers for gene delivery, J. Med. Chem. 49 (2006) 3872-3887.

[21] F. Tang, J.A. Hughes, Synthesis of a single-tailed cationic lipid and investigation of its transfection, J. Control. Release. 62 (1999) 345-358.

[22] S. Kawakami, F. Yamashita, M. Nishikawa, Y. Takakura, M. Hashida, Asialoglycoprotein receptor-mediated gene transfer using novel galactosylated cationic liposomes, Biochem. Biophys. Res. Commun. 252 (1998) 78-83.

[23] C. Tros de Ilarduya, Y. Sun, N. Duzgunes, Gene delivery by lipoplexes and polyplexes, Eur. J. Pharm. Sci. 40 (2010) 159-170.

[24] K. Matsumoto, E. Kikuchi, M. Horinaga, T. Takeda, A. Miyajima, K. Nakagawa, M. Oya, Intravesical Interleukin-15 Gene Therapy in an Orthotopic Bladder Cancer Model, Hum. Gene Ther. (2011).

[25] G. Czibik, J. Gravning, V. Martinov, B. Ishaq, E. Knudsen, H. Attramadal, G. Valen, Gene therapy with hypoxia-inducible factor 1 alpha in skeletal muscle is cardioprotective in vivo, Life Sci. 88 (2011) 543-550.

[26] C. Madeira, R.D. Mendes, S.C. Ribeiro, J.S. Boura, M.R. Aires-Barros, C.L. da Silva, J.M. Cabral, Nonviral gene delivery to mesenchymal stem cells using cationic liposomes for gene and cell therapy, J. Biomed. Biotechnol. 2010 (2010) 735349.

[27] R. Bottega, R.M. Epand, Inhibition of protein kinase C by cationic amphiphiles, Biochemistry. 31 (1992) 9025-9030.

[28] A.J. Bradley, D.E. Brooks, R. Norris-Jones, D.V. Devine, C1q binding to liposomes is surface charge dependent and is inhibited by peptides consisting of residues 14-26 of the human C1qA chain in a sequence independent manner, Biochim. Biophys. Acta. 1418 (1999) 19-30.

[29] S. Li, S.P. Wu, M. Whitmore, E.J. Loeffert, L. Wang, S.C. Watkins, B.R. Pitt, L. Huang, Effect of immune response on gene transfer to the lung via systemic administration of cationic lipidic vectors, Am. J. Physiol. 276 (1999) L796-804.

[30] F. Scherer, M. Anton, U. Schillinger, J. Henke, C. Bergemann, A. Kruger, B. Gansbacher, C. Plank, Magnetofection: enhancing and targeting gene delivery by magnetic force in vitro and in vivo, Gene Ther. 9 (2002) 102-109.

[31] A.K. Salem, P.C. Searson, K.W. Leong, Multifunctional nanorods for gene delivery, Nat. Mater. 2 (2003) 668-671.

[32] F.L. Graham, A.J. van der Eb, A new technique for the assay of infectivity of human adenovirus 5 DNA, Virology. 52 (1973) 456-467.

[33] I. Roy, S. Mitra, A. Maitra, S. Mozumdar, Calcium phosphate nanoparticles as novel non-viral vectors for targeted gene delivery, Int. J. Pharm. 250 (2003) 25-33.

[34] S. Bisht, G. Bhakta, S. Mitra, A. Maitra, pDNA loaded calcium phosphate nanoparticles: highly efficient non-viral vector for gene delivery, Int. J. Pharm. 288 (2005) 157-168.

[35] H. Fu, Y. Hu, T. McNelis, J.O. Hollinger, A calcium phosphate-based gene delivery system, J. Biomed. Mater. Res. A. 74 (2005) 40-48.

[36] M. Urabe, A. Kume, K. Tobita, K. Ozawa, DNA/Calcium phosphate precipitates mixed with medium are stable and maintain high transfection efficiency, Anal. Biochem. 278 (2000) 91-92.

[37] B. Goetze, B. Grunewald, S. Baldassa, M. Kiebler, Chemically controlled formation of a DNA/calcium phosphate coprecipitate: application for transfection of mature hippocampal neurons, J. Neurobiol. 60 (2004) 517-525.

[38] H. Liu, H. Yazici, C. Ergun, T.J. Webster, H. Bermek, An in vitro evaluation of the Ca/P ratio for the cytocompatibility of nano-to-micron particulate calcium phosphates for bone regeneration, Acta Biomater. 4 (2008) 1472-1479.

[39] R. Narayanan, T.Y. Kwon, K.H. Kim, Preparation and characteristics of nano-grained calcium phosphate coatings on titanium from ultrasonated bath at acidic pH, J. Biomed. Mater. Res. B. Appl. Biomater. 85 (2008) 231-239.

[40] Q. Xu, J.T. Czernuszka, Controlled release of amoxicillin from hydroxyapatite-coated poly(lactic-co-glycolic acid) microspheres, J. Control. Release. 127 (2008) 146-153.

[41] S. Deville, E. Saiz, A.P. Tomsia, Freeze casting of hydroxyapatite scaffolds for bone tissue engineering, Biomaterials. 27 (2006) 5480-5489.

[42] N.W. Kam, Z. Liu, H. Dai, Functionalization of carbon nanotubes via cleavable disulfide bonds for efficient intracellular delivery of siRNA and potent gene silencing, J. Am. Chem. Soc. 127 (2005) 12492-12493.

[43] T.D. Corso, G. Torres, C. Goulah, I. Roy, A.S. Gambino, J. Nayda, T. Buckley, E.K. Stachowiak, E.J. Bergey, H. Pudavar, P. Dutta, D.C. Bloom, W.J. Bowers, M.K. Stachowiak, Assessment of viral and non-viral gene transfer into adult rat brains using HSV-1, calcium phosphate and PEI-based methods, Folia. Morphol. (Warsz). 64 (2005) 130-144.

[44] Katagiri, R. Hamasaki, K. Ariga, J. Kikuchi, Layered paving of vesicular nanoparticles formed with cerasome as a bioinspired organic-inorganic hybrid, J. Am. Chem. Soc. 124 (2002) 7892-7893.

[45] K. Matsui, S. Sando, T. Sera, Y. Aoyama, Y. Sasaki, T. Komatsu, T. Terashima, J. Kikuchi, Cerasome as an infusible, cell-friendly, and serum-compatible transfection agent in a viral size, J. Am. Chem. Soc. 128 (2006) 3114-3115.

[46] D.T. Curiel, S. Agarwal, E. Wagner, M. Cotten, Adenovirus enhancement of transferrin-polylysine-mediated gene delivery, Proc. Natl. Acad. Sci. U. S. A. 88 (1991) 8850-8854.

[47] E. Kleemann, N. Jekel, L.A. Dailey, S. Roesler, L. Fink, N. Weissmann, R. Schermuly, T. Gessler, T. Schmehl, C.J. Roberts, W. Seeger, T. Kissel, Enhanced gene expression and reduced toxicity in mice using polyplexes of low-molecular-weight poly(ethylene imine) for pulmonary gene delivery, J. Drug Target. 17 (2009) 638-651.

[48] G. Navarro, G. Maiwald, R. Haase, A.L. Rogach, E. Wagner, C.T. de Ilarduya, M. Ogris, Low generation PAMAM dendrimer and CpG free plasmids allow targeted and extended transgene expression in tumors after systemic delivery, J. Control. Release. 146 (2010) 99-105.

[49] G.F. Lemkine, B.A. Demeneix, Polyethylenimines for in vivo gene delivery, Curr. Opin. Mol. Ther. 3 (2001) 178-182.

[50] C.P. Lollo, M.G. Banaszczyk, P.M. Mullen, C.C. Coffin, D. Wu, A.T. Carlo, D.L. Bassett, E.K. Gouveia, D.J. Carlo, Poly-L-lysine-based gene delivery systems. Synthesis, purification, and application, Methods Mol. Med. 69 (2002) 1-13.

[51] G. Borchard, Chitosans for gene delivery, Adv. Drug Deliv. Rev. 52 (2001) 145-150.

[52] P. Dubruel, E. Schacht, Vinyl polymers as non-viral gene delivery carriers: current status and prospects, Macromol. Biosci. 6 (2006) 789-810.

[53] C. Pfeifer, G. Hasenpusch, S. Uezguen, M.K. Aneja, D. Reinhardt, J. Kirch, M. Schneider, S. Claus, W. Friess, C. Rudolph, Dry powder aerosols of polyethylenimine (PEI)-based gene vectors mediate efficient gene delivery to the lung, J. Control. Release. 154 (2011) 69-76.

[54] S. Abbasi, A. Paul, S. Prakash, Investigation of siRNA-Loaded Polyethylenimine-Coated Human Serum Albumin Nanoparticle Complexes for the Treatment of Breast Cancer, Cell Biochem. Biophys. (2011).

[55] B. Zhang, M. Kanapathipillai, P. Bisso, S. Mallapragada, Novel pentablock copolymers for selective gene delivery to cancer cells, Pharm. Res. 26 (2009) 700-713.

[56] H. Arima, S. Yamashita, Y. Mori, Y. Hayashi, K. Motoyama, K. Hattori, T. Takeuchi, H. Jono, Y. Ando, F. Hirayama, K. Uekama, In vitro and in vivo gene delivery mediated by Lactosylated dendrimer/alpha-cyclodextrin conjugates (G2) into hepatocytes, J. Control. Release. 146 (2010) 106-117.

[57] M. Liu, J. Chen, Y.N. Xue, W.M. Liu, R.X. Zhuo, S.W. Huang, Poly(beta-aminoester)s with pendant primary amines for efficient gene delivery, Bioconjug. Chem. 20 (2009) 2317-2323.

[58] Y. Liu, T.M. Reineke, Degradation of poly(glycoamidoamine) DNA delivery vehicles: polyamide hydrolysis at physiological conditions promotes DNA release, Biomacromolecules. 11 (2010) 316-325.

[59] X. Zhang, M. Oulad-Abdelghani, A.N. Zelkin, Y. Wang, Y. Haikel, D. Mainard, J.C. Voegel, F. Caruso, N. Benkirane-Jessel, Poly(L-lysine) nanostructured particles for gene delivery and hormone stimulation, Biomaterials. 31 (2010) 1699-1706.

[60] M. Thibault, M. Astolfi, N. Tran-Khanh, M. Lavertu, V. Darras, A. Merzouki, M.D. Buschmann, Excess polycation mediates efficient chitosan-based gene transfer by promoting lysosomal release of the polyplexes, Biomaterials. 32 (2011) 4639-4646.

[61] R.E. Vandenbroucke, B.G. De Geest, S. Bonne, M. Vinken, T. Van Haecke, H. Heimberg, E. Wagner, V. Rogiers, S.C. De Smedt, J. Demeester, N.N. Sanders, Prolonged gene silencing in hepatoma cells and primary hepatocytes after small interfering RNA delivery with biodegradable poly(beta-amino esters), J. Gene Med. 10 (2008) 783-794.

[62] E. Salva, J. Akbuga, In vitro silencing effect of chitosan nanoplexes containing siRNA expressing vector targeting VEGF in breast cancer cell lines, Pharmazie. 65 (2010) 896-902.

[63] B.A. Clements, V. Incani, C. Kucharski, A. Lavasanifar, B. Ritchie, H. Uludag, A comparative evaluation of poly-L-lysine-palmitic acid and Lipofectamine 2000 for plasmid delivery to bone marrow stromal cells, Biomaterials. 28 (2007) 4693-4704.

[64] J.C. Fernandes, H. Wang, C. Jreyssaty, M. Benderdour, P. Lavigne, X. Qiu, F.M. Winnik, X. Zhang, K. Dai, Q. Shi, Bone-protective effects of nonviral gene therapy with folate-chitosan DNA nanoparticle containing interleukin-1 receptor antagonist gene in rats with adjuvant-induced arthritis, Mol. Ther. 16 (2008) 1243-1251.

[65] A.S. Hoffman, Hydrogels for biomedical applications, Adv. Drug Deliv. Rev. 54 (2002) 3-12.

[66] B. Jeong, S.W. Kim, Y.H. Bae, Thermosensitive sol-gel reversible hydrogels, Adv. Drug Deliv. Rev. 54 (2002) 37-51.

[67] L. Dong, Z. Huang, X. Cai, J. Xiang, Y.A. Zhu, R. Wang, J. Chen, J. Zhang, Localized Delivery of Antisense Oligonucleotides by Cationic Hydrogel Suppresses TNF-alpha Expression and Endotoxin-Induced Osteolysis, Pharm. Res. (2010).

[68] R.M. Olabisi, Z.W. Lazard, C.L. Franco, M.A. Hall, S.K. Kwon, E.M. Sevick-Muraca, J.A. Hipp, A.R. Davis, E.A. Olmsted-Davis, J.L. West, Hydrogel microsphere encapsulation of a cell-based gene therapy system increases cell survival of injected cells, transgene expression, and bone volume in a model of heterotopic ossification, Tissue Eng. Part A. 16 (2010) 3727-3736.

[69] Y. Lei, Q.K. Ng, T. Segura, Two and three-dimensional gene transfer from enzymatically degradable hydrogel scaffolds, Microsc. Res. Tech. 73 (2010) 910-917.

[70] E.B. Dickerson, W.H. Blackburn, M.H. Smith, L.B. Kapa, L.A. Lyon, J.F. McDonald, Chemosensitization of cancer cells by siRNA using targeted nanogel delivery, BMC Cancer. 10 (2010) 10.

[71] Y. Lei, M. Rahim, Q. Ng, T. Segura, Hyaluronic acid and fibrin hydrogels with concentrated DNA/PEI polyplexes for local gene delivery, J. Control. Release (2011).

[72] J.A. Gustafson, R.A. Price, K. Greish, J. Cappello, H. Ghandehari, Silk-elastin-like hydrogel improves the safety of adenovirus-mediated gene-directed enzyme-prodrug therapy, Mol. Pharm. 7 (2010) 1050-1056.

[73] S. Choi, S. Oh, M. Lee, S.W. Kim, Glucagon-like peptide-1 plasmid construction and delivery for the treatment of type 2 diabetes, Mol. Ther. 12 (2005) 885-891.

[74] M. Jean, M. Alameh, M.D. Buschmann, A. Merzouki, Effective and safe gene-based delivery of GLP-1 using chitosan/plasmid-DNA therapeutic nanocomplexes in an animal model of type 2 diabetes, Gene Ther. (2011).

[75] T. Aoyama, S. Yamamoto, A. Kanematsu, O. Ogawa, Y. Tabata, Local delivery of matrix metalloproteinase gene prevents the onset of renal sclerosis in streptozotocin-induced diabetic mice, Tissue Eng. 9 (2003) 1289-1299.

[76] P.Y. Lee, Z. Li, L. Huang, Thermosensitive hydrogel as a Tgf-beta1 gene delivery vehicle enhances diabetic wound healing, Pharm. Res. 20 (2003) 1995-2000.

[77] I.K. Ko, A. Ziady, S. Lu, Y.J. Kwon, Acid-degradable cationic methacrylamide polymerized in the presence of plasmid DNA as tunable non-viral gene carrier, Biomaterials. 29 (2008) 3872-3881.

SOL-GEL SYNTHESIZED BIO-ACTIVE NANOPOROUS SODIUM ZIRCONATE COATING ON 316L STAINLESS STEEL FOR BIOMEDICAL APPLICATION

K. Bavya Devi, N. Rajendran[*]
Department of Chemistry, Anna University, Chennai – 600 025 INDIA.
[*]Email: nrajendran@annauniv.edu

ABSTRACT

Bio-active sodium substituted zirconium was synthesised by sol-gel method and coated on 316L SS substrate. XRD patterns of as prepared specimen are amorphous in nature. Calcination leads to the formation of crystalline sodium zirconate. The FT-IR spectra showed a broad band between 3300 and 3000 cm^{-1}, which was assigned to fundamental stretching vibrations of different OH^- groups. AFM, SEM-EDAX and TEM micrographs showed the surface morphology of coated sodium zirconate to be nanoporous and uniform. The contact angle value was found to be 19°. The corrosion performance of the coatings has been evaluated by using electrochemical impedance spectroscopy measurements, which proved increased resistance of nano sodium zirconate particles coated 316L SS substrate against corrosion. The polarisation results show a shift in the positive potential associated with a decrease in current density. These results indicated that the sodium zirconate coatings on 316L SS substrate exhibited excellent corrosion resistance.

INTRODUCTION

The short and long term success of metallic implant devices utilized for orthopaedic and dental rehabilitation is dependent on their ability to maintain physical and chemical properties in the dynamic physiologic environment. The implant surface is the first to interact with the host, its biocompatible and osteoconductive properties are key for the modulation of the biofluid, cellular, and tissue level interactions that modulate osseointegration[1]. Traditional medical metallic implants are usually made of one of the three types of materials: 316L stainless steels, cobalt–chromium alloys and titanium and its alloys[2]. Among the various metallic materials 316L stainless steel (SS) is one of the most commonly used, because of its lower cost, acceptable biocompatibility and ease of fabrication. Moreover 316L stainless steel possesses reasonable corrosion resistance, tensile strength, fatigue resistance and suitable density for load-bearing purposes thus making this material a desirable surgical-implant material[3]. However, 316L SS orthopaedic implants suffers localised corrosion and releases iron, chromium and nickel ions to the neighbouring tissue thus inducing fibrosis around the implant[4]. To overcome this problem, the prostheses should have high corrosion resistance and good adhesion to the tissue so that a stable biological bond is formed with the bone.

Several surface engineering methods have been developed in recent times to improve the biocompatibility, osteoconductivity and corrosion resistance of the implants. One of the most effective methods is to deposit a protective bio-ceramic coating layer on the metal surface. Various bio-ceramic coatings such as titanium dioxide, bioglass, calcium phosphate, niobium oxide and sintered hydroxyapatite have been used[5-7]. However due to their poor mechanical properties, these materials cannot be used in load bearing applications, such as in femoral and tibial bones. Therefore, it is necessary to develop new materials with adjustable mechanical properties, as well as adequate biocompatibility and biosafety. In recent years, bioceramic materials based on zirconia have been used in biomedical applications because of their good fracture toughness, high strength, excellent biocompatibility, low cost[8]. Many attempts have been made for the deposition of thin film coatings, such as blast coating, sputtering, physical and chemical vapour deposition and sol-gel method[9,10].

Among the above techniques sol-gel method is the most versatile technique because it requires low temperature and considerably less expensive.

Thus the main aim of the work is to deposit sodium zirconate on 316L SS to reduce the corrosion of the substrate, increase the mechanical properties and to induce the growth of hydroxyapatite on the implant. The bioceramic coated 316L SS was investigated using surface characterisation techniques and the corrosion resistance was evaluated by electrochemical techniques.

MATERIALS AND METHODS

Substrate pre-treatments

Commercially available AISI stainless steel (316L SS) specimen of composition 18% Cr, 12% Ni, 0.5% Mn, 1 % Si, 0.03% S, 0.03%C, 0.045% P, 3% Mo and balance Fe cut into 20 mm x 10 mm x 2 mm were used as the substrate in the present study. The specimens were mechanically polished using 400 grit silicon carbide paper in order to remove coarse scratches and deformations. To perform fine grinding, samples were drawn back to front (one direction only) across the paper. Fine grinding was performed using a continuous water flow for lubrication. Each step was done for approximately 5 min. After grinding, the specimens were thoroughly washed with double distilled water to prevent abrasive particles from being carried. Then the specimens were cleaned with acetone in an ultrasonic cleaner for 20 min. Then it is degreased using 5% NaOH solution at $50 \pm 1^{\circ}$C and then etched in a mixed acid solution of HNO_3 (150 g/L) and HF (50 g/L) for 5 min, at a temperature of $28 \pm {}^{\circ}$C to ensure that the surface was free from any oxide layer, impurities, etc. The substrate were then dipped in 30% NH_4Cl solution for 30 min at $50 \pm 1^{\circ}$C to avoid further surface oxidation and to enhance adhesion of the molten metal onto the substrate during dipping process.

Synthesis of the sodium zirconate

The sodium zirconate was prepared by sol-gel method, all the chemicals were obtained from Merck (India) and used as such without purification. The starting compounds were sodium hydroxide and zirconium oxychloride. The molar ratio of components used was 0.075: 0.0031 (Sodium hydroxide: Zirconium oxychloride). Stochiometric amounts of sodium hydroxide were dissolved in double distilled water and it was slowly added to zirconium oxychloride solution. The resulting suspensions were stirred in a flask fitted with a reflux condenser for 14 h at 80 °C under vigorous condition until a gelatinous solution was obtained. The solution with the milky white precipitate turns to gel after long stirring. The prepared solution was aged in dark for 48 h in order to complete the nucleation process. Then the 316L SS substrates were immersed in the sol solution and withdrawn at a speed of 5 cm/min at room temperature to ensure the elimination of organic residues as well as to develop a thin film. The sol-gel sodium zirconate thus obtained on 316L SS substrate was later dried in an oven at of 70 °C for a period of 10 h for the gelation process. Then the samples were sintered at 700 °C for 1 h.

Characterisation techniques

Thermo gravimetric analysis was carried out in nitrogen atmosphere, using STA 409, Netzch instrument on green powders. The heating rate was fixed at 10°C/min. The contact angle subtended by SBF was measured using an FTA 200 contact angle goniometer. X-ray diffraction patterns were recorded with a PAN Analytical X-pert pro diffractometer using Cu Kα radiation, with 40 kV and 30 mA, at a scan rate of (2θ) 0.02°. FT–IR characterizations were carried out for the apatite layer formed over the surface in the range 400 – 4000 cm^{-1} on a Perkin – Elmer using the KBr tablet technique to confirm the phosphate and carbonate groups. The morphology of synthesised sodium zirconate powder was recorded using transmission electron microscopy (JEOL 2000FX). The surface morphology of the coating before and after immersion in SBF were recorded using scanning electron microscopy on a Hitachi Model-S 3400. The thickness of the coating was measured using elcometer.

Bio-electrochemical measurement

A conventional three-electrode cell was used for all the electrochemical measurements. A saturated calomel electrode (SCE) was used as a reference electrode, platinum foil as a counter electrode and the test material as the working electrode. The preparation of simulated body fluid (SBF) and the procedure for electrochemical experiment and the in vitro studies were carried out according to an earlier report[11]. The chemical compositions of simulated body fluid (SBF) are NaCl 8.035 g/ml, NaHCO$_3$ 0.355 g/ml, KCl 0.225 g/ml, K$_2$HPO$_4$.3H$_2$O 0.231 g/ml, MgCl$_2$.6H$_2$O 0.311 g/ml, 1 Kmol / cm^3 HCl 40 cm^3 g/ml, CaCl$_2$ 0.292 g/ml, Na$_2$SO$_4$ 0.072 g/ml, ((HOCH$_2$)$_3$CNH$_2$) 6.118 g/ml, 1 Kmol / cm^3 HCl appropriate amount for adjusting the pH. Potentiodynamic polarization studies were carried out for the test specimens in SBF solution. The potentiostat (model PGSTAT 12, AUTOLAB, the Netherlands B.V.) controlled with a personal computer was used for conducting the experiments. In order to test the reproducibility of the results, the experiments were performed in triplicate.

RESULTS AND DISCUSSIONS

Surface characterisations

The TG curves of as-prepared sodium zirconate sol are shown in Figure 1. A significant mass loss is measured from room temperature to approximately 600°C. The first curve at 105°C was due to removal of water molecule. The second and the third curve upto 503°C were due to the removal of deposited organics, chloride, solvent etc. No weight loss was observed above 700°C and the total weight loss amount to 75%[12].

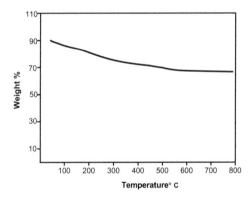

Figure 1. TGA curves of sol-gel sodium zirconate coated 316L SS

The thickness of the sodium zirconate coated 316L SS was found to be 1μ. The TF-XRD pattern of sodium zirconate coated on 316L SS is shown in Figure 2. Based on the thermal analysis, the sintering schedule is established. A slow-heating rate upto 700°C is expected to drive off the organic parts of the precursors gradually and completely and the sodium zirconate crystal nucleons are formed simultaneously. The heating process upto the desired temperature of 700°C with a hold time for 1h is expected to make the sodium zirconate crystallisation and densification. The sodium zirconate coated specimens were spontaneously cooled to room temperature in the furnace to obtain the desired product. The corresponding 2θ values of 18.04, 28.94, 36.80, 38.77, 39.89, 40.01, 44.57 and 61.30 represent the planes of (020), (311), (131), (040), (331) (122), (340) and (260) respectively (JCPDS Card - 35-0770). The crystalline peaks were identified as monoclinic. The average grain size of sodium zirconate was calculated using Scherrer's equation $D = 0.9\lambda/(\beta \cos (2\theta)$[13] where λ is the wavelength of the radiation, θ the diffraction angle, β the corrected half-width of the diffraction peak. The calculated crystalline size of the sodium zirconate is 18.6 nm. The signal observed at 2θ values of 44.60, 53.78 and 74.56 corresponds to the planes (111), (200) and (220) of cubic centric $Cr_{0.19} Fe_{0.7} Ni_{0.11}$, respectively for the 316L SS[14].

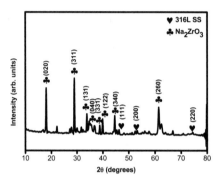

Figure 2. X-ray diffraction pattern of sodium zirconate coated 316L SS

Figure 3 shows the morphology of the synthesized sodium zirconate powder using transmission electron microscopy. Surface coating structures of implants are known to play significant roles in the interactions between living bone tissues and implants. Particularly, nano-scale features have attracted ever increasingly attention due to their unique effects on in vitro cell responses[15]. It is clear from the figure that the sodium zirconate coatings are in the form of rod shape. The shape also depends on the temperature, viscosity of the precursor materials and the rotation speed.

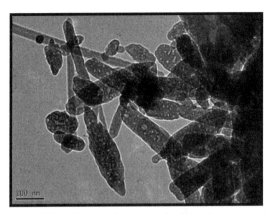

Figure 3.   Transmission electron microscopy of sodium zirconate powder

Mechanical properties

Scratch test

Adhesion of the coating to the substrate is the utmost importance for the implant to function in the body fluid environment. The scratch patterns for sodium zirconate coated on 316L SS taken at loads of 1, 4.2, 7, 16.7, 32.8 and 41.6 N loads are shown in Figure 4 (a-f). When the applied load increased from 1 N to 7 N, there was no change on the surface. The surface of the coating remains undamaged and smooth without any cracks. However, the applied load of 16. 7 N, some cracks are visible. This load could be taken as Lc_1, i.e. the first critical load for cohesive failure. The critical point for the coating failure is detected by the abrupt change with a continual fluctuation in the displacement of the probe. However, pores and chippings were visible at 32.8 N, thus making this load as the load for adhesive failure i.e. Lc_2, which can be used as indication to evaluate the adhesion force between the coating and the substrate. At 41.6 N the coated substrate is found to have the interlayer gaps and detachment from the substrate. The online measurements of frictional force and acoustic emission are shown in Figure 5. Coefficient of friction (μ), increased with the increase in scratch load. The μ value increased from 0.3 to 0.6 when the load was increased from 1 N to 7 N. With the increase in scratch load to 16 N, the μ value increased to 0.7. The increase in the value as the scratch test progressed is due to the rough nature of the coatings. Thus the bioceramic sodium zirconate coatings achieved a structural integrity and the chemical bonding enhances the densification of the coating effectively. Therefore sodium zirconate coated 316L SS substrate have acceptable mechanical properties which can be used as an implant biomaterial in high load-bearing conditions[16].

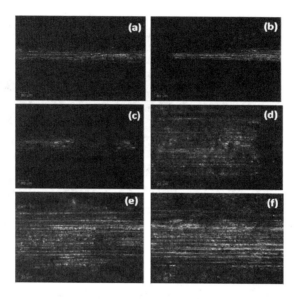

Figure 4. Optical micrograph scratch patterns of sodium zirconate coated 316L SS at (a) 1 N, (b) 4.2 N, (c) 7 N, (d) 16.7 N, (e) 32.8 N and (f) 41.6 N loads.

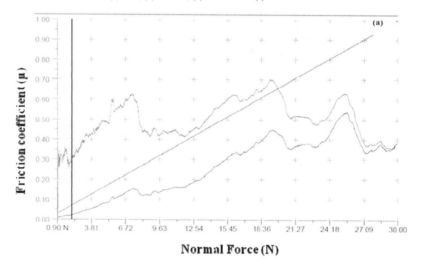

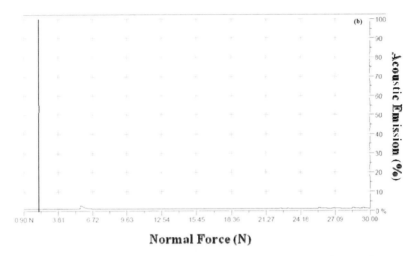

Normal Force (N)

Figure 5. Scratch test results (a) coefficient of friction (b) Acoustic emission during scratch test

Invitro characterisations

The TF-XRD patterns of the sodium zirconate coated 316L SS after immersion in SBF for 7 days are shown in Figure 6. The presence of Zr-OH groups on the sodium zirconate coating combines with the positively charged calcium ions to form calcium zirconate. As the calcium ions are accumulated, the surface is positively charged and hence combines with negatively charged phosphate ions to form an amorphous calcium phosphate. This phase is metastable and hence eventually transforms into crystalline bone-like hydroxyapatite[17].

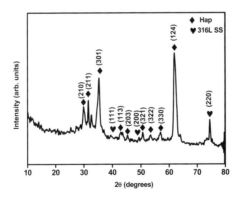

Figure 6. X-ray diffraction pattern of hydroxyapatite over sodium zirconate coated 316L SS after immersion in SBF solution for 7 days

The apatite peaks were seen to occur at 2θ of 29.98, 31.58, 33.52, 35.38, 42.40, 44.56, 51.34, 54.81, 58.21and 61.79 which are consistent with the standard XRD peaks for apatite (JCPDS-74-0566). It is therefore confirmed that the cluster of white spherical shape structure, covering the sodium zirconate coated 316L SS, is a bone-like apatite. The HR-SEM analyses of uncoated, sodium zirconate coated and sodium zirconate coated samples after immersion in SBF solution are shown in Figure 7 (a-c). The uncoated specimen shows the presence of smooth uniform surface. However, a homogenous porous rod shaped crack free morphology was observed for sodium zirconate coating. After immersion in SBF solution, white globules were observed on the surface of the coating, confirming the presence of apatite. Once the apatite nuclei are formed, they can grow spontaneously by consuming the calcium and phosphate ion in the SBF[18].

Figure 7. SEM (a) uncoated 316L SS (b) sodium zirconate coated 316L
SS (c) sodium zirconate coated after immersion in SBF for 7 days

The typical shape of simulated body fluid (SBF) droplet on sodium zirconate coated 316L SS is shown in Figure 8. The SBF droplet on uncoated 316L SS was hydrophobic with contact angle value of 83.2 ± 1.0° (lowest adhesion energy)[19] and the sodium zirconate coated 316L SS was hydrophilic with the value of 19 ± 1.0° (highest adhesion energy). The SBF solution spreads evenly over the coated surface

Figure 8. Contact angle measurements for sodium zirconate coated 316L SS in the
presence of SBF drops

making it hydrophilic. The hydrophilic nature of the nano sized coating facilitates the ion exchange behaviour from the SBF solution, which favour effective apatite growth. The FTIR spectrum of sodium zirconate coated 316L SS before and after immersion in SBF solution for seven days is shown in Figure 9. Before immersion, a broad band between 3500-3300 cm^{-1} was assigned to fundamental stretching vibration of hydroxyl groups. Another peak related to hydroxyl group was found at 1650 cm^{-1}.

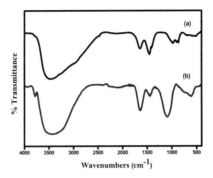

Figure. 9. FTIR spectra of sodium zirconate coated 316L SS (a) before immersion (b) after immersion in SBF solution for 7 days

The peaks in the range of 810 - 520 cm^{-1} are related to Zr-O and Zr-O-Zr groups. From the above results it is indicated the presence of sodium zirconate on 316L SS substrate was confirmed[20]. However after immersion in SBF, the peak at 3500 cm^{-1} clearly indicated the presence of the OH group of the hydroxyapatite. The band with small shoulder over the region 1630 cm^{-1} together with the absorption band at 1319 cm^{-1} is related to the CO_3^{2-} group from the carbonate apatite. The peak at 1030 cm^{-1} corresponds to the stretching of the P=O bond on the hydroxyapatite. The peaks between 500 and 710 cm^{-1} are related to Zr-O and Zr-O-Zr vibrations which might overlay the bands of CO_3^{2-} or P-O and O-P-O vibrations.

Bio-electrochemical characterisations

The potentiodynamic cyclic polarisation curves of uncoated and sodium zirconate coated 316L SS substrate are shown in Figure 10. The sodium zirconate coated 316L SS with elevated open circuit potential and lower corrosion current density indicated the corrosion resistance of the coatings. The corrosion parameters obtained from the electrochemical measurements are given in Table 1.

Table 1 Corrosion parameters of uncoated and sodium zirconate coated 316L SS in SBF solution

Sample	OCP (mV)	E_b (mV)	E_p (mV)	ΔE	Corrosion rate mmpy X 10^{-2}
Uncoated 316L SS	-510	400	-450	850	3.75
Sodium zirconate coated 316L SS	-320	930	650	280	0.82

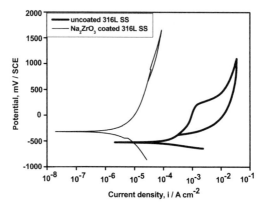

Figure 10. Potentiodynamic polarization curves for uncoated and sodium zirconate coated 316L SS in SBF solution

The coated substrate showed a considerable ennoblement in the corrosion potential and reduction in the current density. This showed that there was no detachment of the coatings from the substrate. Thus the nano sodium zirconate coating serves as a considerably effective barrier against corrosion electrolyte.

The electrochemical impedance spectra of uncoated, sodium zirconate coated specimen during immediate immersion in SBF solution are shown Figure 11a. The presence of two time constants confirms the sodium zirconate coatings on the 316L SS substrate. The Bode resistance plot for the sodium zirconate coated substrate showed higher resistance when compared to that of uncoated substrate. The uncoated and sodium zirconate coated 316L SS immersed in SBF solution for 7 days are shown in (Figure 11b). The sodium zirconate coated specimen showed three time constant with distinct phase angle behaviour which confirms the formation of new apatite layer over the porous surface, can be attributed to a growth of apatite as inferred from the SEM analysis (Figure 7c). The coated substrate indicated that the barrier effect on the film remains unaltered, which shows the ingress of anions in the electrolyte from attacking the metal surface[21].

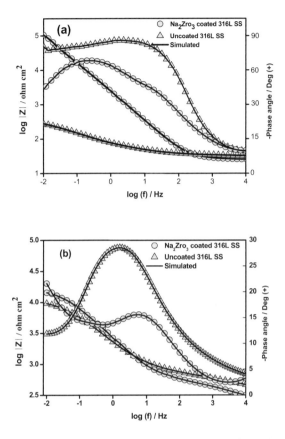

Figure 11. Impedance spectra of uncoated and sodium zirconate coated 316L SS (a) immediate immersion (b) Immersion for 7 days in SBF solution

CONCLUSIONS

In conclusion, cheap and reproducible nano bioceramic sodium zirconate coatings were prepared by sol-gel method and coated on 316L SS implant material by dip-coating method. The XRD patterns revealed that the coated sodium zirconate exhibit monoclinic phase. The surface morphological studies revealed that the coatings obtained were rod shaped, nanoporous and uniform. The scratch test revealed that the coatings possess good adhesion to the substrate. In vitro response of coated samples showed the formation of apatite over the coated specimen. The contact angle measurements showed the hydrophilic nature of the coatings with increased bio-activity. The bio-electrochemical studies confirmed the lower passive current density and nobler corrosion potential for sodium zirconate coating compared to uncoated 316L SS in SBF solution. These results showed that sodium zirconate coating on 316L SS is an interesting material which offers the required bio-activity and corrosion resistance (a twin role) in the simulated body fluid making it applicable for biomaterials.

ACKNOWLEDGEMENT
      One of the authors Ms. K. Bavya devi is thankful to All India Council of Technical Education (AICTE-NDF), New Delhi, India for the financial assistance.

REFERENCES
[1]J. E Lemons, Biomaterials, biomechanics, tissue healing, and immediate-function dental implants. *J. Oral Implantol.*, **30**, 318-324 (2004).

[2]R. M. Hall, A. Unsworth, Friction in hip prostheses, *Biomaterials,* **18**, 1017-26 (1997).

[3]Laura Indolfi, Filippo Causa, Paolo Antonio Netti. Coating process and early stage adhesion evaluation of poly (2-hydroxy-ethyl-methacrylate) hydrogel coating of 316L steel surface for stent applications. *J. Mater. Sci: Mater Med.,* **20**, 1541-1551 (2009).

[4]R. B. Tracana, J. Sousa, G. S. Carvalho, Mouse inflammatory response to stainless steel corrosion products, *J. Mater. Sci. Mater. Med.*, **5**, 596-600 (1994).

[5]J. Lausmaa, B. Kasemo, H. Mattsson, surface spectroscopic characterisation of titanium implant materials, *App. Sur. Sci.*, **44**, 133-146 (1990).

[6]P. Li, and P. Dueheyne, Quasi-biological apatite film induced by titanium in a simulated body fluid, *J. Biomed. Mater. Res*, **41**, 341-348 (1998).

[7]L. L. Hench, The story of Bioglass, *J. Mater. Sci: Mater. Med.*, **17**, 967-978 (2006).

[8]Guocheng Wang, Fanhao Meng, Chuanxian Ding, Paul K Chu, and J. Xuanyong Liu, Microstructure, bioactivity and osteoblast behavior of monoclinic zirconia coating with nanostructured surface, *Acta Biomater.*, **6**, 990-1000 (2010).

[9]S. Nagarajan, V. Raman, and N. Rajendran, Synthesis and characterization of porous niobium oxide coated 316L SS for orthopedic application, *Mat. Chem. Phy.*, **119**, 363-366 (2010).

[10]Chun Wang, Jurgen Meinhardt, Peer Lobmann, Growth mechanism of Nb-doped TiO_2 sol-gel multilayer films characterized by SEM and focus/defocus TEM, *J. Sol-Gel. Sci. Technol.* DOI 10.1007/S10971-009-2070-7

[11]Ines becker, Ingo Hofmann, Frank A, Muller. Preparation of bioactive sodium titanate ceramics, *J. Europ. Ceram. Soc*, **27**, 4547-4553 (2007).

[12]Bussarin Ksapabutr, Erdogan Gulari, Sujitra Wongkasemjit. Preparation of zirconia powders by sol-gel route of sodium glycozirconate complex, *Powder technol.* , **148**, 11-14 (2004).

[13]T. Lindgren, J. H. Muabora, E. Avendeno, J. Jonsson, A. Hoel, C. G Granquist, S. E Lindquist. Photoelectrochemical and optical properties of Nitrogen doped titanium dioxide films prepared by reactive DC magnetron sputtering *J. Phys. Chem. B* **107**, 5709-5716 (2003).

[14]G. Teufer, The crystal structure of tetragonal ZrO_2. *Acta Cryst.*, **15**, 1187 (1962).

[15]R. G. Flemming, C. J Murphy, G. A. Abrams, S. L. Goodman, and P. F. Nealey, Effects of synthetic micro and nano structured surfaces on cell behaviour, *Biomaterials,* **20**, 573-588 (1999).

[16]K Bavya Devi, Kulwant Singh, and N. Rajendran, Synthesis and characterization of nanoporous sodium-substituted hydrophilic titania ceramics coated on 316L SS for biomedical applications, *J. Coat. Technol. Res.,* DOI 10.1007/s11998-011-9344-z.

[17]Tadashi Kokubo, Design of bioactive bone substitutes based on biomineralization process, *Mater. Sci. Eng C,* **25**, 97-104 (2005).

[18]Chikara Ohtsukic, Tadashi Kokubo , Takao Yamamuro, Mechanism of apatite formation on CaO-SiO_2-P_2O_5 glasses in a simulated body fluid, *J. Non-Cryst. Solids,* **143**, 84-92 (1992).

[19]K. Bavya Devi, Kulwant Singh, and N. Rajendran, Sol-gel synthesis and characterization of nanoporous zirconium titanate coated on 316L SS for biomedical applications, *J. Sol-Gel. Sci. Technol.,* DOI 10.1007/s1097-011-2520-x

[20]S. Nagarajan, and N. Rajendran, Sol-gel derived porous zirconium dioxide coated on 316L SS for orthopedic applications, *J. Sol-Gel. Sci. Technol.,* **52**, 188-196 (2009).

[21]S. Nagarajan, and N. Rajendran, Surface characterisation and electrochemical behaviour of porous titanium dioxide coated 316L stainless steel for orthopaedic applications, *Appl. Surf. Sci.,* **255**, 3927-3932 (2009).

INFLUENCES OF SR, ZN AND MG DOPANTS ON OSTEOCLAST DIFFERENTIATION AND RESORPTION

Mangal Roy, Gary Fielding, and Susmita Bose*

W. M. Keck Biomedical Materials Research Laboratory
School of Mechanical and Materials Engineering
Washington State University, Pullman, WA 99164, USA

*Corresponding Author: e-mail: sbose@wsu.edu, Tel. 509-335-7461, Fax. 509-335-4662

ABSTRACT

Osteoclasts, the bone resorbing cells, are specialized macrophages generating from differentiation of mononuclear cells. In this research strontium (Sr), magnesium (Mg) and zinc (Zn) substituted beta-tricalcium phosphate (β-TCP) on differentiation of mononuclear cells into osteoclasts and its resorptive activity were studied. 1.0 wt% Sr, 1.0 wt% Mg, and 0.25 wt% Zn doped β-TCP were prepared by sintering at 1250 °C. Osteoclast differentiation and resorption was studied *in vitro* using RAW 264.7 precursor cell, supplemented with receptor activator of nuclear factor $\kappa\beta$ ligand (RANKL). Formation of osteoclast cells were monitored by field emission scanning electron microscope (FESEM) and immunohistochemical analysis. Formation of osteoclasts were noticed on pure, Sr doped and Zn doped β-TCP samples after 8 days of culture. However, no such osteoclast formation was found on Mg doped β-TCP samples. Cells on the substrates surface expressed specific osteoclast markers such as; actin ring, multiple nucleus, and vitronectin receptor. Immunohistochemical analysis indicated that matured osteoclasts were formed on β-TCP, Sr, and Zn doped samples after 14 days of culture whereas it took 21 days to form matured osteoclasts on Mg doped samples. Differentiated osteoclasts successfully resorbed β-TCP samples which increased for Zn doped samples and decreased for Sr doped samples. Mg doped β-TCP samples initially restricted osteoclast differentiation, however, once formed showed prominent resorption lacunae formation.

1. INTRODUCTION

Osteoclasts are large multinucleated cells whose function is to resorb bone. The activities of osteoclasts are controlled by cytokines and colony stimulating factors[1,2]. Among the cytokines, receptor activator for nuclear factor $\kappa\beta$ ligand (RANKL) plays an important role in osteoclast formation. The process of osteoclast formation is known as osteoclastogenesis[2]. During bone remodeling, osteoblasts generally produce RANKL and other cytokines, if not externally controlled, to induce osteoclastogenesis.

Bone remodeling is a lifelong active and coordinated biochemical process of osteoclastic resorption and osteoblastic synthesis of new bone. Any disruption in the equilibrium of these two processes results in either osteoporosis (increased osteoclastic activity-loss of bone density) or osteopetrosis and fragile bone (dysfunction of osteoclastic -increase in bone density)[1]. The bone mass can also increase by stimulated osteoblast activity, generally known as osteosclerosis[1]. Increase in bone density decreases bone toughness and eventually increase the chances of bone fracture. Therefore, bone replacement materials must be able to control both osteoblast and osteoclast functions in order to maintain successful bone remodeling. Calcium phosphate ceramics, especially β-TCP, found wide application in reconstructive bone surgery due to its close proximity in chemical composition to natural bone and the ability to get resorbed by natural biological processes without creating immunogenic responses[3,4]. It has also been reported that mechanical, *in vitro* and *in vivo* bioactivity of β-TCP, including osteoblast and osteoclast activity, can be controlled by metal ion doping[5-7].

In this paper, we report the effects of Sr, Zn and Mg doping in β-TCP on the formation of osteoclast and its resorptive activity. Osteoclast formation was studied *in vitro* by using RAW 264.7 mononuclear osteoclast precursor cells, supplemented with RANKL. Osteoclast cell formation was monitored by cellular morphology, acting ring formation and multiple nucleuses. Activity of the osteoclast cells was studied by the resorption lacunae formation.

2. MATERIALS AND METHODS

2.1 Materials processing

β-tricalcium phosphate nanopowder (β-TCP, Berkeley Advanced Biomaterials Inc. Berkeley, CA, USA) with an average particle size of 550 nm was used for this study. Strontium oxide (SrO, 99.9% purity) and zinc oxide (ZnO, 99.9% purity) was purchased from Sigma-Aldrich (MO, USA) and magnesium oxide (MgO, 99.998%) was procured from Alfa Aesar (MA, USA). Four compositions of β-TCP were prepared for this study: β-TCP, β-TCP doped with 1.0 wt% Mg, β-TCP doped with 1.0 wt% Sr and β-TCP doped with 0.25 wt% Zn (hereafter refereed as TCP, Mg-TCP, Sr-TCP Zb-TCP, respectively). The dopant concentrations were selected based on our previous optimization studies for excellent osteoblast-TCP interaction and superior mechanical properties[7, 8]. Samples were prepared by mixing β-TCP powder and appropriate amounts of dopants in polypropylene Nalgene bottles containing 75 mL of anhydrous ethanol and 100 g zirconia milling media. The mixtures were then milled for 6h at 70 rpm to minimize agglomeration, and increase homogeneity. Milled powder was dried in an oven at 60°C for 72h and pressed to disc shapes (12 mm diameter and 2.5 mm thickness) at 145 MPa using a uniaxial press. Green compacts were then cold isostatically pressed at 414 MPa for 5 min followed by sintering at 1250 °C for 2h in a muffle furnace.

2.2 Phase and microstructural analysis

Siemens D500 Krystalloflex X-ray diffractometer using Cu Kα radiation at 35 kV and 30 mA at room temperature was used to determine different phases over the 2θ range between 25° and 40°. Apparent density of each composition was determined by the Archimede's method and reported as normalized density to the theoretical density of β-TCP, i.e., 3.07 g/cm3. Average grain size of the sintered disks were determined from FESEM images via a linear intercept method using the equation $G = (L/N) \, C$, where G is the average grain size (μm), L is the test line length (cm), N is the number of intersections with grain boundaries along test line L; and C is the conversion factor (μm/cm) of the picture on which the test lines were drawn as obtained from the scale bar. 30 lines from 3 replicates were used from each composition to calculate the grain size. The data is reported as mean± standard deviation.

2.3 Contact angle measurement

Contact angles were determined by using deionized distilled water (D.I.) and cell culture media (Dulbecco's modified Eagle's medium/10 vol. % fetal bovine serum (DMEM) at pH 7.4) on polished samples. Contact angles were measured using the static sessile drop method using a face contact angle set-up equipped with a camera (VCA Optima, AST Products Inc., MA, USA)[9].

2.4 Osteoclast cell culture

The monocyte-like cell line RAW 264.7 (ATCC, USA) was cultured in DMEM media (ATCC, USA) complemented with 10 vol. % fetal bovine serum (FBS, Sigma, Germany) and 1 vol. % penicillin/streptomycin (Invitrogen, Germany) at 37°C in an atmosphere of 5% CO_2. β-TCP disks were autoclaved at 121°C for 20 min and then RAW 264.7 cells were added at a concentration of 10^5 cells/ml and incubated at 37°C in an atmosphere of 5% CO_2. At day 1, 50 ng/ml RANKL (Biolegend, CA,

USA) was added to the culture media. Cell media, containing 50 ng/ml, was changed every 2 days duration rest of the experiment.

2.5 Osteoclast cell morphology

Cellular morphology was assessed by field emission scanning electron microscope (FESEM, FEI 200F, FEI Inc., OR, USA). Samples were removed from culture after 5 and 8 days of incubation and were rinsed with 0.1M phosphate-buffered saline (PBS). Cells were subsequently fixed with 2% paraformaldehyde/2% glutaraldehyde in 0.1 M cacodylate buffer overnight at 4°C. After washing with 0.1M cacodylate buffer, each sample was post-fixed in 2% osmium tetroxide (OsO4) for 2h at room temperature. Fixed samples were dehydrated in an ethanol series (30%, 50%, 70%, 95% and 100% three times), followed by hexamethyl-disililane (HMDS) drying. Dried samples were gold coated and observed under FESEM for cell morphologies.

2.6 Immunofluorescence and confocal laser-microscopy

At specific culture days, osteoclast cells were fixed in 3.7% paraformaldehyde/ phosphate buffered solution, pH 7.4 and at ambient temperature for 10 min. After washing with PBS for 3 times, cells were permeabilized with 0.1% Triton X-100 (in PBS) for 4 min at room temperature. Samples were then washed with PBS and incubated in TBST-BSA (Tris-buffered saline with 1% bovine serum albumin, 250 mM NaCl, pH 8.3) blocking solution for 1h at room temperature. Actin staining was completed by incubating the samples in Rhodamine-phalloidine (molecular probes, invitrogen), with 1:40 dilution in PBS, for 30 min in the dark. Samples were then rinsed in PBS and incubated in the primary antibody (mouse-anti-vitronectin, abcam), with 1:50 in TBST solution for 2h and kept at 4 °C overnight. Samples were then washed with TBST-BSA for 10 min for 3 times. The secondary antibody (Alex Fluor 488 anti-mouse) was diluted 1:100 in TBST and was used to incubate the cells for 1h. After washing with TBST-BSA and with PBS, the samples were then mounted on glass coverslips with Vectashield mounting medium (Vector Labs, Burlingame, CA) with 4′,6-diamidino-2-phenonylindole (DAPI) and kept at 4°C for future imaging. The microscopical examinations were performed on a Zeiss 510 laser scanning microscope (LSM 510 META, Carl Zeiss MicroImaging, Inc., NY, USA)[10].

2.7 Resorption pit assay

Osteoclast cells were removed from sample surface after 14 and 28 days of culture. Samples were ultrasocicated in 1M NaCl solution with 0.2% Triton X-100 to remove the osteoclast cells and then washed with PBS[11]. After gold coating, samples were observed in FESEM for resorption lacuna.

3. RESULTS AND DISCUSSIONS

3.1. Physico-chemical properties

Figure 1 shows the X-ray diffraction (XRD) patterns of the doped and undoped β-TCP ceramics, sintered at 1250 °C for 2h. β-TCP (JCPDS # 09-169) was identified as the major phase in the doped and undoped samples. Few peaks related to α-TCP (JCPDS # 09-0348) phase were also identified in the XRD spectra. Presence of α-TCP indicated that some of the β-TCP transformed to α-TCP phase during sintering. No such α-TCP peaks were detected in Mg-TCP samples. Recently, we have shown that the phase stability of β-TCP can be increased by Sr and Mg doping[7]. The relative density of TCP, Sr-TCP, Mg-TCP and Zn-TCP were found to be 96.17 ±0.89%, 95.02 ±1.39%, 98.61 ±0.97%, and 98.07±0.52%, respectively. Addition of dopants resulted in slight increase in density of Mg-TCP and Zn-TCP compacts. Influence of dopants on grain size of sintered TCP compacts are shown in Table 1. The grain size was found to reduce for Mg-TCP and Zn-TCP samples while it increased for Sr-TCP samples.

Table 1 shows the contact angles of two different liquids (D.I. water and cell culture media) measured on TCP, Sr-TCP, Mg-TCP, and Zn-TCP samples. For all samples, contact angle significantly dropped when measured in cell culture media compared to that of D.I. water. However, the difference in contact angles between TCP, Sr-TCP, Mg-TCP, and Zn-TCP samples were insignificant when measured in D.I. water as well as cell culture media.

3.2. Osteoclast cell morphology

Cellular morphology and its interaction with doped and undoped β-TCP samples were analyzed using FE-SEM. Figure 2 shows RAW 264.7 cell morphology on TCP, Sr-TCP, Mg-TCP, and Zn-TCP samples after 5 days of culture in RANKL supplemented media. The osteoclast precursor monocytes spontaneously adhered to all β-TCP substrates. Cellular attachment is related to the extensions of plasma membrane, such as filopodia and lamellopodia. Existence of cellular micro-extensions indicated good cellular adhesion on all samples. At day 5, some large cells with round and flat morphology were noticed on TCP, Sr-TCP, and Zn-TCP samples which indicated onset of osteoclast differentiation. The characteristic "blebs" can be noticed on the cell surfaces[12]. In comparison, cellular agglomeration was found on Mg-TCP samples. Similar monocyte attachment after 5 days of culture on doped and undoped β-TCP samples could be explained by the contact angle results. Lower contact angle generally favors cellular attachment[9]. Insignificant difference in contact angles among the samples resulted in similar monocyte attachment on all the substrates.Osteoclast cellular morphology after 8 days of culture is shown in Figure 3. After 8 days, cells on TCP, Sr-TCP, and Zn-TCP samples grew in size with morphological features resembling to osteoclast. The cells showed highly oriented pseudopodia all around the plasma membrane. It is remarkable that monocytes only proliferated on Mg-TCP samples without differentiation, i.e. the formation of osteoclast cells was appear to be inhibited by the presence of Mg in β-TCP samples. In a recent article it has been reported that *in vivo* conditions, magnesium hydroxide temporarily reduce the osteoclast formation around the implant[13].

Osteoclast formation was also monitored by immunohistochemistry. The cells were analyzed for multinuclearity, actin ring formation and positivity for vitronectin receptor $\alpha_v\beta_3$ integrin. Figure 4 shows the fluorescence microscopy images of osteoclasts after specific culture days. After 14 days, cells were found to be multinuclear with continuous actin ring on TCP, Zn-TCP, and Sr-TCP samples. However, matured osteoclast cells were noticed on Mg-TCP samples only after 21 days. It is well known that multinuclearity and formation of actin ring are the essential characteristic markers of osteoclasts[2]. Presence of multiple nucleus, actin ring, and vitronectin receptor $\alpha_v\beta_3$ integrin confirmed the giant cells to be osteoclasts. The $\alpha_v\beta_3$ integrin, a vitronectin receptor protein, was highly expressed in the cells plasma membrane and cytoplasm. Although osteoclast formation was scarce on Mg-TCP samples on 8th day of culture, multinucleated osteoclasts were found after 21 days of culture on these sample surfaces which indicated that the osteoclastogenesis was slowed down in Mg-TCP samples, however, was not completely restricted. Matured osteoclasts were also found in TCP, Sr-TCP, and Zn-TCP samples. Therefore, our results indicated that Mg does not inhibit cellular attachment and proliferation on β-TCP substrates, however, plays an important role in osteoclast differentiation.

3.3. Osteoclastic surface resorption

Surface resorption characteristics of the doped and undoped β-TCP samples were studied by removing the osteoclast cells from the sample surfaces after 14 and 28 days of culture. Figure 5 shows the surface degradation of pure and doped β-TCP samples after 14 days. After 14 days of culture, formation of osteoclast imprints on the TCP, Sr-TCP, and Zn-TCP sample surfaces which indicated cellular attachment and beginning of cell mediated resorption. Resorption pits on TCP, Zn-TCP and Sr-TCP surfaces were mostly surface phenomenon. The resorption lacunae were similar in size to the matured osteoclasts found at day 8 on TCP, Zn-TCP and Sr-TCP which indicated that lacunae formation was primarily due to the differentiated osteoclasts at 8th day of culture. These results clearly indicated that most of the osteoclasts formed at day 8 started the resorption process. Due to restricted osteoclast cell differentiation, resorption lacunae were not found on Mg-TCP sample surfaces even after 14 days of culture. After 28 days of culture, significant increase in depth of the resorption pits was noticed in TCP, Zn-TCP and Mg-TCP samples. Formation of resorption lacuna on Mg-TCP samples after 28 days of culture could be attributed to the osteoclast-like cells that were found at 21st day of culture.

Degradation of TCP ceramic is dependent on many factors including pH, grain size, crystallinity, presence of secondary phases etc. However, decreased resorption rate for Sr-TCP samples could be due to the larger grain size. In comparison, Zn-TCP and Mg-TCP, owing to their smaller grain size, got readily resorbed by osteoclasts. It has been reported that decrease in hydroxyapatite (similar bone substitute material) grain size can significantly increase the osteoclast mediated resorption[14]. The increase in depth of resorption lacunae on Zn-TCP samples indicated that grain size plays an important role in osteoclastic resorption kinetics.

4. CONCLUSIONS

This study showed a surface chemistry mediated control of osteoclast formation on β-TCP and its activity. Monocytes were readily adhered to all of the TCP substrates, however; differentiated to osteoclasts only on TCP, Sr-TCP, Zn-TCP samples. Formations of osteoclasts were confirmed by both cellular morphology and Immunohistochemical analysis. Our results indicated that in presence of sufficient amount of RANKL, osteoclastogenesis can be drastically reduced on TCP ceramics by Mg doping. Moreover the cell mediated resorption of TCP ceramics can be significantly controlled by Zn, Sr and Mg doping. The present results can be used for adjusting bone graft substitutes by metal ion doping in order to affect their biological and degradation properties.

ACKNOWLEDGEMENTS

Authors acknowledge financial support from the National Institute of Health (N1H-RO1-EB-007351). The authors also like to acknowledge the financial support from the Office of Naval Research and the W. M. Keck Foundation to establish a Biomedical Materials Research Laboratory at WSU. Authors thank Dr. Ted S Gross and L. Worton from University of Washington for technical help and Dr. Christine Davitt and Dr. Valerie Lynch- Holm, of Franceschi Microscopy and Imaging Center, Washington State University for their helpful assistance in the immunohistochemical and FESEM analyses.

REFERENCES

1. S.L. Teitelbaum, Bone Resorption by Osteoclasts. *Science* **289**, 1504-1508 (2000).

2. W.J. Boyle, W.S. Simonet, D. Lacey, Osteoclast Differentiation and Activation. *Jature* **423**, 337-342 (2003).

3. S.V. Dorozhkin, M. Epple, Biological and Medical Significance of Calcium Phosphates, *Angewandte Chemie-International Edition*. **41**, 3130-3146 (2002).

4. T. Okuda, K. Ioku, I. Yonezawa, H. Minagi, G. Kawachi, Y. Gonda, H. Murayama, Y. Shibata, S. Minami, S. Kamihira, H. Kurosawa, T. Ikeda, The Effect of the Microstructure of β-Tricalcium Phosphate on the Metabolism of Subsequently Formed Bone Tissue, *Biomaterials* **28**, 2612- 2621 (2007).

5. L. Yang, S. Perez-Amodio, F. Groot, V. Everts, C.A. Blitterswijk, P. Habibovic, The Effects of Inorganic Additives to Calcium Phosphate on in Vitro Behavior of Osteoblasts and Osteoclasts, *Biomaterials* **31**, 2976-2989 (2010).

6. A. Bandyopadhyay, S. Bernard, W. Xue, S. Bose, Calcium Phosphate-Based Resorbable Ceramics: Influence of MgO, ZnO, and SiO_2 dopants, *J Am Ceram Soc.* **89**, 2675-2688 (2006).

7. S.S. Banerjee, M.S. Tarafder, N.M. Davies, A. Bandyopadhyay, S. Bose, Understanding the Influence of MgO and SrO Binary Doping on Mechanical and Biological Properties of β-TCP Ceramics, *Acta Biomaterialia*, **6**, 4167–4174 (2010).

8. S. Bose, M.S. Tarafder, S.S. Banerjee, N.M. Davies, A. Bandyopadhyay, Understanding in vivo response and mechanical property variation in MgO, SrO and SiO2 doped β-TCP, *Bone*, **48**, 1282-1290 (2011).

9. M. Roy, B.V. Krishna, S. Bose, A. Bandyopadhyay, Comparison of Tantalum and Hydroxyapatite Coatings on Titanium for Applications in Load Bearing Implants, *Advanced Biomaterials (Advanced Engineering Materials)* **12[11]**, B637-B641: (2010).

10. A.F. Schilling, W. Linhart, S. Filke, M. Gebauer, T. Schinke, J.M. Rueger, M. Amling. Resorbability of Bone Substitute Biomaterials by Human Osteoclasts. *Biomaterials*, **25**, 3963-3972 (2004).

11. S. Patntirapong, P. Habibovic, P.V. Hauschka, Effects of Soluble Cobalt and Cobalt Incorporated into Calcium Phosphate Layers on Osteoclast Differentiation and Activation, *Biomaterials*, **30**, 548–555 (2009).

12. R. Detsch, H. Mayr, G. Ziehler, Formation of Osteoclast-Like Cells on HA and TCP Ceramics. *Acta Biomaterialia*, **4**, 139-148 (2008).

13. C. Janning, E. Willbold, C. Vogt, J. Nellesen, A. Meyer-Lindenberg, H. Windhagen, F. Thorey, F. Witte. Magnesium Hydroxide Temporarily Enhancing Osteoblast Activity and Decreasing the Osteoclast Number in Peri-Implant Bone Remodeling. *Acta Biomaterialia*, **6**, 1861–1868 (2010).

14. R. Detsch, D. Hagmeyer, M. Neumann, S. Schaefer, A. Vortkamp, M. Wuelling, G. Ziegler, M. Epple. The Resorption of Nanocrystalline Calcium Phosphates by Osteoclast-Like Cells. *Acta Biomaterialia*, **6**, 3223-3233 (2010).

Table 1. Relative density and grain size of undoped and doped samples.

	Composition	Relative density (%)	Grain size (μm)	Contact angle	
				DI water	Cell media
TCP	β-TCP	96.17 ±0.89	4.09±0.69	57.83±7.44	45.41±4.38
Zn-TCP	β-TCP+0.25 wt% Zn	98.07±0.52	3.05±0.33	54.79±6.17	49.76±5.93
Sr-TCP	β-TCP+1.0 wt% Sr	95.02 ±1.39	4.86±0.84	52.35±2.65	47.10±4.63
Mg-TCP	β-TCP+1.0 wt% Mg	98.61 ±0.97	3.21±0.71	55.72±2.9	50.61±5.3

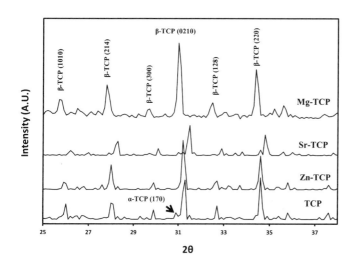

Figure 1. X-ray powder diffraction patterns of pure and doped β-TCP samples.

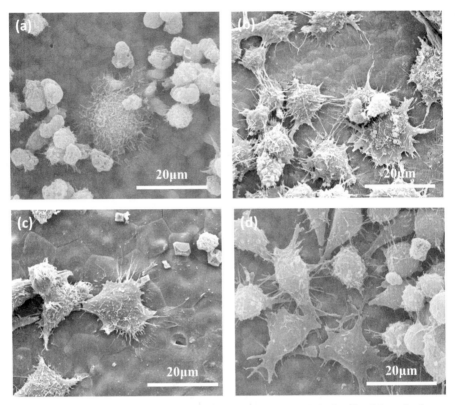

Figure 2: FESEM micrographs illustrating the RAW 264.7 cell morphologies after 5 days of culture with RANKL (a) TCP, (b) Zn-TCP, (c) Sr-TCP, and (d) Mg-TCP.

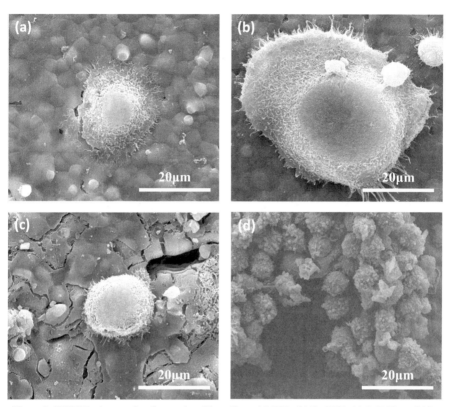

Figure 3: FESEM micrographs of osteoclast-like cells on (a) TCP, (b) Zn-TCP, (c) Sr-TCP, and (d) Mg-TCP samples after 8 days of culture. Well defined pseudopodia can be noted around the cell membane attached to TCP, Zn-TCP, and Sr-TCP.

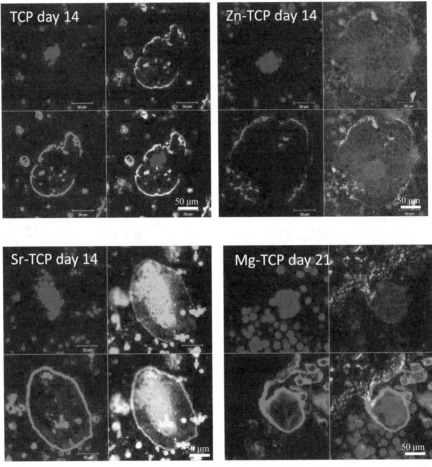

Figure 4: Fluorescence microscopy images of cells cultured for 21 days. The red represents the actin cytoskeleton, green indicates vitronectin receptor $\alpha_v\beta_3$ integrin and the blue represents the nucleolus. Osteoclast-like cells were identified by the presence of multiple nucleuses, acting ring and were vitronectin receptor $\alpha_v\beta_3$ integrin.

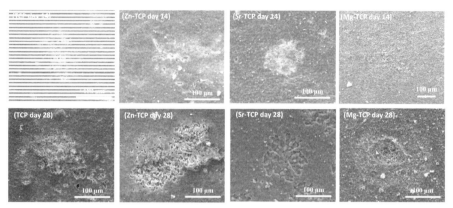

Figure 5: Surface morphology of pure and doped TCP samples after culturing with RAW 264.7 cells at day 14 and 28.

A COMPARITIVE STUDY OF CELL BEHAVIORS OF HYDROXYAPATITE AND Ti-6Al-4V

Ling Li, Kyle Crosby, Monica Sawicki, Leon L. Shaw, Yong Wang
Department of Chemical, Materials, and Biomolecular Engineering
University of Connecticut, Storrs, CT 06269

ABSTRACT

In this study, we have investigated the cell attachment and cell proliferation properties on the surfaces of hydroxyapatite (HA) and Ti-6Al-4V (Ti-6-4). The HA and Ti-6-4 samples are prepared via sintering and the sintering temperature is used to control grain sizes and the level of porosity. Cell attachment and proliferation properties are investigated with the aid of ROS17/2.8 cells and MTT assay, while cell morphologies on different surfaces are studied using scanning electron microscopy. It is found that when pore sizes are smaller than 100 μm for Ti-6-4 and 1 μm for HA, denser HA and Ti-6-4 are better in enhancing cell attachment and proliferation than porous counterparts.

INTRODUCTION

Current methods for bone implants include autologous, allogeneic graft, metallic implant and calcium phosphate-based bone graft. For autologous bone graft, it is regarded as "gold standard" in the field. Although it has many advantages such as osteoinductivity, osteoconductivity and no rejection problems, it is restricted by autologous resource limitation and potential donor site morbidity [1]. For allogeneic graft, the advantage is the retaining of proteins and growth factors in the structure, but the disadvantage is the potential transmission of disease and variation in different batches. In the case of calcium phosphate-based bone graft, they have compositional and structural similarities to natural bone inorganic phase, which result in osteoinductivity and osteoconductivity. However, they are brittle and lack the shape flexibility. What is more, most of the commercially available synthetic bone grafts lack bioactive molecules which are critical for osteoinductivity. Although there are some companies who provide scaffolds with bioactive proteins, there are no effectors in it to control the release of bioactive molecules. They use sponge to absorb bone morphogenetic proteins (BMPs) and release these proteins while sponge is dissolving. As for metallic implants [2], the problems are the limited life expectancy of metallic implants, inability to bond to living bone in vivo [3,4], movement at the implant-tissue interface [5], and higher elastic moduli than that of natural bones, all of which eventually result in loosening of the implant and the need for revision surgery to replace the loose implant.

Chang et al. [6] found that in cylindrical-type HAs, the 50-μm sized pore was enough for osteoconduction, and the 300-μm sized pore was optimal. Tsuruga et al. [7] reported that the optimal pore size for attachment, differentiation and growth of osteoblasts and vascularization was approximately 300 ~ 400 μm. Flatley et al. [8] reported the 500-μm sized pore was the most compatible for osteoconduction. Hubert et al. [9] concluded that the 100-μm sized pore was the smallest pore that can be used for osteoconduction, and the 150-μm sized pore was the optimal one. However, there is no research focusing on the effect of small pore sizes on cell behaviors. Thus, in this study we have investigated the effect of HA with pore sizes < 1 μm and Ti-6-4 with pore sizes < 100 μm on the cell attachment and proliferation. The results from this study are described below.

MATERIALS

ROS17/2.8 cell line was used as a cell model to study cell/HA and cell/Ti-6-4 interactions. ROS17/2.8 cells were cultured in the incubator and maintained in Dulbecco's modified Eagle's medium (DMEM) in a 5% CO_2 atmosphere at 37°C. The medium was supplemented with FBS (10%; v/v), 2mM L-glutamine and 100 U/mL penicillin–streptomycin (Hyclone, Logan, UT). LIVE/DEAD cell staining kit was obtained from Invitrogen. Alcohol, glutaraldehyde and paraformaldehyde were from Electron Microscopy Sciences. Other reagents were purchased from Fisher Scientific and Sigma.

Two conditions of HA discs were produced via sintering of HA nano-rods which were synthesized using the procedure described in a previous study [10]. The HA nano-rod powder was uniaxially pressed (300 MPa) into discs using a ½" diameter steel die. The pressed discs were sintered in air in a box furnace at 900°C or 1100°C for 2 hours. The 900°C-sintered discs will be designated as HA900 hereafter, whereas the 1100°C-sintered discs will be denoted as HA1100.

There were also two conditions for Ti-6-4 samples. One was Ti-6-4 discs sintered at 1250°C under vacuum for 2 hours using the as-received coarse-grained Ti-6-4 powder. This set of discs will be termed as the as-received Ti-6-4 hereafter. The other condition was Ti-6-4 discs sintered for 2 hours also at 1250°C under vacuum, but using Ti-6-4 powder ball milled at room temperature for 4 hours with the aid of a SPEX mill. This set of Ti-6-4 discs will be called SPEX-milled Ti-6-4 hereafter. Note that the surfaces of both Ti-6-4 and HA discs were polished up to 0.05 μm SiO_2 colloidal suspension before being subjected to bioactivity studies.

CHARACTERIZATION

Fluorescence Microscopy Imaging of Cell Attachment and Proliferation on HA and Ti-6-4 Discs

The initial seeding density of ROS17/2.8 cells was 2×10^4 cells/well in a 24-multiwell plate. Cells were directly seeded on the surface of glass coverslips, HA900, HA1100, SPEX-milled Ti-6-4 and as-received Ti-6-4 discs. At 1, 3, 5 and 8 days, cells on the surfaces of these discs were directly stained with LIVE/DEAD cell staining kit [11, 12]. Cell images were captured under an inverted fluorescence microscope (Zeiss Axiovert 40 CFL microscope, Thornwood, NY). The experiment was performed in triplicate.

Field Emission Scanning Electron Microscopy Imaging of Cell Morphology on HA and Ti-6-4 Discs

For SEM observations, the ROS17/2.8 cells (2×10^4 cells/well) were cultured on the glass coverslips, HA900, HA1100, SPEX-milled Ti-6-4 and as-received Ti-6-4 discs, and fixed in 1.5% glutaraldehyde and 1.5% paraformaldehyde in a 0.12M phosphate buffer solution (PBS) at room temperature for 2 hours. The ROS17/2.8 cells on the disc were rinsed 3 times with PBS for 20 min, and then dehydrated by increasing the concentration of alcohol (15%, 30%, 50%, 70%, 85%, 95% and $3 \times 100\%$). The critical point drying of specimens was undertaken with liquid CO_2. The specimens were sputter-coated with gold and examined by SEM [11, 13]. The cell morphology on the surfaces of different samples was observed at 1, 3, 5 and 8 days. The experiment was performed in triplicate.

Evaluation of Cell Activity on HA Discs by MTT Assay

We have also studied cell adhesion on the surface of different materials using MTT assay [11, 14]. MTT assay has been used to characterize cell activity. It is also a method to indicate the number of living cells. The materials were naked well (polystyrene), glass coverslips and HA discs. Glass coverslips and HA discs were placed in a 24-multiwell plate. The initial cell seeding density was 2×10^4 cells/well. At 1, 3, 5 and 8 days, cell activity was analyzed with MTT assay. Briefly, 1 mg/ml 3-(4, 5-dimethylthiazol-2-yl)-2, 5-diphenyltetrazolium bromide (MTT) solution was added into the culture medium. After 4 hours of incubation at 37°C in a 5% CO_2 atmosphere, the medium was removed and the formed purple formazan crystals were then dissolved in dimethyl sulfoxide (DMSO). The UV absorbance was measured using a microplate reader (BioTek). The experiment was performed in triplicate.

RESULTS

Microstructure of Sintered HA and Ti-6-4

Figure 1 shows the polished surface of HA900 and HA1100 discs. It is obvious that HA1100 is much denser than HA900. Indeed, based on the weight and volume of the discs, it is found that the density of HA900 is ~60% of the theoretical, whereas the density of HA1100 is close to 90% of the theoretical. Most of the pores for HA1100 are less than 1 μm and so is the case for HA900, as shown in Figure 1. However, the number of pores in HA1100 is much less than that in HA900.

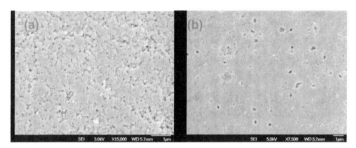

Figure 1. SEM images of (a) polished HA900 and (b) polished HA1100 discs.

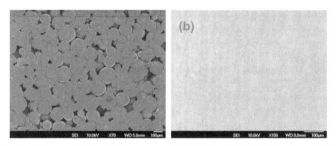

Figure 2. SEM images of (a) polished as-received Ti-6-4 and (b) SPEX-milled Ti-6-4 discs.

Figure 2 shows the polished surface of Ti-6-4 discs. It is clear that the SPEX-milled sample is almost 100% dense, whereas the as-received counterpart contains porosity. Furthermore, the pore sizes in the as-received sample are about 100 μm. The density for SPEX-milled samples is 97% of the theoretical, while the density for as-received samples is ~90% of the theoretical.

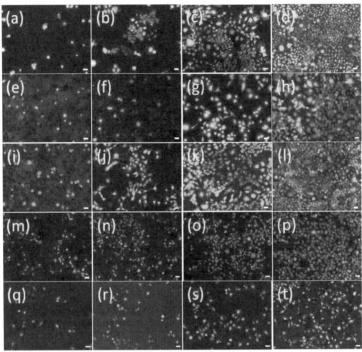

Figure 3. Comparison of the cell adhesion and proliferation on different materials at different time points. (a-d) 1, 3, 5 and 8 days on glass coverslips; (e-h) 1, 3, 5 and 8 days on HA900 discs; (i-l) 1, 3, 5 and 8 days on HA1100 discs; (m-p) 1, 3, 5 and 8 days on SPEX-milled Ti-6-4 discs; (q-t) 1, 3, 5 and 8 days on as-received Ti-6-4 discs. Green fluorescence indicates live cells, while red fluorescence indicates dead cells. The scale bar represents 10 μm.

Fluorescence Microscopy Imaging of Cell Attachment and Proliferation
Figure 3(a), (e), (i), (m) and (q) show that ROS17/2.8 cells attach to the surface of glass coverslips, HA900, HA1100, SPEX-milled Ti-6-4 and as-received Ti-6-4 discs after 1-day culture. The number of cells increases with time, and cells grow confluence after 8 days on glass coverslips and HA1100 discs (Figures 3d and 3l), but not on other discs. Cells grow the best on HA1100, followed by SPEX-milled Ti-6-4, and then HA900. As-received Ti-6-4 discs have the worst result. Clearly, HA has better bioactivity than Ti-6-4, and dense HA and Ti-6-4 are better than their porous counterparts.

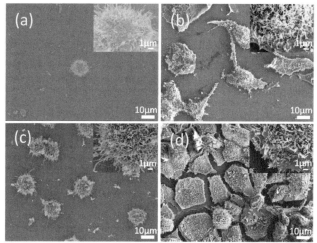

Figure 4. FESEM imaging of cell morphology on the surfaces of glass coverslips after incubation with cells at different time points. (a) 1 day; (b) 3 days; (c) 5 days; (d) 8 days. Inserts are higher magnification views of cell morphology.

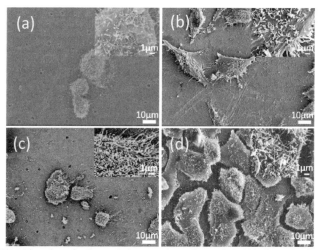

Figure 5. FESEM imaging of cell morphology on the surfaces of HA900 discs after incubation with cells at different time points. (a) 1 day; (b) 3 days; (c) 5 days; (d) 8 days. Inserts are higher magnification views of cell morphology.

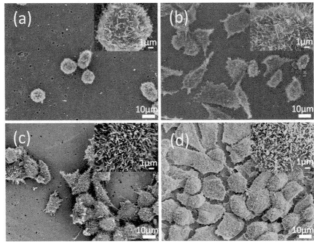

Figure 6. FESEM imaging of cell morphology on the surfaces of HA1100 discs after incubation with cells at different time points. (a) 1 day; (c) 3 days; (e) 5 days; (g) 8 days. Inserts are higher magnification views of cell morphology.

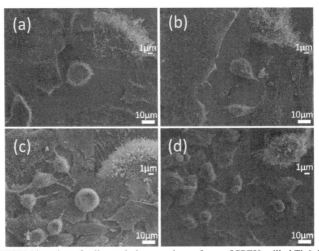

Figure 7. FESEM imaging of cell morphology on the surfaces of SPEX-milled Ti-6-4 discs after incubation with cells at different time points. (a) 1 day; (b) 3 days; (c) 5 days; (d) 8 days. Inserts are higher magnification views of cell morphology.

Morphology of ROS17/2.8

The morphology of cells on the different materials was observed at 1, 3, 5 and 8 days. Figure 4 shows FESEM images of ROS17/2.8 cells adhered on glass coverslips, while Figure 5 shows the ROS17/2.8 cell morphology on HA900 discs for the indicated period. Figure 6 shows FESEM images of ROS17/2.8 cells adhered on HA1100 discs. The cells on HA1100 discs are more than that on HA900 discs at the same indicated period. At 8 days (Figures 4d and 6d), the glass coverslips and HA1100 discs were found to be totally covered with ROS17/2.8 cells. The ROS17/2.8 cell morphology on SPEX-milled Ti-6-4 discs at the indicated period is shown in Figure 7, while Figure 8 shows cell morphology of ROS 17/2.8 cells adhered on as-received Ti-6-4 discs. The cells on SPEX-milled Ti-6-4 discs grow better than that on as-received Ti-6-4 discs. These results are consistent with the observation of fluorescence microscopy (Figure 3).

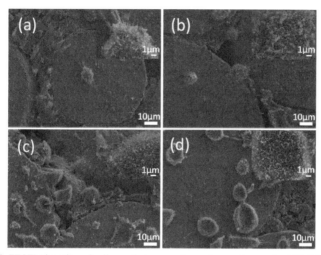

Figure 8. FESEM imaging of cell morphology on the surfaces of as-received Ti-6-4 discs after incubation with cells at different time points. (a) 1 day; (b) 3 days; (c) 5 days; (d) 8 days. Inserts are higher magnification views of cell morphology.

MTT Assay

Consistent with the microscopy observation, the MTT assay indicates that the number of cells on the surface of different materials increases with time (Figure 9). In addition, glass and HA1100 exhibit similar capability of enhancing cell adhesion. However, the numbers of cells on glass and HA1100 are lower than that of cells growing on the naked polystyrene surface (well in Figure 9). Comparing HA900 with HA1100, HA1100 discs have more cell adhesion. This result is consistent with the studies of SEM imaging (Figures 5 and 6) and fluorescence microscopy imaging (Figure 3).

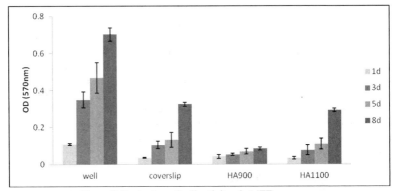

Figure 9. Evaluation of cell activity via MTT assay.

DISCUSSION

The studies of fluorescence microscopy imaging, SEM imaging and MTT assay all reveal that ROS17/2.8 cells grow better on HA1100 than on HA900. These results indicate that denser surface is more effective for cell attachment and proliferation when the pore size of HA discs is less than 1 μm. Similar results are obtained for Ti-6-4 when the pore size of Ti-6-4 discs is less than 100 μm. The present finding should not be confused with the previous notion, showing that the introduction of macro-porosity (with pore sizes in the range of 300 ~ 400 μm) to HA can enhance bone proliferation and fixation [6,7,15,16]. It has been shown that for bone formation and remodeling, a porous form of HA is more suitable for implantation than non-porous HA. Furthermore, HA with pore sizes of 300 ~ 400 μm has the best result for attachment, differentiation and growth of osteoblasts and vascularization [6,7,15]. The discrepancy between the present and previous studies lies in the pore size. In the present study the pore size of HA is less than 1 μm and it is less than 100 μm for Ti-6-4. In contrast, the beneficial effect is demonstrated with pores in 300 ~ 400 μm [6,7,15,16]. The mechanism with which small pores impede cell attachment and proliferation is not clear at this stage and will be the topic of future studies.

CONCLUSION

The present study reveals, for the first time, that cells exhibit better growth on the surface of denser HA and Ti-6-4 samples when the pore size of HA is smaller than 1 μm and the pore size of Ti-6-4 smaller than 100 μm. Furthermore, HA has better activity in enhancing cell attachment and proliferation than Ti-6-4.

ACKNOWLEDGMENTS

This research was sponsored by the U.S. National Science Foundation (NSF) under the contract number CBET-0930365. The support and vision of Ted A. Conway is greatly appreciated.

REFERENCES

[1]Finkemeier, C.G., Bone-grafting and bone-graft substitutes, *J. Bone Joint Surg. Am.*, **84**, 454-464 (2002).

[2]Hench, L.L., Bioceramics - from Concept to Clinic, *Journal of the American Ceramic Society*, **74**, 1487-1510 (1991).

[3]Kokubo, T., Novel bioactive materials, *Anales De Quimica*, **93**, S49-S55 (1997).

[4]Bobyn, J.D., et al., The susceptibility of smooth implant surfaces to periimplant fibrosis and migration of polyethylene wear debris, *Clin Orthop Relat Res*, **311**, 21-39 (1995).

[5]Moreland, J.R. and M.A. Moreno, Cementless femoral revision arthroplasty of the hip: minimum 5 years followup, *Clin Orthop Relat Res*, **393**, 194-201 (2001).

[6]Lee, C.K., et al., Osteoconduction at porous hydroxyapatite with various pore configurations, *Biomaterials*, **21**, 1291-1298 (2000).

[7]Tsuruga, E., et al., Pore size of porous hydroxyapatite as the cell-substratum controls BMP-induced osteogenesis, *Journal of Biochemistry*, **121**, 317-324 (1997).

[8]Flatley, T.J., K.L. Lynch, and M. Benson, Tissue-Response to Implants of Calcium-Phosphate Ceramic in the Rabbit Spine, *Clin Orthop Relat Res*, **179**, 246-252 (1983).

[9]Hulbert, S.F., S.J. Morrison, and J.J. Klawitter, Tissue reaction to three ceramics of porous and non-porous structures, *J Biomed Mater Res*, **6**, 347-74 (1972).

[10]J. Wang, L.L. Shaw, Morphology-Enhanced Low-Temperature Sintering of Nanocrystalline Hydroxyapatite, *Adv. Mater.*, **19**, 2364-2369 (2007).

[11]Zhao, H., et al., The structural and biological properties of hydroxyapatite-modified titanate nanowire scaffolds, *Biomaterials*, **32**, 5837-46 (2011).

[12]Wang, C., et al., Enhancing cell affinity of nonadhesive hydrogel substrate: the role of silica hybridization, *Biotechnol Prog*, **24**,1142-6 (2008).

[13]Tan, F., et al., In vitro and in vivo bioactivity of CoBlast hydroxyapatite coating and the effect of impaction on its osteoconductivity, *Biotechnol Adv*, 2011, in press.

[14]Roy, M., A. Bandyopadhyay, and S. Bose, Induction plasma sprayed Sr and Mg doped nano hydroxyapatite coatings on Ti for bone implant, *J Biomed Mater Res B Appl Biomater*, 2011, in press.

[15]Q.-M. Jin, H. Takia, T. Kohgo, K. Atsumi, H. Itoh and Y. Kuboki, Effects of Geometry of Hydroxyapatite as a Cell Substratum in BMP-Induced Ectopic Bone Formation, *J. Biomed. Mater. Res.*, **51**, 491-499 (2000).

[16]A. Magan and U. Ripamonti, Geometry of Porous Hydroxyapatite Implants Influences Osteogenesis in Baboons (Papio ursinus), *J. Craniofacial Surg.*, **7**, 71-78 (1996).

COMPARATIVE STUDIES OF COLD AND THERMAL SPRAYED HYDROXYAPATITE
COATINGS FOR BIOMEDICAL APPLICATIONS- A REVIEW

Singh, Ravinder Pal[1], Bala, Niraj[2]

[1]Chitkara University, Punjab, India-140401
[2] BBSBEC, Fatehgarh Sahib, Punjab, India- 140407

ABSTRACT
    Various conventional spray techniques have been exploited for more than two decades to coat
bio-ceramics such as Hydroxyapatite (HAP) and its composites on metallic substrates for biomedical
applications besides many limitations viz. residual stresses, evaporation, phase alterations and de-
bonding etc. Cold spray technology overcome number of traditional thermal spray shortcomings and
has been studied for engineering metallic materials. Deposition of HAP and its composites using cold
spraying is still unexploited and holds tremendous potential in dental and orthopedic implant
applications. This review will enumerate the research works and other crucial aspects of HAP
deposition using thermal spraying. Present R&D efforts of cold spraying will be discussed and
compared with the conventional spraying techniques. Studies and their conclusions related with
synthesis, characterization, electrochemical evaluation and other valuable information of HAP will be
extensively focused. In light of relative merits and demerits, research gap of using cold spraying for
HAP deposition will be proposed and concluded.

INTRODUCTION
    Biomaterial by definition is "a non-drug substance suitable for inclusion in systems which
augment or replace the functions of bodily tissues or organs" [1]. Many researchers defined
biomaterials differently [1,2]. Williams defined biomaterials as "nonviable materials used in medical
devices, intended to interact with the biological systems" [3]. Biomaterials have vast applications in
human body [4], but are widely used in repair, replacement or augmentation of diseased/damaged parts
of the musco-skeletal systems i.e. bones, joints and teeth [1,2,5,6]. Biomaterials are broadly classified
as Polymers, ceramics, composites, metals and alloys [1]. Class of ceramics used for repair and
replacement of diseased and damaged parts of musco-skeletal systems are termed as bio-ceramics [6].
Calcium phosphate group of ceramics are categorized as bio-resorbable and have become an accepted
group of materials for medical applications, mainly for implants in orthopedics, maxillofacial surgery
and dental implants [6]. Most widely used calcium phosphate based bioceramics are Hydroxyapatite
(HAP) and β- tricalcium phosphates [6]. In the past few decades, HAP was the subject of many
scientific researches due to its high chemical and thermal stability, unique sorption properties and
biocompatibility [7,8,9]. HAP has the chemical formula: $Ca_{10}(PO_4)_6(OH)_2$, the Ca/P ratio being 1.67
and possess a hexagonal structure. It is the most stable phase of various calcium phosphates [6]. It is
stable in body fluid and in dry/moist air up to $1200^0 C$ and does not decompose and has shown to be
bioactive due to its resorbable behavior [6]. Its usage is limited to non load bearing applications
because of its inferior mechanical properties due to poor sinter-ability [10], lower fracture toughness
compared to human bone [1, 11] and very low reliability under tensile loads [12]. Various techniques
have been developed to synthesize HAP powder such as acid-base reaction, precipitation, sol-gel
method, hydrothermal method, mechano-chemical synthesis, combustion method and various
techniques of wet chemistry [13-20]. There are other alternative techniques for preparation of HAP
powders such as flux method, electro-crystallization, spray pyrolysis, freeze-drying, microwave
irradiation etc. [6].

SURFACE COATING TECHNOLOGY

To coat HAP on substrates, many surface treatment techniques like plasma spraying [21-23], immersion in physiological fluid [24-26], sol-gel method [27-29], electrophoretic deposition [30-32], cathodic deposition [33-38], ion-beam techniques [39-41] and sputtering techniques [42-43] have been used. In general, method for the production of HAP prosthesis can be classified into two categories- thin coating and thick coating techniques [44]. Among the coating techniques, plasma spraying is by far the most widely adopted process [45-47]. This is due to advantages that include high deposition efficiency, improved technology and equipments [48]. However, this potency is limited by the poor cohesion strength and adhesion to the metallic substrate and most importantly, large quantities of phase transformations [49-51]. High Velocity Oxy-fuel spray technique is also highly exploited HAP coating technique because it can deposit HAP with similar set of advantages that plasma spraying can with small amount of phase transformation following thermal decomposition. Higher adhesion strength due to high particle velocities in HVOF has been also reported [48] over plasma spraying. Li and Khor [58] reported partial melt state of bigger HAP particles during HVOF spraying, while the smallest particles were subjected to a full melt state. Plasma sprayed HAP particles were reported to achieve a full melt state. Partial melt state during HVOF and overheating state during Plasma spraying of HAP particles could result in a faster cooling rate of the molten splats and causes deformation of particle/grain shapes than feedstock. In addition, different melt states influence both the microstructure and phases in resultant coatings. Additional, when HAP is plasma/HVOF sprayed, it may be converted into other calcium phosphate phases i.e. α-or β-tricalcium phosphate, tetra-calcium phosphates or calcium oxide and the crystallinity of HAP may also be lowered due to rapid solidification [52-54]. These alterations in chemistry and crystallinity often deteriorate the novel bioactive properties of HAP as well as its adhesion to the implant [55]. Furthermore, the presence of these phases other than HAP may also account for the low modulus of the coating. Achieving the appropriate phase and crystallinity in HAP coatings has been an actively pursued research area and lead to the development of new thermal spray process, known as cold spraying (CS) to produce metal, alloy and composite coating with superior qualities [56]. In contrast to conventional thermal spraying, deposition in cold spraying occurs without melting only by the deformation of the solid particles. This can be realized due to high kinetic energies of the particles upon impact on the substrate. A highly pressurized gas, typically nitrogen, helium or hydrogen is preheated and then expanded in a converging-diverging De-Laval nozzle. Through a separate gas line, powder feedstock is fed into or upstream of the nozzle. The powder is accelerated by the gas stream and impacts the substrate with high velocities. Depending on the choice of process gas, spray parameters and particle size (1-50 μm) [102], impact velocities between 200 and 1200 m/s can be realized [101]. The particle temperature upon impact also depends on various factors such as gas temperature, nozzle design and heat capacity of the particles-and can be in the range between room temperature and 800^0 C [101]. Coating is formed by only those particles which achieve the velocity more than the material-dependent critical velocity [102]. As CS process uses high velocity of spraying particles rather than high temperature to produce coatings, and thereby avoid/ minimize many deleterious high temperature reactions, which are characteristics of typical thermal sprayed coatings [57]. Compared to the other thermal spray coating processes, CS shows superiority due to the fact that the coating is formed from particles in the solid state. It's typical advantages include compressive rather than tensile stresses, wrought like microstructure, near theoretical density, oxides and other inclusions-free coatings etc. [57]. To the best of our knowledge, report of successful fabrication of HAP coating by cold spray technology is not available in open literature; hence needs attention for research. HAP being a ceramic lacks deformability and difficult to coat using cold spraying [55] and arises a need for composite coating of mixture of ductile metallic and hard ceramic powders. Choudhari and Mohanty (2009) deposited Ti-HAP composite coatings on titanium and aluminum coupons [55]. Cold sprayed HAP composite coatings revealed average adhesion strength of 47.4 MPa

[55] more than Plasma sprayed (average 10-15 MPa) [55] and HVOF (average 28-32 MPa) [58] respectively, revealing dense nature of cold sprayed HAP composite coatings.

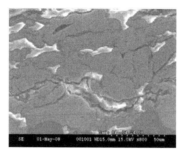

Figure 1: Cross-section view of Sponge-Ti +20% HAP cold sprayed at 35 Bar & 600^{0}C [55]

The microstructure of sponge Ti and HAP coating is shown in Figure 1 [55] which reveals very dense coating with well distributed HAP particles. The encapsulation of HAP particles (white portion) within the Ti matrix (dark black portion) is shown in Figure 2.

Figure 2: High magnification view of Sponge-Ti + 20% HAP [55]

The corresponding XRD (Figure 3) of the coating as well as the precursor powders show that the phases of both the powders were unaltered in the coating, which bears tremendous potential especially for HAP [55].

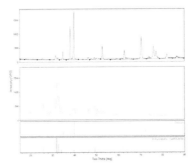

Figure 3: XRD of Sponge-Ti + 20% HAP composite coating made at 35 Bar & 600^0 C [55]

Lot of research studying various metallic elements i.e. titanium [59], copper [60], SS 316L [61], WC-Co [62,63], Aluminum [64] and composites i.e. iron alloy [65], Tungsten/Copper [66] and Al-Al$_2$O$_3$ [67] etc. have been successfully coated and investigated using cold spraying. So, pure HAP or its composite coatings using cold spraying is still unexploited and requires more research investigations.

The magnitude of the critical velocity can be estimated through the use of empirical relationships, which generally depend on particle material characteristics such as density, ultimate strength and melting point as well as the particle temperature immediately before impact [103]. The progress in the computational fluid dynamics (CFD) has made it possible to simulate gas-solid two phase flow precisely. Several reports have shown the feasibility to use CFD approach to obtain reasonable results [104]. Li and Li [105] made use of commercial CFD code to optimize nozzle geometry for maximum particle velocity. Pardhasaradhi et al. [106] compared the laser illuminated, time of flight velocity measurements with the empirical models. Jodoin et al. [107] utilized Reynolds average Navier Stokes equations within a computational platform to model flow with boundary conditions and documented the effects of gas type and stagnation conditions. Samareh et al. [108] used computational fluid mechanics to describe the effect of particle concentration on gas velocity for two nozzle geometries.

Therefore CFD approach has become popular to estimate particle velocity at certain conditions [107]. FLUENT software package has been proven to be reliable for modeling the gas flow in a Laval nozzle through experimental validation [107,108]. Author has extensively evaluated the individual effect of various cold spraying parameters on HAP particle velocity using CFD and mathematical approaches [109].

HAP SYNTHESIS AND CHARACTERIZATION

Synthetic Hydroxyapatite is a representative material for a bone substitute because of its chemical similarities with the inorganic phase of bone [68-70]. As a consequence, many investigations have been already carried out to prepare HAP [13] and most of them can be classified into two groups according to the processing method: one is wet [71-75] and the other is dry process [76-79]. The advantages of the wet process are that the by-product is almost water [80-81] and the probability of contamination during processing is very low [13]. However, the disadvantages of wet precipitation technique are the composition of the resulting product is greatly affected by even a slight difference in the reaction conditions and the time needed for obtaining the hydroxyapatite of a stiochiometric composition takes around 20 days, which is inconceivable in an industrial scale production [13]. Process is complicated due to poor reproducibility and high processing cost. It also needs highly qualified and controlled parameters such as pH and temperature of the solutions, nature and composition of the starting materials, reagents concentration and addition rate, stirring techniques and stirring speed, maturation and presence of impurities to obtain HAP mono-phase [82-83].

The calcium and phosphorus compounds used as the stirring materials in the dry process are di-calcium phosphate anhydrous [78], di-calcium phosphate dehydrate [78], mono-calcium phosphate monohydrate [79], calcium pyrophosphate [76], calcium carbonate [78,79], calcium oxide [78,79] and calcium hydroxide [78] etc.

Some authors also suggested the cheaper methods to produce Hydroxyapatite from other sources i.e. seashells [12], eggshells [84,5,85].

After synthesis, characterization of HAP powder is vital to ensure the presence of desirable chemical constituents, phase various functional groups and evaluation of morphology. Extensive research works have been conducted by various authors and different spectroscopic and thermal techniques i.e. X-Ray Diffraction (XRD), Scanning Electron Microscopy (SEM), Fourier Transform Infrared Spectroscopy (FTIR) and Thermo gravimetric Analysis (TGA).

HAP and other phases present in synthesized HAP powder was characterized using XRD and peak data and their angle (2θ) were compared with standard Joint Committee on Powder Diffraction (JCPDS) data. Crystallinity/amorphous nature of powder were also ascertained using XRD spectroscopy [86,87,13,12, 10,88,9,11,5,85] and Rietveld analysis [89].

Various functional groups present in the prepared HAP powder using FTIR [10,12,13,86,87,88,85], morphology of powder particles [88, 85,12,13,87,11] i.e. shape and size of particles were evaluated using SEM techniques by various authors. Mass loss during heating can be calculated by TGA method at different temperatures [86]. Some authors also analyzed the Ca/P molar ratio of the precipitated powder by quantitative chemical analysis via. EDTA titration technique [5,85] and Atomic Absorption Spectroscopy (AAS) [88].

ELECTROCHEMICAL EVALUATION

Corrosion, the gradual degradation of materials by electrochemical attack, is a concern particularly when a metallic implant is placed in the hostile electrolytic environment provided by the human body [90]. Human body environment consists of oxy-genated saline solution with salt content of about 0.9% at pH 7.4 and temperature of 37 ± 1^0 C [1,91, 92]. When an orthopedic implant is surgically implanted into human body, it is constantly bathed in extra-cellular tissue fluid [93] lead to electrochemical dissolution at some finite rate, causes the release of metal ions into the tissues and increases the suspected role in induction of tumors e.g. malignant fibrous histiocytoma [94]. Various types of corrosion in orthopedic implants i.e. uniform attack [94], galvanic corrosion [95], fretting corrosion [90], crevice corrosion [96], pitting corrosion [94], inter-granular corrosion [94], leaching [94] and stress-corrosion cracking [97] were studied. Therefore, the most important requirement for an implantable material is that it must have corrosion resistance to physiological environment. In-vitro corrosion studies have been carried out in saline/physiological solutions, using electrochemical technique e.g. potential-time curves (E-t) for isolated specimens, potential-current density (E-log j) curves for anodes and current density-time (j-t) curves for anodes maintained electronically at constant potential [98]. While there are several test methods used to determine corrosion resistance i.e. ASTM G5, G61, F86.5, B117, one standard ASTM F 2129 has been developed specifically to assess a small implant's resistance to pitting and crevice corrosion [99]. An acceptance criterion is not addressed in any of the standards [99]. It is up to the device manufacturer to interpret the results and justify the conclusions [99]. If a device is found to be acceptable in service through clinical trials, ASTM F 2129 test can be performed and a mean breakdown potential can be identified [99]. Even Food and Drug Administration (FDA) does not specify a breakdown potential that must be met to consider a device acceptable [100]. According to Corbett, if a breakdown potential exceeds +600mv (SCE) in PBS at 37^0 C, then the alloy is in its optimum corrosion resistant condition [100]. If the breakdown potential is more electronegative than +300 mV, then the alloy is not in an optimal corrosion resistant condition [100]. At breakdown potentials between +300 and +600 mV, the alloy is marginal and a high confidence level is needed for acceptance that the alloy will exhibit stable passive behavior [100].

Hence it is important to realize that corrosion of orthopedic biomaterials is not just an exercise in physics and chemistry, but is a pertinent clinical issue confronting all orthopedic surgeons. While the need for stringent quality control on the part of implant manufacturers remains of paramount importance in the context of rising demands, orthopedic surgeons and other theatre personnel need to be made more aware to ensure that implants do not corrode and fail due to carelessness or ignorance.

CONCLUSIONS

In the light of above reviewed literature, it can be concluded that extensive research work has been conducted in thermal spraying technique of HAP coating and there is a need of exploration of other methods of spraying especially cold spraying method. HAP has been synthesized from various natural as well as artificial resources. Electrochemical evaluation is a latest technique of HAP characterization having definite procedures and serves as an effective technique of measuring corrosion resistance of bio implants.

REFERENCES

[1]G. Henes and B. Nissan, Innovative Bioceramics, *Materials Forum*, **27**, 104-14 (2004).

[2]U.K. Mudali, T.M. Sridhar and B. Raj, Corrosion of Bio Implants, *Sadhana*, **28**, 3&4, 601-7 (2003).

[3]D.F. Williams, Electrochemical aspects of Corrosion in the Physiological Environment, *Fundamental aspects of Biocompatibility (ed)*, **1**, 11-20 (1981).

[4]L.L. Hench, Inorganic Biomaterials, Advances in Chemistry series 245: *Materials Chemistry-an emerging discipline (ed)*, **523** (1985).

[5]P. Dasgupta, A. Singh, S. Adak and K.M. Purohit, Synthesis and Characterization of Hydroxyapatite produced from Eggshells, *International Symposium of Research students on Materials Science and Engineering*, December 20-22, Chennai, India (2004).

[6]T.V. Thamaraiselvi and S. Rajeswari, Biological Evaluation of Bioceramic Materials-A Review, *Trends in Biomaterials and Artificial Organs,* **18**(1), 9-17 (2004).

[7]I. Smiciklas, A. Onjia and S. Raicevic, Experimental Design approach in the synthesis of Hydroxyapatite by neutralization method, *Separation and Purification Technology*, **44** (2), 97-102 (2005).

[8]T.S.B. Narasaraju and D.E. Phebe, Review of some Physico-chemical aspects of Hydroxylapatite, *Journal of Material Science*, **31**(1), 1-21 (1996).

[9]K.K. Saju, N.K. George and P.S. Sreejith, Bioactive Coatings of Hydroxyapatite on titanium substrates for body implants, *International conference on Advanced Materials and composites*, 729-734 (2007).

[10]S. Sasikumar and R. Vijayaraghavan, Low Temperature Synthesis of Nanocrystalline Hydroxyapatite from Egg shells by combustion method, *Trends in Biomaterials and Artificial Organs*, **19**(2), 70-73 (2006).

[11]S. Ramesh, Grain size-Properties correlation in Polycrystalline Hydroxyapatite Bioceramics, *Malaysian Journal of Chemistry*, **3**(1), 35-40 (2001).

[12]R. Narayan, S. Dutta and S.K. Seshadri, Hydroxyapatite coatings on Ti-6Al-4V from Seashell, *Surface & Coatings Technology*, **200**, 4720-4730 (2006).

[13]S.H. Rhee, Synthesis of Hydroxyapatite via Mechanochemical treatment, *Biomaterials*, **23**, 1147-1152 (2002).

[14]K.C.B. Yeong, J. Wang and S.C. Ng, Mechanochemical Synthesis of Nanocrystalline Hydroxyapatite from CaO and $CaHPO_4$, *Biomaterials*, **22** (20), 2705-2712 (2001).

[15]C. Tas, Synthesis of Hydroxyapatite and application in bone and dental regeneration in human body, *Journal of European Ceramic Society*, **20** (14-15), 2389-2394 (2000).

[16]http://Gralib.Hcmuns.Edu.Vn/Gsdl/Collect/Hnkhbk/index/Assoc/Hashd89f.Dir/doc.pdf.

[17]A. Afshar, M. Ghorbani, N. Ehsani, M.R. Saeri and C.C. Soroell, Some important factors in the wet chemical process of Hydroxyapatite, *Materials and Design*, **24**, 197-202 (2003).

[18]M.R. Saeri, A. Afshar, M. Ghorbani, N. Ehsani and C.C. Sorrell, The wet precipitation process of Hydroxyapatite, *Materials Letters*, **57** (24-25) 4064-69 (2003).

[19]Y. Liu, D. Hou and G. Wang, A simple wet chemical synthesis and characterization of Hydroxyapatite nanorods, *Materials Chemistry and Physics*, **86** (1), 69-73 (2004).

[20]L.B. Kong, J. Ma and F. Boey, Nanosized Hydroxyapatite powders derived from co-precipitation process, *Journal of Materials Science*, **37**(6), 1131-34 (2002).

[21]H.C. Gledhill, I.G. Turner and C. Doyle, In vitro dissolution behavior of two morphologically different thermally sprayed Hydroxyapatite coatings, *Biomaterials*, **22**(7), 695-700 (2001).

[22]O. Grabmann and R.B. Heimann, Compositional and Microstructural changes of Engineered Plasma-sprayed Hydroxyapatite on Ti6Al4V substrates during incubation in protein free simulated body fluids, *Journal of Biomedical Materials Research*, **53**, 685-693 (2000).

[23]K.A. Khor, Z.L. Dong, C.H. Quek and P. Cheang, Microstructure investigation of Plasma sprayed HA/Ti6Al4V composites by TEM, *Materials Science and Engineering A*, **281** (1-2), 221-228 (2000).

[24]W.H. Song, Y.K. Jun, Y. Han and S.H. Hong, Biomimetic apatite coatings on micro-arc oxidized titania, *Biomaterials*, **25**, 3341 (2004).

[25]D.M. Liu and H.M. Chou, Formation of a New Bioactive Glass-Ceramic, *Journal of Material Science: Materials in Medicine*, **5**, 7-10 (1994).

[26]L. Jonasova, F.A. Muller, A. Helebrant, J. Strnad and P. Greil, Hydroxyapatite formation on alkali treated titanium with different contents of Na^+ in the surface layer, *Biomaterials*, **23**, 3095-3101 (2002).

[27]A. Montenero, G. Gnappi, F. Ferrari, M. Cesari, E. Salvioli, l. Mattogno, S. Kaciulis and M. Fini, Sol–gel derived hydroxyapatite coatings on titanium substrate, *Journal of Materials Science*, **35**, 2791-97 (2000).

[28]H.W. Kim, Y. H. Koh, L.H. Li, S. Lee and H.E. Kim, Hydroxyapatite coating on titanium substrate with titania buffer layer processed by sol-gel method, *Biomaterials*, **25**(13), 2533-38 (2004).

[29]D.M. Liu, T. Troczynski and W.J. Tseng, Water-based sol-gel synthesis of hydroxyapatite: process development, *Biomaterials*, **22**, 1721-30 (2001).

[30]L.A.D. Sena, M. Andrade, A. Rossi, and G. Soares, Hydroxyapatite deposition by electrophoresis on titanium sheets with different surface finishing, *Journal of Biomedical Materials Research*, **60**, 1-7 (2002).

[31]A. Stoch., G. Kmita, A. Brozek, J. Stoch, W. Jastizebski and A. Rakowska, FTIR study of copper patinas in the urban atmosphere, *Journal of Molecular Structure*, **596**, 201-6 (2001).

[32]X. Nie, A. Leyland, A. Matthews, J.C. Jiang and E.I. Meletis, Effects of solution pH and electrical parameters on Hydroxyapatite coatings deposited by a plasma-assisted electrophoresis technique, *Journal of Biomedical Materials Research*, **57**, 612-18 (2001).

[33]B. Seiji and M. Shiego, Effect of temperature on electrochemical deposition of calcium phosphate coatings in a simulated body fluid, *Biomaterials*, **16**(13), 977-81 (1995).

[34]M.C. Kuo, S.K. Yen, The process of electrochemical deposited Hydroxyapatite on titanium at room temperature, *Materials Science Engineering*, **C20** (1-2), 153-160 (2002).

[35]S.K. Yen and C.M. Lin, Cathodic reactions of electrolytic hydroxyapatite coating on pure titanium, *Materials Chemistry and Physics*, **77**, 70-6 (2002).

[36]M. Shirkhanzadeh, Bioactive calcium phosphate coatings prepared by electrodeposition, *Journal of Materials Science*, **10**, 1415 (1991).

[37]J. Wang, P. Layrolle, M. Stigter and K.D. Groot, Biomimetic and electrolytic calcium phosphate coatings on titanium alloy: physicochemical characteristics and cell attachment, *Biomaterials*, **25**, 583-592 (2004).

[38]H. Silva, G. Soares and C. Elias, The cellular response to titanium electrochemically-coated with Hydroxyapatite compared to titanium with three different levels of surface roughness, *Journal of Material Science: Materials in Medicine*, **14**, 511-19 (2003).

[39]M.F. Maitz, M.T. Pham, W. Matz, H. Reuther and G. Steiner, Promoted calcium-phosphate precipitation from solution on titanium for improved biocompatibility by ion implantation, *Surface Coating Technology*, **158-159C**, 151-6 (2002).

[40]M.T. Pham, W. Matz, H. Reuther, E. Richter, G. Steiner and S. Oswald, Ion beam sensitizing of titanium surfaces to hydroxyapatite formation, *Surface Coating Technology*, **128-129**, 313-319 (2000).

[41]Baszkiewicz, D. Krupa, J.A. Kozubowski, B. Rajchel and M. Mitura, Influence of the Ca- and P-enriched oxide layers produced on titanium and the Ti6Al4V alloy by the IBAD method upon the corrosion resistance of these materials, *Vacuum*, **70**, 163-7 (2003).

[42]J.L. Ong, K. Bessho, R. Cawn and D.L. Carnes, Bone response to radio frequency sputtered calcium phosphate implants and titanium implants in vivo, *Biomaterials*, **59**, 184-190 (2002).

[43]Y. Yang, K.H. Kim, C.M. Agarwal, and J.L. Ong, Effect of post-deposition heating temperature and the presence of water vapor during heat treatment on crystallinity of calcium phosphate coatings, *Biomaterials*, **24**, 5131-37 (2003).

[44]S.J. Yankee, B.J. Pletha, H.P. Luckey and W. Johnson, Process for fabricating Hydroxyapatite coatings for biomedical applications, *Thermal Spray Research and applications*, proceedings of the Third NTSC, Long Beach, CA, USA, 433-38 (1990).

[45]V. Sergo, O. Sbaizero and D.R. Clarke, Mechanical and chemical consequences of the residual stresses in plasma sprayed hydroxyapatite coatings, *Biomaterials*, **18**(6), 477-82 (1997).

[46]W. Tong, Z. Yang and X. Zhang, Patterns of Plasma sprayed HAP coatings, *Biomedical Materials Research*, **40**, 407-13 (1998).

[47]C.Y. Yang, B.C Wang and E. Chang, Bond degradation at the plasma sprayed HA coating/Ti6Al4V alloy interface: an in vitro study, *Journal of Materials Science: Materials in Medicine*, **6**, 249-57 (1995).

[48]H. Li, K.A. Khor and P. Cheang, Effects of the powders melting state on the properties of HVOF sprayed Hydroxyapatite coatings, *Materials Science and Engineering*, **2936**, 71-80 (2000).

[49]R.L. Reis, F.J. Monteiro, Crystallinity and Structural Changes in HÁ Plasma-Sprayed Coatings Induced by Cyclic Loading, *Journal of Materials Science: Materials in Medicine*, **7**, 407-411 (1996).

[50]K.A. Gross, V. Gross and C.C. Berndt, Thermal analysis of Amorphous phases in Hydroxyapatite coatings, *Journal of American Ceramics Society*, **81**(1), 106-12 (1998).

[51]D.M. Liu, H.M. Chou and J.D. Wu, Plasma-Spray Hydroxyapatite Coating: Effect of Different Calcium Phosphate Ceramics, *Journal of Materials Science: Materials in Medicine*, **51**, 147-53 (1994).

[52]Y.C. Tsui, C. Doyle and T.W. Clyne, Plasma sprayed Hydroxyapatite coatings on titanium substrates Part I: Mechanical Properties and Residual stress levels, *Biomaterials*, **19**(22), 2015-29 (1998).

[53]L. Sun, C.C Berndt and K.A. Gross, Material Fundamentals and Clinical performance of Plasma sprayed Hydroxyapatite Coatings: A Review, *Journal of Biomedical Materials Research: Part B*, **58** (5), 570-92 (2001).

[54]P.K. Stephenson, M.A. Freeman, P.A. Revell, J. Germain, M. Tuke and C.J. Pirie, The effect of Hydroxyapatite coating on ingrowth of bone into cavities in an implant, *Journal of Arthroplasty*, **6**(1), 51-58 (1991).

[55]A. Choudhuri and P.S. Mohanty, Bio-Ceramics composite coatings by cold spray technology, *Thermal Spray 2009*: Proceedings of the International Thermal Spray Conference, 391-96 (2009).

[56]P. Alkhimov, A.N. Papyrin, V.P. Dosarev, N.J. Nesterovich and M.M Shuspanov, US patent **5**, 302, 414 (1994).

[57]J. Karthikeyan, Cold Spray Technology: International Status and USA efforts (2004).

[58]H. Li and K.A. Khor, Characteristics of the nanostructures in thermal sprayed Hydroxyapatite coatings and their influence on coating properties, *Surface & Coatings technology*, **201**, 2147-54 (2006).

[59]R.S. Lima, A. Kucuk and C.C. Berndt, Deposition Efficieny, Mechanical Properties and coating roughness in cold-sprayed titanium, *Journal of Materials Science Letters*, **21**, 1687-89 (2002).

[60]T. Kairet, M. Dogrez, F. Campana and J.P. Janssen, Influence of the powder size distribution on the microstructure of cold-sprayed copper studied by X-Ray Diffraction, *Journal of Thermal Spray Technology*, **16** (5-6), 610-18 (2007).

[61]W.Y. Li, W.Y. X Liao, G. Douchy, , and C. Coddet, Optimal design of a cold spray nozzle by numerical analysis of particle velocity and experimental validation with 316L S.S. powder, *Materials and Design*, **28**, 2129-37 (2006).

[62]H.J. Kim, C.H. Lee and S.Y. Hwang, Fabrication of WC-Co coatings by Cold spray Deposition, *Surface & Coating Technology*, **191**, 335-40 (2004).

[63]H.J. Kim, C.H. Lee and S.Y. Hwang, Superhard nano WC-12% Co coating by cold spray deposition, *Materials Science and Engineering*, **391**, 243-248 (2004).

[64]K. Balani, A. Agarwal, S. Seal and J. Karthikeyan, Transmission Electron Microscopy of Cold Sprayed 1100 Aluminium coating, *Scripta Materialia*, **53**, 845-50 (2005).

[65]T.V. Steenkiste, Kinetic sprayed rare earth Iron alloy composite coatings, *Journal of Thermal Spray Technology*, **15**(4), 501-06 (2006).

[66]H.K Kang and S.B. Kang, Tungsten/Copper composite deposits produced by a cold spray, *Scripta Materialia*, **49**(12), 1169-74 (2003).

[67]E. Irissou, J.G. Legoux, B. Arsenault and C. Moreau, Investigation of Al-Al$_2$O$_3$ cold spray coating formation and properties, *Journal of Thermal Spray Technology*, **16**(5-6), 661-68 (2007).

[68]W.F. DeJong, La Substance Minerals das Les Os, *Tec. Trav. Chim.*, **45**, 445-58 (1926).

[69]R.Z. LeGerus, J.P. Legeros, Dense Hydroxyapatite, An Introduction to Bio-ceramics, Singapore: World Scientific Publishing Co. Pvt. Ltd., pp. 139-180 (1998).

[70]H. Aoki, Science and Medical Applications of Hydroxyapatite, Takayama Press System Center Co. Inc., 1-10 (1991).

[71]J.G.C. Peclen, B.V. Rejda and G. De, Preparation and Properties of sintered Hydroxyapatite, *Cermurgia Int*, **4**, 71-3(1980).

[72]H. Akao, J. Aoki and K. Kato, Mechanical Properties of Sintered Hydroxyapatite for Prosthetic Applications, *Journal of Materials Science*, **16**, 809-12 (1981).

[73]M. Jarchu, C.H. Bolen, M.B. Thomas, J. Bobick, J.F. Kay and R.H. Doremus, Hydroxyapatite Synthesis and Characterization in sense Polycrystalline, *Journal of Materials Science*, **11**, 2027-35 (1976).

[74]E.C. Moreno, T.M. Gregory and W.E. Brown, Preparation and Solubility of Hydroxyapatite, *J. Res. Mat. Bureau standards*, **72**, 773-782 (1968).

[75]M. Asada, Y. Miura, A. Osaka, K. Oukami and S. Nakamura, Hydroxyapatite crystal Growth on Calcium Hydroxyapatite Ceramics, *Journal of Materials Science*, **23**, 3202-05 (1988).

[76]B.O. Fowler, Infrared Studies of Apatite II. Preparation of Normal and Isotopically substituted calcium, strontium and Barium Hydroxyapatite and Spectra-structure-composition correlations, *Inorganic Chemistry*, **13**, 207-14 (1974).

[77]S.J. Joris and C.H. Amberg, The nature of deficiency in nonstiochiometric Hydroxyapatite II: Spectroscopic studies of calcium and strontium Hydroxyapatite, *Journal of Physical Chemistry*, **75**, 3172-78 (1971).

[78]M. Otsuka, Y. Matsuda, J. Hsu, J.L. Fox and W.I. Higuchi, Mechanochemical synthesis of Bioactive materials: Effect of environmental conditions on the phase transformation of calcium phosphates during grinding, *Bio-Medical Materials and Engineering*, **4**, 357-62 (1994).

[79]M. Chaikina, Mechanochemical synthesis of Phosphates and apatites- new way of preparation of complex materials, *Phosphorus Research Bulletin*, **7**, 35-8 (1997).

[80]J.A.S. Bett, L.G. Cristner and W.K. Hall, Studies of the hydrogen held by solids, XII Hydroxyapatite catalysts, *Journal of American Chemical Society*, **89**, 5535-41 (1967).

[81]A. Slosarczyk, E. Stobierska, Z. Paszkiewicz and M. Gawlicki, Calcium Phosphate materials from precipitates with various calcium: phosphorus molar ratio, *Journal of American Ceramics Society*, **79**, 2539-44 (1996).

[82]S. Lazic, S. Zec, N. Miljevic, and S. Milonjic, The effect of Hydroxyapatite precipitated from Calcium Hydroxide and Phosphoric acid, *Thermochimica. Acta*, **374**, 13-22 (2001).

[83]L. Bernard, M. Freche, J.L. Laccout and B. Biscans, Modeling of the dissolution of Calcium Hydroxide in the preparation of Hydroxyapatite by neutralization, *Chemical Engineering Science*, **55**, 5683-92 (2000).

[84]S. Sasikumar and R. Vijayaraghavan, Low temperature synthesis of nanocrystalline Hydroxyapatite from Egg shell by Combustion method, *Trends in Biomaterials and Artificial Organs*, **19**(2),70-73 (2006).

[85]S.J. Lee and S.H. Oh, Fabrication of Calcium Phosphate Bioceramics by using eggshell and phosphoric acid, *Materials Letters*, **57**, 4570-74 (2003).

[86]D.K. Pattanayak, P. Divya, S. Upadhayay, R.C. Prasad, B.T. Rao and T.R. Mohan, Synthesis and Evaluation of Hydroxyapatite Ceramics, *Trends in Biomaterials and Artificial Organs*, **18**(2), 87-92 (2005).

[87]A.B.H. Yoruc and Y. Koca, Double Step Stirring: A Novel method for precipitation of nano-sized Hydroxyapatite powder, *Digest Journal of Ianoma terials and Biostructures*, **4**(1), 73-81 (2009).

[88]H. Eslami, M.S Hashjin and M. Tahriri, Synthesis and Characterization of nanocrystalline Fluorinated Hydroxyapatite powder by a modified wet-chemical process, *Journal of ceramic processing research*, **9**, 3, 224-29 (2008).

[89]J.C. Knowles, K. Gross, C.C. Berndt, and W. Bonfiels, Structural changes of thermally sprayed Hydroxyapatite investigated by Rietveld analysis, *Biomaterials*, **17**, 639-45 (1996).

[90]A.S. Litsky and M. Spector, Biomaterials in Simon SR (Ed): Orthopedic basic science, *American Academy of Orthopedic Surgeons*, 470-73 (1994).

[91]D. Sharan, The Problem of corrosion in Orthopedic implant materials, *Orthopedic Update (India)*, **9**(1), 1-5 (1999).

[92]D.C. Hansen, Metal corrosion in the Human Body: The ultimate Bio-Corrosion Scenario, *The Electrochemical Society Interface*.

[93]O.E.M. Pholer, Failure of orthopedic Metallic Implants, *ASM handbook on failure analysis and prevention*, 9th edn, **11**, 670-79 (1986).

[94]J. Black, Corrosion and degradation in Orthopedic Biomaterials in research and practice, New York, Churchill, Livingstone, 235-266 (1988).

[95]J.J. Jacobs, J.L. Gilbert and R.M. Urban, Corrosion of Metal Orthopedic Implants, *Journal of Bone and Joint Surgery*, **80**, 268-82 (1998).

[96]F.P. Bowden, J.C.P. Williamson and P.G. Laing, The significance of Metal transfer Orthopaedic Surgery, *Journal of Bone and Joint Surgery*, **37**, 676-90 (1955).

[97]E. Greener and S.E. Lauten, Materials for Bio-engineering applications, Biomedical Engineering, Philadelphia, FA Davis (1971).

[98]R.J. Solar, Corrosion resistance of titanium surgical Implant Alloys: A Review, *Corrosion and Degradation of Implant Materials*, 259-73 (1979).

[99]S.N. Rusenbloom and R.A. Corbett, An assessment of ASTM-F 2129 electrochemical testing of small medical implants-Lessons Learned, *IACE Corrosion 2007 Conference & Expo*, Paper No. 07674, 1-10 (2007).

[100]R.A. Corbett, Laboratory Corrosion Testing of Medical Implants. Medical Device Materials: Proceedings of the Materials & Processes for Medical Devices Conference 2003 (ASM International) (2004).

[101]T. Schmidt, H. Assadi, F. Gartner, H. Richter, T. Stoltenhoff, H. Kreye and T. Klassen, From Particle Acceleration to Impact and Bonding in Cold Spraying, *Journal of Thermal Spray Technology*, 794-08 (2009).

[102]S. Yin, X.F. Wang, W. Y. Li and X.P. Guo, Examination on Substrate Preheating Process in Cold Gas Dynamic Spraying, *Journal of Thermal Spray Technology*, 852-59 (2011).

[103]V.K. Champagne, D.J. Helfritch, S.P.G. Dinavahi, P.F. Leyman, Theoretical and Experimental Particle Velocity in Cold Spray, *Journal of Thermal Spray Technology* (2010).

[104]B. Jodoin, F. Raletz and M. Vardelle, Cold Spray Modeling and Validation using an Optical Diagnostic Method, *Surface and Coating Technology*, 200, 4424-32 (2006).

[105]W.Y. Li and C.J. Li, Optimal Design of a Novel Cold Spray Gun Nozzle at a Limited Space, *Journal of Thermal Spray Technology*, 14(3), 391-96 (2004).

[106]S.P. Pardhasaradhi, V. Venkatachalapathy, S.V. Joshi and S. Govindanet, Optical Diagnostic study of Gas particle Transport Phenomena in Cold Gas Dynamic Spraying and Comparison with Model Predictions, *Journal of Thermal Spray technology*, 17(4), 551-63 (2008).

[107]B. Jodoin, F. Raletz and M.Vardelle, Cold Spray Modeling and Validation using an Optical Diagnostic Method, *Surface and Coating Technology*, 200, 4424-32 (2006).

[108]B. Samareh, O. Steir, V. Luthen and A. Dolatabadi, Assessment of CFD Modeling via Flow Visualization in Cold Spray Process, *Journal of Thermal Spray Technology*, Published online 2009.

[109]R.P. Singh, Numerical Evaluation, optimization and Mathematical Validation of Cold Spraying of Hydroxyapatite using Taguchi Approach, *International Journal of Engineering Science and Technology*, 3(9) 2011.

INJECTABLE BIOMIMETIC HYDROGELS WITH CARBON NANOFIBERS AND NOVEL SELF ASSEMBLED CHEMISTRIES FOR MYOCARDIAL APPLICATIONS

Xiangling Meng[1], David Stout[3], Linlin Sun[3], Hicham Fenniri[4] and Thomas Webster [2,3]

[1]Department of Chemistry, Brown University, Providence, RI, 02912
[2]Department of Orthopedics, Brown University,
[3]School of Engineering, Brown University, Providence, RI, 02917
[4]National Institute for Nanotechnology and Department of Chemistry, University of Alberta

Corresponding author: Professor Thomas J. Webster, School of Engineering, Brown University, 184 Hope Street, Providence, RI, 02917. Tel: +1 401-863-2318. Fax: +1 401-863-9107
Email Addresses: xiangling_meng@brown.edu (Xiangling Meng), david_stout@brown.edu (David Stout), linlin_sun@brown.edu (Linlin Sun), hicham.fenniri@ualberta.ca (Hicham Fenniri), thomas_webster@brown.edu (Thomas Webster)

ABSTRACT

The objective of the present in vitro study was to investigate cardiomyocyte functions, such as adhesion and proliferation, on injectable polyHEMA (poly (2-hydroxyethyl methacrylate)), HRN (helical rosette nanotube) and CNF (carbon nanofiber) composites to determine their potential for myocardial tissue engineering applications. HRNs are novel biocompatible nanomaterials that are formed from synthetic DNA base analog building blocks that self-assemble within minutes when placed in aqueous solutions at body temperatures. Thus, HRNs were used in this study as a material that could potentially improve cardiomyocyte functions and solidification time of polyHEMA and CNF composites. Moreover, since heart tissue is conductive, CNFs were added to polyHEMA to increase the composite conductivity. PolyHEMA was chosen here since it has been one of the most widely used injectable biocompatible hydrogels for tissue repair applications. Contact angle, conductivity and scanning electron microscopy studies characterized the structure and surface energy of the composites. More importantly, the results showed that cardiomyocyte density increased with greater amounts of CNF and greater amounts of HRNs in polyHEMA (up to 10mg/ml CNFs and 0.05mg/ml HRNs).The contact angles of the samples decreased with more CNFs and HRNs. Adding CNFs increased the conductivity of the samples. In summary, the presently observed properties of these injectable composites make them promising candidates for further study for myocardial tissue engineering applications.

INTRODUCTION

Heart disease or cardiovascular disease is a leading cause for death today. There is a significantly high and increasing rate of cardiovascular disease, which clearly has led to a public health problem. For example, each year, an estimated 785, 000 Americans will have a new coronary attack and around 470, 000 will have a recurrent heart attack[1]. Myocardial infarctions, also known as heart attacks, will result in the large scale loss of cardiac muscle, and are usually caused by a sudden block by an obstruction, in a major blood vessel [2-5]. After the heart attack, dead heart tissue remains and the scared cardiac muscle may lead to another myocardial infarction in the future.

Various techniques have been developed to regenerate cardiomyocytes after a myocardial infarction. Injectable scaffolds have attracted much attention due to their outstanding properties compared with other techniques, like cardiac patches and 3D printing techniques. Specifically, injectable scaffolds offer the unparalleled advantage of avoiding invasive surgical procedures since they are able to take the shape of a tissue defect, avoiding the need for patient specific scaffold prefabrication and highly invasive surgery [10]. A variety of injectable materials are being explored to date for such applications, including synthetic materials and naturally derived or biologically inspired materials.

Variations of PNIPAAM (Poly (*I* -isopropylacrylamide)) have shown improved cardiac function in small animal infarct models [16-17]. Commercially available collagen has also been used as an injectable naturally derived scaffold and has shown positive effects for cardiac tissue engineering [18-20]. However, synthetic materials offer the advantage of ease of manufacture allowing the design of a material with appropriate porosity, mechanical stability, and degradation properties [21]. Unfortunately, most of injectable scaffolds developed for myocardial applications to date are non-conductive, such as those mentioned above, yet heart tissue is conductive. So the objective of this in vitro study was to develop a novel injectable biomimetic hydrogel scaffold for myocardial applications based on helical rosette nanotubes (HRNs), carbon nanofibers (CNFs), and poly (2-hydroxyethyl methacrylate) (pHEMA, a biocompatible hydrogel).

HRNs are soft nanotubes which self-assemble from a synthetic DNA base analog when placed in aqueous solutions (Figure 1). These tubes have a 1.1 nm inner diameter and can extend from several hundred nanometers up to several micrometers in length as controlled by non-covalent interactions [6]. Moreover, HRNs can be functionalized by different amino acids and peptide side chains (such as lysine, RGD (Arg-Gly-Asp), KRSR (Lys-Arg-Ser-Arg)) in order to change the chemical and physical properties of these nanotubes for specific tissue engineering applications[7]. Previous studies have highlighted the exceptional cytocompatibility properties of HRNs for orthopedic (bone forming cell) applications [7-8], yet they remain to be tested for myocardial applications. For example, the cytocompatibility properties of HRNs have been improved when functionalized with RGD [7]. In addition, endothelial cell functions can be enhanced when placed on HRN coated titanium vascular stents [9]. Besides the cytocompatibility properties, HRN can also improve other properties of hydrogels. Adding HRN into hydrogels can decrease the time to polymerize and increase the adhesive strength. The adhesive strength of PEG on collagen casings mimicking skin will also increase with higher concentrations of HRNs in the PEG solution [22]. Based on these exceptional properties of HRNs, the purpose of present study was to explore its potential for the first time for injectable myocardial applications.

pHEMA was chosen here since it has been one of the most widely used biocompatible hydrogels for corneal implants, drug delivery devices, tissue repair surgeries, and orthopedic applications due to its biomimetic properties[11-14]. CNFs were added to the composite to increase conductivity which has been shown to promote cardiomyocyte functions. CNFs were modified with – OH and – COOH on the surface in order to enhance their dispersion in HRN and HEMA solutions. Thus, an injectable

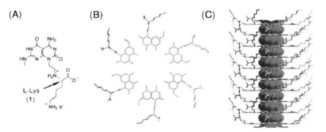

Figure 1. The self-assembling process of helical rosette nanotube (HRN) in water. (A) A DNA base pair building block with a lysine side chain (G^C-K), (B) a rosette self-assembled from six building blocks by forming 18 H-bonds, and (C) a stacking helical rosette nanotube (HRN) 3.5 nm in diameter.

myocardial scaffold was prepared by dispersing different concentrations of HRNs (from 0.01mg/ml to 0.05mg/ml) and CNFs (from 5mg/ml to 10mg/ml) within polyHEMA hydrogels and tested for myocardial applications.

MATERIALS AND METHODS

Material Fabrication

HRNs were synthesized using our previously reported synthetic strategy [6]. 1mg /ml of an HRN solution was prepared by directly assembling respective amounts of G˘C-K in water solutions. CNFs were modified with –OH and –COOH groups on their surfaces (Pyrograf Products Inc.) then added into 9ml of HEMA monomer (Polysciences Inc) separately. The solutions were sonicated in a water bath (VWR B3500A-DTH) at level 9 below 30℃ for 30 min. 300mg of 2, 2-azobisisobutyronitrile (free radical initiator, Sigma-Aldrich) was then added into 10ml of HEMA monomer to make a solution with concentration of 30mg/ ml. At last, 1ml of 2, 2-azobisisobutyronitrile solution, 9ml of HEMA with different weight ratios of CNFs and different volumes of 1mg/ml HRN solutions were mixed together to obtain the following composites: polyHEMA, polyHEMA+0.01 mg/ml HRN, poly HEMA+5mg/ml CNF, polyHEMA+5mg/ml CNF+0.01mg/ml HRN, polyHEMA+10mg/ml CNF+0.01mg/ml HRN and polyHEMA+10mg/ml CNF+0.05mg/ml HRN. Materials were then coated on the top of a 12 mm diameter circle microscope cover glass (Fisher Science circles No. 1, 0.13–0.17 mm thick, diameter 12 mm, catalog No. 12-545-80) using a disposable pipette (Fisherbrand No. 13-711-9AM), where around 0.3ml of the solution was placed on the cover glass and polymerized in an oven at 75°C for 2 h [15].

Before coating the material, silanization of the glass surface was completed in order to enhance the interaction of the glass surface and the samples. Silanization occurred by the following procedure: first soak the cover glasses in 1M NaOH solution for 10 min, and then wash the cover glass using dH$_2$O several times, air dry and put the cover glass in a 2% 3-Aminopropyl-trimethoxysilane (ACROS, CAS No. 13822-56-5) solution, whose solvent is acetone for 30min, then wash the cover glass using acetone several times and air dry.

All the samples were sterilized in 70% ethanol and 100% ethanol for 5 min. Then, the samples were evaporated and sterilized under UV light in the biohood overnight prior to cell seeding.

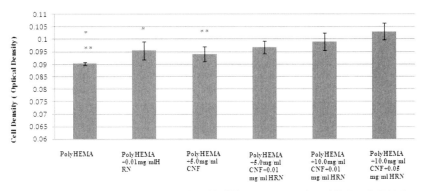

Figure 2: Cardiomyocyte adhesion on composite with different concentration of CNF and HRN. Data are mean ±SEM. n=3, Time=4h.*p < 0.05, **p < 0.05 compared to PolyHEMS + 10.0 mg CNF + 0.05 mg/ml HRN.

Material Characterization

A scanning electron microscope (SEM, LEO 1530-VP) operating at a 10 kV accelerating voltage was used to image CNFs in the pHEMA composite. The composite substrates were coated with a thin layer of gold–palladium using a PS-2 coating unit (International Scientific Instruments) under argon for 2 min. SEM images of the top were taken after contact angles on the surface were measured by a drop shape analyzer (Easy Drop FM40, Kruss, Germany). Before measurements, 3 µl of deionized water was dropped onto the surfaces. Contact angles on nine different samples were measured and averaged.

The electrical resistance of the samples was determined using a multimeter (HP 34401) by touching the samples surface with two probes, via alligator clips. The resistance was tested dry at room temperature and the probes were 10mm apart. Three measurements were taken for each sample.

Human Cardiomyocyte in vitro Assays

Human cardiomyocytes (Celprogen catalog,Cat#36044-15-T25) were seeded at a cell density of 3500 cells/cm^2 for the cell adhesion assay on the materials of interest in complete growth medium containing 10% fetal bovine serum (Celprogen catalog No.M36044-15S). The samples were incubated for 4 h for the cell adhesion assay under standard incubator conditions (5% CO_2, 95% humidified air and 37°C). Alive cells on the top of materials were counted following the well established MTT (Promega Inc.) assay.

Statistical Analysis

Data are expressed as the standard error mean+/ − deviation. Statistical analyses were performed using a Student's one-tailed t-test, with $p < 0.05$ considered statistically significant.

RESULTS AND DISCUSSION

Cardiomyocyte Adhesion

Results of cardiomyocyte adhesion on the various substrates are shown in Figure 2. It was observed that both CNFs and HRNs promoted cell adhesion. Samples with highest concentration of CNF and HRN (polyHEMA+10mg/mlCNF+0.05mg/ml HRN) had the highest cell density.

Materials Characterization

Some representative SEM pictures of the surface of the composites are shown in Figure 3. CNFs were well dispersed in polyHEMA. More CNFs were seen in the samples with higher concentrations of CNFs. Contact angle results are shown in Figure 4. When the concentration of the HRNs increased, the contact angle of the samples decreased, which means that HRNs increased the hydrophilicity of the materials. The same result was observed when the concentration of the CNFs increased. The reason may be that the surface of the CNFs used was modified with –COOH and –OH, which helped to increase hydrophilicity.

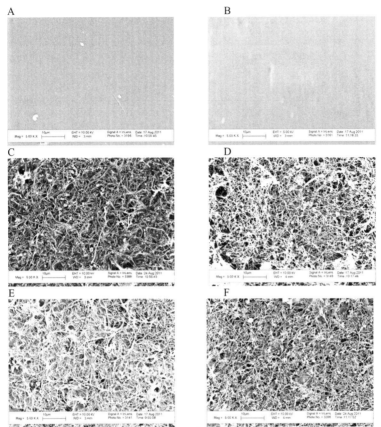

Figure 3. SEM images at 5K magnification showing the distribution of CNFs in the hydrogels.
(A) PolyHEMA, (B) PolyHEMA+0.01mg/ml HRN, (C) PolyHEMA+5mg/ml CNF,
(D) PolyHEMA+5mg/ml CNF+0.01mg/ml HRN, (E) PolyHEMA+10mg/ml CNF+0.01mg/ml HRN,
(F) PolyHEMA+10mg/ml CNF+0.05mg/ml HRN

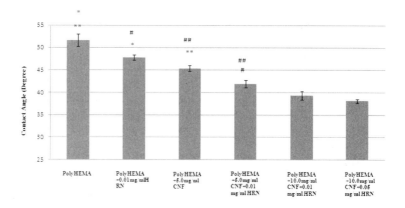

Figure 4: The water contact angle on the materials of interest. Data are mean ± SEM. n=3. *p < 0.05,**p < 0.005, # p < 0.005, ##p < 0.005 compared to polyHEMS + 10 mg/ml CNF + 0.05 mg/ml HRNs.

The result of conductivity measurements are shown in Figure 5. Significantly greater conductivity was observed when the concentration of CNFs increased. The pure polyHEMA and PolyHEMA+0.01ml/ml HRN had negligible conductivity. From this result, it was observed that HRNs did not play any positive role in increasing the conductivity of the composites.

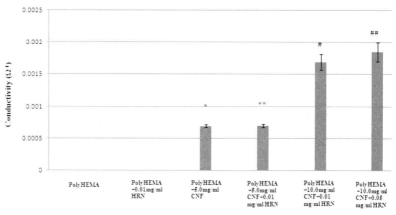

Figure 5: Conductivity of the materials of interest. Data are mean ± SEM (n=3).*p < 0.005, ** p < 0.005, # p < 0.005 compared to polyHEMA.

An important issue from the results of the present study is the toxicity of CNFs after the matrix degrades. The toxicity of nanoparticles is currently a hotly debated issue[23-27], including inhalation and dermal exposure. Importantly, Price et al. studied the toxicity of the same CNFs used here when exposed to osteoblasts (bone forming cells) and determined that CNFs are less toxic than conventional carbon fibers[28]. Moreover, Stout et al. demonstrated that PLGA: CNF composites can also promote the growth of cardiomyocytes and neurons, especially composites with 75 wt% CNFs [29]. Collectively, although requiring further investigation, such results indicate a pro-cell interaction with CNFs, not toxicity at the concentrations used in the present study.

CONCLUSIONS

HRNs and CNFs were successfully used to design a novel biomimetic composite with improved cytocompatibility properties for cardiomyocytes. HRNs and CNFs were well dispersed in polyHEMA via sonication. This composite showed increased cytocompatibility, especially for the composite with highest concentration of HRNs and CNFs (polyHEMA+0.05mg/ml HRN+10mg/ml CNF). This study also showed that HRNs can increase the hydrophilicity of polyHEMA. Thus, HRNs and CNFs provided a supportive matrix for myocardial applications.

ACKNOWLEDGEMENT
This study was supported by Hermann Foundation.

REFERENCES
[1] American Heart Association Statistics Committee and Stroke Statistics Subcommittee. Heart disease and stroke statistics – 2011 update. Circulation 2011; 123: e18–209.

[2] Laflamme M A, Murry C E, Regeneration the heart, *Jature Biotechnology* , 2005;23;845-856.

[3] Locca D, Bucciarelli-Ducci C, Ferrante G, La Manna A, Keenan NG, Grasso A, et al. New universal definition of myocardial infarction: applicable after complex percutaneouscoronary interventions, *JACC Cardiovasc Interv*, 2010; 3: 950–8.

[4] Thygesen K, Alpert JS, White HD, Universal definition of myocardial infarction. *J Am Coll Cardiol*, 2007; 50:2173–95. 445

[5] Jain AK, Knight C, Myocardial infarction – new definitions and implications. 2010; 38: 421–423.

[6] Fenniri H, Mathivanan P, Vidale K L, Sherman D M, Hallenga K, Wood K V and Stowell J G, Helical rosette nanotubes: design, self-assembly and characterization, *J. Am. Chem. Soc.* 2001, 123; 3854-5

[7] Zhang L, Rakotondradany F, Myles A J, Fenniri H and Webster T J, 2009, Arginine-glycine-aspartic acid modified rosette nanotube-hydrogel composites for bone tissue engineering, *Biomaterials*, 30; 1309–20

[8] Zhang L, Ramsaywack S, Fenniri H and Webster T J, 2008, Enhanced osteoblast adhesion on self-assembled nanostructured hydrogel scaffolds, *Tissue Eng.* 14: 1353–64

[9] Fine E, Zhang L, Fenniri H andWebster T J, 2009, Enhance endothelial cell functions on rosette nanotube coated titanium vascular stents, *Int. J. Janomed* . 4: 91–97.

[10] Kretlow J D, Klouda L, Mikos A G, Injectable matrices and scaffolds for drug delivery in tissue engineering, *Advanced Drug Delivery Reviews,* 59 (2007) 263–273

[11] Filmon R, Basl'e M F, Atmani H and Chappard D, 2002, Adherence of osteoblast-like cells on calcospherites developed on a biomaterial combining poly (2-hydroxyethyl) methacrylate and alkaline phosphatase, *Bone*, 30:152–8

[12] Hoffman A S, 2002, Hydrogels for biomedical applications, *Adv Drug Deliv Rev*. 43: 3–12

[13] Lowman A M and Peppas N A, 1999, H*ydrogels Encyclopedia of Controlled Drug Delivery ed: E Mathiowitz (Jew Jersey: Wiley)* 397–418

[14]Zhang L, Rodriguez J, Raez J, Myles A J, Fenniri H and Webster T J, Biologically inspired rosette nanotubes and nanocrystalline hydroxyapatite hydrogel nanocomposites as improved bone substitutes, *Ianotechnology,* 20 (2009) 175101 (12pp)

[15] Zhang L, Ramsaywack S, Fenniri H and Webster T J, 2008, Enhanced osteoblast adhesion on self-assembled nanostructured hydrogel scaffolds, *Tissue Eng*, 14:1353–64

[16] Wang T, Wu DQ, Jiang XJ, Zhang XZ, Li XY, Zhang JF, et al. Novel thermosensitive hydrogel injection inhibits post-infarct ventricle remodelling. *European Journal of Heart Failure.*2009; 11(1):14–19

[17] Li XY, Wang T, Jiang XJ, Lin T, Wu DQ, Zhang XZ, et al. Injectable hydrogel helps bone marrow-derived mononuclear cells restore infracted myocardium. *Cardiology.* 2010; 115(3):194–199.

[18] Dai W, Wold LE, Dow JS, Kloner RA. Thickening of the infarcted wall by collagen injection improves left ventricular function in rats: A novel approach to preserve cardiac function after myocardial infarction. *Journal of the American College of Cardiology.* 2005; 46(4):714–719.

[19] Huang NF, Yu J, Sievers R, Li S, Lee RJ. Injectable biopolymers enhance angiogenesis after myocardial infarction, *Tissue Engineering.* 2005; 11(11–12):1860–1866.

[20] Thompson CA, Nasseri BA, Makower J, Houser S, McGarry M, Lamson T, et al. Percutaneous transvenous cellular cardiomyoplastyA novel nonsurgical approach for myocardial cell transplantation, *Journal of the American College of Cardiology.* 2003; 41(11):1964–1971.

[21] Lutolf MP, Hubbell JA. Synthetic biomaterials as instructive extracellular microenvironments for morphogenesis in tissue engineering, *Iature Biotechnology* . 2005; 23(1):47–55.

[22] Chen Y, Bilgen B, Pareta R.A, Myles A.J, Fenniri H, Ciombor D.M, Aaron R.K and Webster T.J. Self-Assembled Rosette Nanotube/Hydrogel Composites for Cartilage Tissue Engineering. *TISSUE EIGIIEERIIG: Part C* . Volume 16, Number 6, 2010

[23] Brown D.M., Kinloch I.A., Bangert U., Windle A., Walter D.M., Walker G.S., Scotchford C.A., Donaldson K. and Stone V., An in vitro study of the potential of carbon nanotubes and nanofibres to induce inflammatory mediators and frustrated phagocytosis. *Carbon,* 45 9 (2007), pp. 1743–1756.

[24] Grabinski C., Hussain S., Lafdi K., Braydich-Stolle L. and Schlager J., Effect of particle dimension on biocompatibility of carbon nanomaterials. *Carbon,* 45 (2007), pp. 2828–2835.

[25] Lindberg H.K., Falck G.C., Suhonen S., Vippola M., Vanhala E., Catalán J., Savolainen K. and Norppa H., Genotoxicity of nanomaterials: DNA damage and micronuclei induced by carbon nanotubes and graphite nanofibres in human bronchial epithelial cells in vitro. *Toxicol. Lett.,* 186 3 (2009), pp. 166–173.

[26] Magrez A., Kasas S., Salicio V., Pasquier N., Won Seo J., Celio M., Catsicas S., Schwaller B. and Forró L., Cellular toxicity of carbon-based nanomaterials. *Iano Lett.* , 6 6 (2006), pp. 1121–1125.

[27] Price R.L., Haberstroh K.M. and Webster T.J., Improved osteoblast viability in the presence of smaller nanometre dimensioned carbon fibres. *Ianotechnology* , 15 8 (2004), p. 92.

[28] Kisin ER, Murray AR, Sargent L, Lowry D, Chirila M, Siegrist KJ, Schwegler-Berry D, Leonard S, Castranova V, Fadeel B, Kagan VE, Shvedova AA. Genotoxicity of carbon nanofibers: Are they potentially more or less dangerous than carbon nanotubes or asbestos? *Toxicol Appl Pharmacol.* 2011 Apr 1;252(1):1-10.

[29] Stout D. and Webster T.J., Poly(lactic-co-glycolic acid): carbon nanofiber composites for myocardial tissue engineering applications. *Acta Biomater.* 2011 Aug;7(8):3101-12.

A QUANTITATIVE METHOD TO ASSESS IRON CONTAMINATION REMOVAL FROM A NON-FERROUS METAL SURFACE AFTER PASSIVATION

Sophie X. Yang, Lakshmi Sharma, Bernice Aboud
DePuy Othopaedics, Inc. Warsaw, IN, U.S.A.

ABSTRACT

Removal of exogenous iron or iron contaminants from machined or polished surfaces of metallic medical devices is done by chemical dissolution in an acid solution. Test methods used to determine if the iron has been removed from a non-ferrous metal surface are subjective, as it is unknown how much, or if any iron is even present prior to testing. A quantitative method is developed in order to assess the effectiveness of iron removal between nitric acid passivation and the more environmentally friendly citric acid passivation. Polished disks (Ti6Al4V, CoCrMo) were intentionally contaminated with steel shot beads tumbled in a Turbula shaker mixer for a set time. X-ray Photoelectron Spectroscopy (XPS) and Energy Dispersive Spectroscopy (EDS) were used to quantify iron in each uniquely identified disk before and after passivation. The 2 factor and 2 level Design of Experiments (DoE) was used to optimize concentration of citric or nitric acid and solution temperature that was more effective at removing iron contamination for the surfaces.

INTRODUCTION

Although nonferrous medical implant alloys like Ti6Al4V and CoCrMo are inherently passive and corrosion resistant materials, the practice of passivation by means of an acid solution treatment is common in order to remove exogenous iron resulting from manufacturing operations. Iron contaminants may attach to metallic surfaces when the part contacts ferrous fixtures or tools in manufacturing. Any iron contaminant must be removed properly.

The challenge in assessing how effective a passivation process is in removing iron contaminants; one must know how much iron is present to begin with. A method is developed by intentionally contaminating a known surface area, quantifying the amount of iron present and evaluating the remaining contaminant after passivation. The standard method for detecting whether free iron resides on the surfaces is a copper sulfate test in accordance to the Practice D in ASTM A967 "Standard Specification for Chemical Passivation Treatments for Stainless Steel Parts" ([1]). In the copper sulfate test, the sample is immersed in $CuSO_4$ solution at least 6 minutes. If there is free iron on the surface, copper will overplate onto iron.

Of interest here is not only to examine the efficiency of iron removal between nitric acid and citric acid passivation, but also to challenge acid concentration and temperature in the efficiency of iron removal. With the characterization method developed in this work, the function of the trending citric acid passivation versus nitric acid passivation on Ti6Al4V and CoCrMo is quantitatively analyzed and equally compared.

MATERIALS AND METHODS

Test Specimen Preparation

Eight Ti6Al4V wrought and eight CoCrMo cast disks were used for this study. Disks measured 19 mm in diameter and 7 mm thick. Disk surfaces were grounded to 1200 grit to present an equal surface finish across all samples. In order to impinge iron on the disks, 1 disk was tumbled in 80ml of 1mm diameter steel shot beads (>96% Fe) in a Turbula shaker mixer (Turbular System Schaze, Willy A Bachofen AG Maschinenfabrik Basel, Switzerland) for a set time. The Ti6Al4V disks were tumbled 3 minutes and the CoCrMo disks were tumbled 10 minutes. CoCrMo needed longer tumbling time because the material was harder than Ti6Al4V and needed longer tumbling time to get enough iron on the surfaces. After tumbling, the disks were dipped in Reverse Osmosis (RO) water twice to remove loose iron particles and dried at 60°C.

Quantification Procedure

Each disk was divided into four quadrants and uniquely indentified (Figure 1). Each quadrant of each disk was analyzed initially in order to establish a quantitative baseline of iron contamination for each disk. After the disks were passivated under the various passivation conditions, they were analyzed again to quantify the amount of iron removed. Each quadrant was uniquely tracked before and after passivation in order to get an accurate percentage of iron removed.

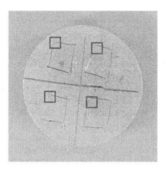

Figure 1. Shown is an example of the four quadrants on each disk which were analyzed before and after passivation. The sample size was 19mm in diameter.

XPS used in this study was a Quantera SXM Scanning X-ray Microprobe with an Al anode (Ulvac-PHI. Inc., MN, USA). It was used to analyze iron removal on Ti6Al4V surfaces. XPS survey scans from 0-1200eV were performed in four areas of each disk for a baseline and those areas were reanalyzed after passivation to establish a percent reduction in iron. XPS could not be used for analyzing iron on CoCrMo due to the Fe2p peaks overlapping the Co LMM Auger peak in the XPS spectrum.

EDS used in this study was an INCA Microanalysis (Oxford Instruments, MA, U.S.A). It attached to a Quanta 600 Field Emission Scanning Electron Microscope (FEI, OR, U.S.A.) and was used to analyze iron removal on both Ti6Al4V and CoCrMo surfaces. In each quadrant, an EDS spectrum was collected in a 100x image with 20KV and 10mm working distance.

Passivation Treatments

Two 2 factor – 2 level Design of Experiment (DoE) matrices were designed to challenge the acid concentration and temperatures of citric acid (CA) passivation (Table 1) and nitric acid (NA) passivation (Table 2). Nitric acid passivation and citric acid passivation parameters conformed to ASTM A967. Nitric acid passivation solutions varied in acid concentration (20 volume% and 55 volume%) and temperature (70°F and 140°F). Immersion time was kept constant at 30 minutes. Citric acid passivation solutions were mixed using citric acid powder (J.T. Baker, NJ) and water for 4 weight% and 10 weight% of concentration. Processing temperature was 70°F and 160°F. Immersion time was constant at 20 minutes.

Each disk was passivated in 100 ml of passivation solution under each condition. After passivation, the disks were rinsed in 131°F RO water 3 times and dried.

Table 1. DoE matrix for citric acid (CA) passivation

Acid Concentration (wt%)	Temperature (°F)	Emersion Time (minute)
4	70	20
10	70	20
4	160	20
10	160	20

Table 2. DoE matrix for nitric acid (NA) passivation

Acid Concentration (v%)	Temperature (°F)	Emersion Time (minute)
20	70	30
55	70	30
20	140	30
55	140	30

RESULTS AND DISCUSSION

XPS Results

The XPS technique used in this analysis could not be used to assess the iron removal of CoCrMo samples due to the overlap of the Fe2p peaks and the Co LMM Auger peak. On the Ti6Al4V disks, XPS consistently showed a reduction in iron after passivation regardless of acid solution, concentration or temperature (Figure 2A). It was noted that samples for the citric acid population exhibited a wider range in contamination from 39% (sample 4%CA70°F20min) to 53% (sample 10%CA160°F20min). It is speculated the variation may be due to the variation of residual iron oxide on the individual steel shot beads.

Figure 2B shows the higher percentage of iron reduction was seen in the higher temperatures for both citric and nitric acid passivation with the greatest reduction in samples treated in nitric acid at 140°F. Therefore, the passivation temperature was a critical factor in increasing the iron removal effectiveness of passivation. The above comparison also showed that nitric acid passivation had significantly higher iron removal effectiveness than citric acid passivation.

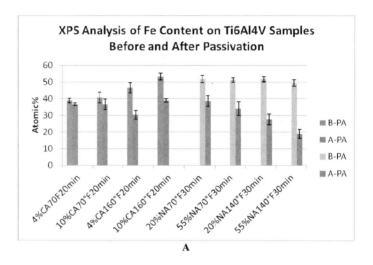

A

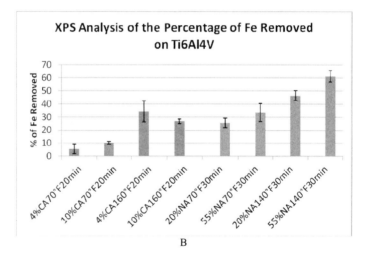

B

Figure 2. A) XPS analysis of iron contents on Ti6Al4V samples before and after citric/nitric acid passivation, B) percentage of iron removed after passivation. Legend: B-PA: before passivation and A-PA: after passivation.

EDS Results

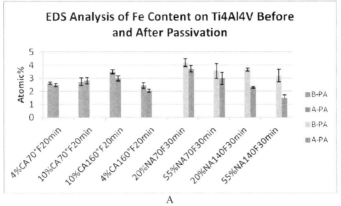

A

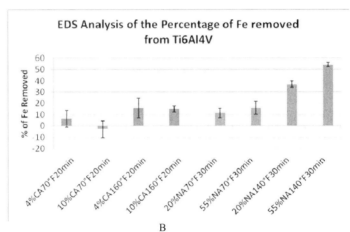

B

Figure 3. A) EDS analysis of iron content on Ti6Al4V before and after passivation, B) Quantification of precentage of iron removed from Ti6Al4V after passivation.

X-rays in the EDS technique typically penetrate the sample to 1 μm or more which is much deeper than the surface analysis technique XPS. Therefore the atomic % of iron pickup using EDS was lower than detected using XPS, since the iron contaminant was concentrated on the surface. EDS of the Ti6Al4V samples detected around 2.5% to 4.2% of iron within the quadrant areas and iron was reduced after passivation in most conditions (Figure 3A). The low temperature citric acid passivation exhibited the least iron removal function (Figure 3B). The negative iron removal value in the graph might be caused by the iron migrating from neighboring regions during passivation or rinsing. The iron removal results obtained from both citric acid passivation and nitric acid passivation indicated that the passivation temperature was a critical factor in improving iron contamination, the high temperature citric acid passivation (160°F) removed more iron (15%) than the low temperature passivation (0%) and high temperature (140°F) nitric acid removed more iron (55%) than low temperature (70°F) nitric acid passivation (15%).

Above results showed that both techniques, XPS and EDS, showed a quantitative reduction in iron on the Ti6Al4V samples, albeit the XPS provided a greater percentage in reduction given the X-rays concentrate mostly on the surface of the sample.

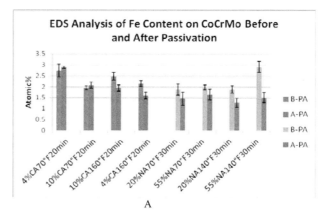

A

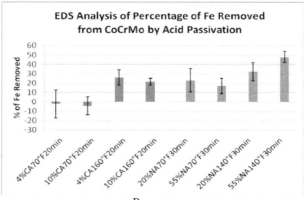

B

Figure 4. A) EDS analysis of iron content on CoCrMo before and after passivation, B) percentage of Fe removed by passivation, Legend B-PA: before passivation and A-PA: after passivation.

EDS analysis on CoCrMo disks detected a range of 1.9 to 2.9% of iron before passivation. Iron was reduced after passivation under most conditions except after low temperature citric acid passivation (Figure 4A).

Similar to the results obtained from Ti6Al4V disks, the percentage of the iron removed from CoCrMo by passivation (Figure 4B) showed that passivation temperature positively impacted iron removal effectiveness. High temperature (160°F) citric acid passivation removed 21-26% of iron while low temperature (70°F) citric acid passivation was less effective. High temperature (140°F) nitric acid passivation which removed up to 48% of iron while low temperature (70°F) nitric acid passivation only removed 17-23% of iron. The results also indicated that nitric acid passivation was more effective to remove iron than citric acid passivation from the CoCrMo surface.

CONCLUSIONS

1. A quantitative method is possible to assess iron removal effectiveness of acid passivation on a non-ferrous surface. The method included putting iron on the non-ferrous surface by tumbling the sample with ferrous metal beads in a shaker mixer, quantifying iron on the surface by analyzing specified areas with XPS and/or EDS before and after passivation, and calculating the percentage of the iron removed.
2. In this study, nitric acid passivation was more effective at iron removal than citric acid passivation and temperature had a greater effect than acid concentration in either solution.

ACKNOWLEDGEMENT
The authors wish to acknowledge Amy Craft and Kori Rivard for their collaboration.

---

[1]. ASTM A 967, 2005e1, "Standard Specification for Chemical Passivation Treatments for Stainless Steel Parts", ASTM International, West Conshohocken, PA, U.S.A, 2005.

Author Index